USE OF LOW ENRICHED URANIUM AND APPLICATIONS OF ACCELERATOR DRIVEN SYSTEMS

IAEA-TECDOC-2107

USE OF LOW ENRICHED URANIUM AND APPLICATIONS OF ACCELERATOR DRIVEN SYSTEMS

FINAL REPORT OF A COORDINATED RESEARCH PROJECT

INTERNATIONAL ATOMIC ENERGY AGENCY
VIENNA, 2025

For further information on this publication, please contact:

Research Reactor Section
International Atomic Energy Agency
Vienna International Centre
PO Box 100
1400 Vienna, Austria
Email: Official.Mail@iaea.org

© IAEA, 2025
Printed by the IAEA in Austria
December 2025
https://doi.org/10.61092/iaea.9ekn-rqsr

IAEA Library Cataloguing in Publication Data

Names: International Atomic Energy Agency.
Title: Use of low enriched uranium and applications of accelerator driven systems / International Atomic Energy Agency.
Description: Vienna : International Atomic Energy Agency, 2025. | Series: IAEA TECDOC series, ISSN 1011-4289 ; no. 2107 | Includes bibliographical references.
Identifiers: IAEAL 25-01795 | ISBN 978-92-0-124825-1 (paperback : alk. paper) | ISBN 978-92-0-124725-4 (pdf)
Subjects: LCSH: Accelerator-driven systems. | Uranium as fuel. | Particle accelerators. | Nuclear physics.

FOREWORD

An accelerator driven system (ADS) integrates a particle accelerator with a subcritical nuclear reactor. The accelerator generates a high energy particle beam that strikes a target, resulting in the emission of neutrons and other particles. These neutrons then induce fission reactions in the surrounding subcritical reactor core, enabling the transmutation of long lived radioactive waste and the generation of energy, as well as other applications.

Some ADS designs have used, or have been proposed to use, high enriched uranium (HEU) fuel. However, there is a growing effort to transition to low enriched uranium (LEU) fuel, given the proliferation and security concerns associated with HEU. Researchers worldwide are working to convert existing systems to use LEU fuel or are developing new ADS designs that can operate with LEU.

From 2005 to 2010, a coordinated research project (CRP) on Analytical and Experimental Benchmark Analyses of Accelerator Driven Systems, along with a parallel international research collaboration, focused on exploring various technical options for conducting ADS research using LEU fuel. The CRP involved participation from 17 Member States, and this cooperative effort resulted in the resolution of a number of discrepancies initially observed for some of the results obtained during the investigations. The results of the CRP were published in IAEA-TECDOC-1821.

Building on the outcomes of the 2005–2010 CRP, the IAEA organized a second CRP to further support Member States, organizations, researchers, regulatory bodies, designers and other interested parties involved in ADS operation and use. This second CRP, on Accelerator Driven System Applications and Use of Low Enriched Uranium in Accelerator Driven Systems, was carried out from 2016 to 2020 and focused on the development of innovative applications for ADS facilities, as well as on the further development and validation of ADS technologies. It also aimed to assist Member States still operating ADS facilities with HEU to convert their facilities to LEU, thereby reducing nuclear proliferation risks. This publication presents the collective efforts and findings resulting from the 2016–2020 CRP.

The IAEA is grateful to all those who participated in the CRP and contributed to the drafting and review of this publication. The IAEA officers responsible for this publication were F. Marshall, S. Geupel, J. Dewes and K. Dunn of the Division of Nuclear Fuel Cycle and Waste Technology.

TABLE OF CONTENTS

Chapter 1

INTRODUCTION

1.1. BACKGROUND

An accelerator driven system (ADS) consists of a subcritical assembly and an external neutron source based on a particle accelerator. The subcritical assembly contains an array of nuclear fuel elements that cannot sustain a nuclear chain reaction on their own. The accelerator generates a high energy particle beam that strikes a target, resulting in the emission of neutrons through processes such as spallation[1], which then drive fission reactions in the subcritical assembly. Various ADS designs have been proposed for the transmutation of radioactive wastes and for energy production because of their advantages of high flexibility of fuel compositions and enhanced safety features [1.1], [1.2], [1.3].

Although some ADS concepts use Th, Th-U or MOX fuel, highly enriched uranium (HEU) or low enriched uranium (LEU) is predominantly used for subcritical assembly fuel. However, the civilian use of HEU raises proliferation concerns which have been the focus of several international undertakings supported by the International Atomic Energy Agency (IAEA). These undertakings include the Reduced Enrichment for Research and Test Reactors (RERTR) programme and the Russian Research Reactor Fuel Return (RRRFR) programme. Accordingly, significant efforts have been undertaken to convert existing research reactors to LEU and to encourage the use of LEU in the design of new research reactors and accelerator driven systems. Ongoing research is focused on adapting ADS technology to effectively utilize LEU while maintaining or improving performance.

In recent years, the number of States investigating ADS has grown. In some States, several universities and institutions have ADS research and development (R&D) programmes. Some States have ADS facilities in operation, and others have plans to construct new ADS facilities. To reduce nuclear proliferation risks, it is desirable that all these facilities use LEU fuel.

This IAEA publication documents the results of the Coordinated Research Project (CRP) T33002 titled "Accelerator Driven Systems (ADS) Applications and Use of Low Enriched Uranium in ADS", which was carried out from 2016 to 2020. In total 22 institutions from 17 IAEA Member States (Argentina, Belarus, China, Germany, Hungary, India, Indonesia, Islamic Republic of Iran, Italy, Japan, Poland, Republic of Korea, Russian Federation, Spain, Ukraine, United States of America, and Uzbekistan) participated in the CRP.

The CRP focused on the development of innovative applications for ADS facilities as well as on further development and validation of ADS technologies and the physics of subcritical systems applicable to the deployment of LEU fuelled ADS. An essential aspect of the CRP was the fostering of a spirit of international collaboration and exchange of information and data acquired from measurements and analytical studies. In the course of the CRP, three research coordination meetings and two consultancy meetings were held. The results of the CRP are particularly useful for Member States that are still using ADS with HEU fuel, to help convert their facilities to LEU fuel, thereby reducing nuclear proliferation risks.

This CRP, coordinated by the IAEA, follows the CRP I32006 on "Analytical and Experimental Benchmark Analysis of Accelerator Driven Systems", which took place from 2005 to 2010 with the participation of 17 IAEA Member States. CRP I32006 was devoted to gaining a better understanding of the physics of accelerator-driven subcritical nuclear systems, and its results were published in IAEA-TECDOC-1821 [1.4].

---

[1] Spallation is the breaking up of a heavy target nucleus into several parts as the result of bombardment by an incident particle, typically a high energy proton. The emitted particles can include neutrons, protons and various others, leaving behind a residual nucleus that is lighter than the original. Although conventional spallation systems often use proton energies above 150 MeV, the term "spallation neutrons" as used in this publication also encompasses neutrons generated by a 100 MeV proton beam.

1.2. OBJECTIVE

The objective of this publication is to present the final consolidated results of the research studies performed by the participants of the CRP T33002 titled "Accelerator Driven Systems (ADS) Applications and Use of Low Enriched Uranium in ADS".

This publication is intended as an information resource for use by operating organizations, research institutions, regulatory bodies, designers, technical support organizations and other interested parties involved in the design, operation and utilization of ADS.

1.3. SCOPE

The publication provides a detailed description of various ADS facilities, the experiments conducted there, and the resulting experimental or computational outcomes. It explores the development of innovative detectors designed for high neutron flux environments and instrumentation for experimental nuclear data measurement. The document also delves into the application of ADS technology for power production and the transmutation of transuranics. Additionally, it presents the methods developed for analysing the physics of subcritical systems, which were utilized for several evaluations within the document.

The following subject matters were within the scope of the CRP:

1. Online subcriticality monitoring techniques

   - Concept development;
   - Experimental verifications;

2. ADS thorium utilization

   - Experimental measurements;
   - ^{233}U ADS breeder design concept development;
   - Energy amplifier concept development;

3. Production of radioactive isotopes

   - Studies;
   - Experimental measurements;

4. Transmutation

   - Experimental measurements;
   - Burner concept development;

5. ADS shielding studies

   - Design studies;
   - Experimental measurements;

6. ADS burnup analyses and subcriticality compensation for the near-term experimental facilities;

7. Solid target design concept development and operational experience.

Conceptual high power ADS designs were out of the scope of this CRP and consequently beyond the scope of this publication. Information on such designs can be found in IAEA-TECDOC-1766, Status of Accelerator Driven Systems Research and Technology Development [1.5].

1.4. STRUCTURE

This publication contains six Chapters (including this introductory Chapter).

Chapter 2 presents the description of the different ADS facilities and the experiments carried out there, the experimental and/or computational results, and a discussion and conclusions.

Chapter 3 describes the development of innovative diamond detectors for use in high neutron fluxes and a description of an experimental facility for cross-section measurement.

Chapter 4 describes studies on the application of ADS technology for power production and transmutation of transuranics.

Chapter 5 presents the methods for analysis of the physics of subcritical systems developed during the CRP which were used for some of the evaluations presented in this document.

Chapter 6 provides an overarching discussion and conclusions, and suggestions for future work.

1.5. REFERENCES TO CHAPTER 1

[1.1]   ABDERRAHIM, H.A., et al., "Accelerator and Target Technology for Accelerator Driven Transmutation and Energy Production", DOE white paper on ADS 1[1], (2010) 1–23.

[1.2]   DE BRUYN, D., et al., "Recent Developments in the Design of the Belgian MYRRHA ADS Facility", ICAPP 2016, San Francisco, CA (2016).

[1.3]   YAN, X., et al., Concept of an Accelerator-Driven Advanced Nuclear Energy System, Energies (Paris) 10[7] (2017) 944.

[1.4]   INTERNATIONAL ATOMIC ENERGY AGENCY, Use of Low Enriched Uranium Fuel in Accelerator Driven Subcritical Systems, IAEA-TECDOC-1821, IAEA, Vienna (2017).

[1.5]   INTERNATIONAL ATOMIC ENERGY AGENCY, Status of Accelerator Driven Systems Research and Technology Development, IAEA-TECDOC-1766, IAEA, Vienna (2015).

Chapter 2

FACILITY EXPERIMENTS AND ANALYSES FOR ADVANCING THE DEVELOPMENT OF A LOW ENRICHED URANIUM ADS

This chapter presents the experimental activities conducted in connection with the project at various facilities and the results of the analyses performed on the available experimental data.

Within Chapter 2, Sections 2.1 and 2.2 describe the work carried out at the Kyoto University Critical Assembly (KUCA) facility in Japan. Although the KUCA facility used HEU fuel for these experiments, the experimental conclusions, analysis and the resulting understanding of the physics of sub-critical systems presented in Section 2 and Section 5 are an important resource for the designers of LEU ADS.

Section 2.1 focuses on the experimental work carried out at the Kyoto University Critical Assembly (KUCA) facility in Japan using the Pb-Bi experimental configuration and on the related computational work. The experimental activity was subdivided into four successive phases as set out in Sections 2.1.1 to 2.1.4. respectively: of the effect of the target, measurements of subcriticality, determination of reaction rates, and the effective delayed neutron fraction studies.

The results for each phase and their analysis are set out in Sections 2.1.5 to 2.1.8. These sections also describe the approaches used for the analysis of the various system configurations and for the interpretation of the experimental data. Further information on the innovative models specifically developed for the analyses can be found in Section 5 of this publication.

Section 2.2 presents the work at the KUCA facility as configured for U–Pb experiments. The benchmark specifications are given in Section 2.2.1 and the two experimental phases are described in Sections 2.2.2 (kinetic parameters) and 2.2.3 (reaction rate distributions). The results of the analyses of these configurations and experimental data are presented in Sections 2.2.4 and 2.2.5 for the kinetic parameters and in Section 2.2.6 for the reaction rate distributions.

Section 2.3 presents the work at the GIACINT critical facility at the Joint Institute for Power and Nuclear Research (JIPNR) in Belarus. The subsections provide a description of the facility (Section 2.3.1), the system composition (Section 2.3.2), the computational models (Section 2.3.3), the experimental results and interpretations (Section 2.3.4). Some concluding remarks are given in Section 2.3.5.

Section 2.4 presents the reaction rate measurements carried out at the QUINTA facility and at the Joint Institute for Nuclear Research (JINR) in the Russian Federation. The experiments are described in Section 2.4.1, and the analysis of the results for the actinide transmutation rates is discussed in Section 2.4.2.

Section 2.5 provides a description of the Beryllium oxide Reflected And HDPe Moderated Multiplying Assembly (BRAHMMA) at the Bhabha Atomic Research Centre in India, together with the results of pulsed neutron experiments that are interpreted for the estimation of the subcriticality level.

2.1. KUCA ADS PB-BI EXPERIMENTS

The neutronic characteristics of the subcritical reactor has been extensively studied theoretically [2.1], [2.2] and experimentally [2.3], [2.4], [2.5]. Since 2009, ADS physics experiments have been carried out at KUCA using spallation neutrons generated by 100 MeV protons from a fixed-field alternating gradient (FFAG) accelerator or 14 MeV neutrons produced by the deuterium–tritium (D–T) reaction in a pulsed neutron generator.

Two experimental benchmarks conducted in the KUCA ADS were published for core configurations embedding solid lead–bismuth plates [2.6] (Pb-Bi benchmark hereafter) and solid lead plates by Ref. [2.7] (U–Pb benchmark hereafter). Sections 2.1 and 2.2 of the report compare neutronics parameters calculated by the

Monte Carlo and deterministic codes for the Pb-Bi and U–Pb benchmarks with the experimental results. The numerical calculations cover both the static and kinetic parameters such as the effective multiplication factor, excess reactivity, control rod worth, reaction rate distribution, and the prompt neutron decay constant. The benchmark specifications for Phase I, II, III and IV of the Pb-Bi benchmark are given in Sections 2.1.1 to 2.1.4, and their analyses are provided in Sections 2.1.5 to 2.1.8.

2.1.1. Experiments: Phase I – Setup of the facility and study of the effect of the target (W, W–Be and Pb-Bi)

2.1.1.1. Introduction

The objective of Phase I was to investigate experimentally the neutron characteristics of solid ADS targets, such as tungsten (W), uranium (U), lead (Pb) or lead–bismuth (Pb-Bi), used with 100 MeV protons.

Prior work included the investigation of the neutronics of the Pb-Bi eutectic spallation target by the MEGAwatt PIlot Experiment (MEGAPIE) [2.8], [2.9], studies at the Transmutation Experimental Facility (TEF) [2.10]–[2.12] at JAEA and the Multi-purpose hYbrid Research Reactor for High-tech Applications (MYRRHA) [2.13]–[2.15] at SCK•CEN, and reactor physics experiments using the KUCA core coupled to the FFAG accelerator [2.5] and [2.16]–[2.25]. The 100 MeV protons from the FFAG produced spallation neutrons which were injected into uranium-loaded [2.5], [2.18], [2.19], [2.20], [2.21], and thorium-loaded [2.17], [2.23] cores.

The measured reactor physics parameters included reaction rates, neutron spectrum, neutron multiplication [2.26], subcriticality and prompt neutron decay constant. The neutron multiplication is a measure of the number of fission neutrons induced in the core by the external neutron source. Earlier studies at KUCA showed that the neutron multiplication was small when the target was located outside the core at position (15, A') in Fig. 2.1. The neutron multiplication was increased by moving the target inside the core [2.18] and by using two-layer W–Be and Pb-Bi eutectic targets (44.5% Pb and 55.5% Bi) [2.21]. The influence of the target material on the high-energy neutron spectrum, neutron yield, neutron multiplication and subcritical multiplication factor, and on the kinetic parameters, the prompt neutron decay constant and subcriticality, was investigated.

2.1.1.2. Description of the KUCA core

The KUCA facility has solid-moderated and solid-reflected type-A and type-B cores, and a water-moderated and water-reflected type-C core. For this series of experiments, the type-A core was combined with a Cockcroft-Walton pulsed neutron generator and the FFAG accelerator.

The A-core configuration with 25 fuel elements was used to measure the reaction rates. Each fuel rod contained 60 fuel cells that combined a high-enriched uranium-aluminium (U-Al) alloy fuel plate and a 3.2 mm thick polyethylene plate and had polyethylene reflectors above and below the fuel region, as shown in Figs 2.2 and 2.3. All these components had a cross-section of 50.8 mm x 50.8 mm. The functional height of the core was approximately 400 mm. The polyethylene reflector and control and safety rods are shown in Figs 2.4 and 2.5. A plan view of the core is shown in Fig. 2.6. The subsequent figures show details of the experimental configuration, the indium wire (Fig. 2.7), control rods (Fig. 2.8), and target and activation foils (Fig. 2.9).

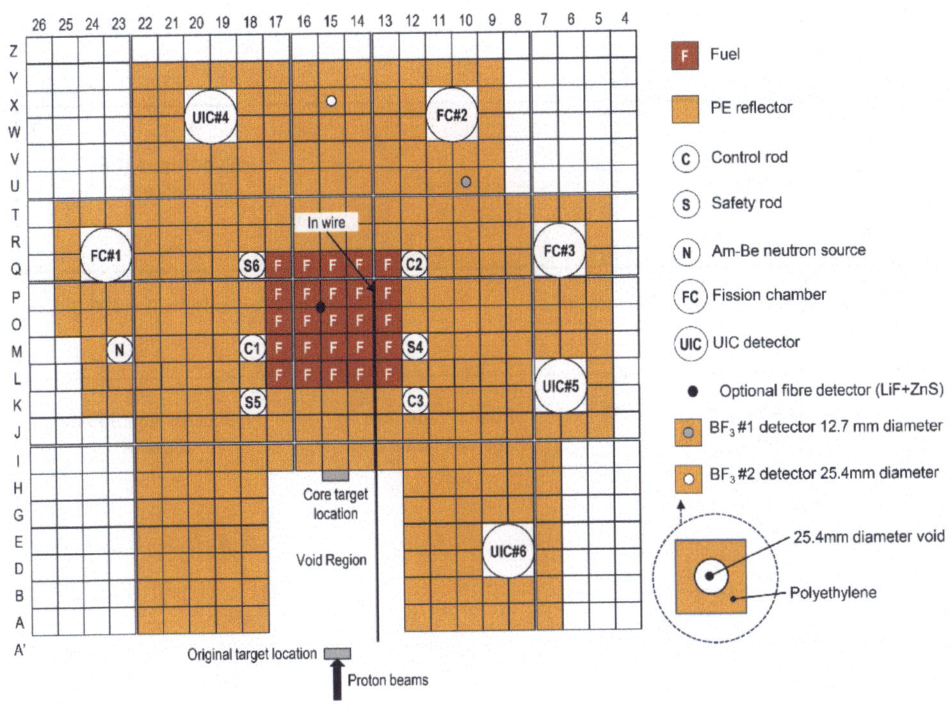

FIG. 2.1. The general view of the KUCA configuration (Adapted from [2.27] with permission courtesy of Taylor & Francis Online).

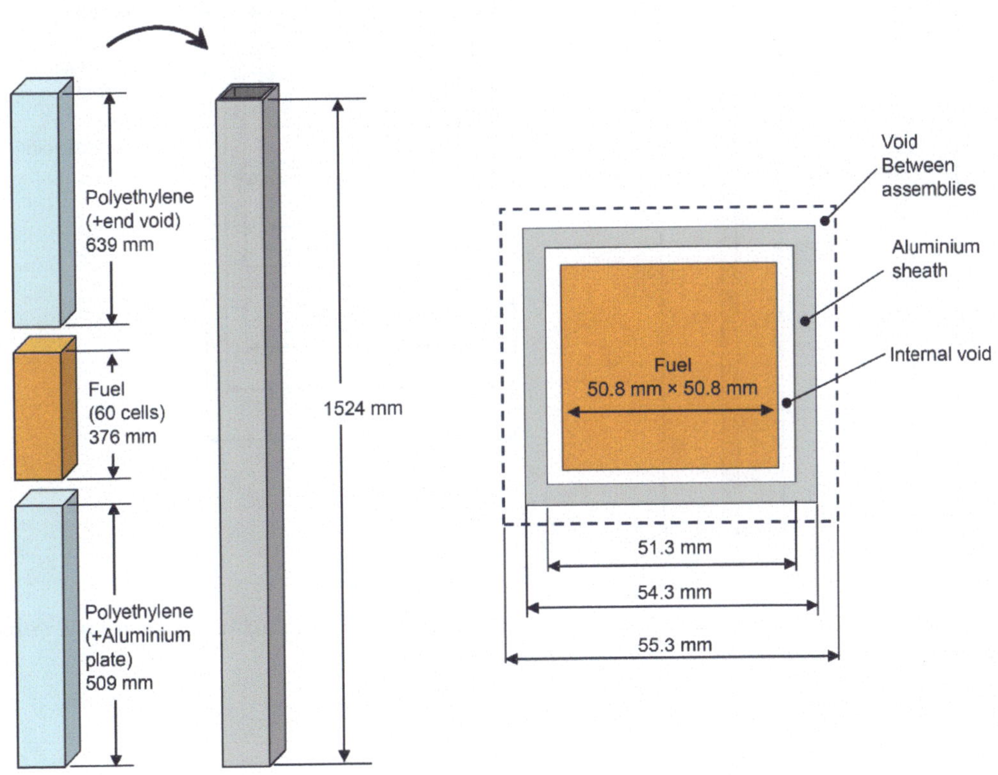

FIG. 2.2. Schematic of KUCA fuel assembly (Reproduced from Ref. [2.6] with permission courtesy of Kyoto University).

7

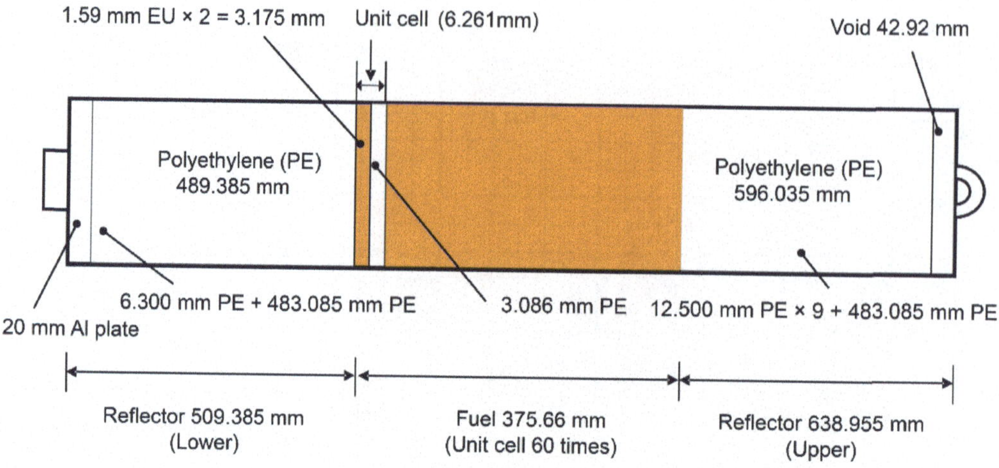

FIG. 2.3. Side view of KUCA fuel assembly (Reproduced from Ref. [2.6] with permission courtesy of Kyoto University).

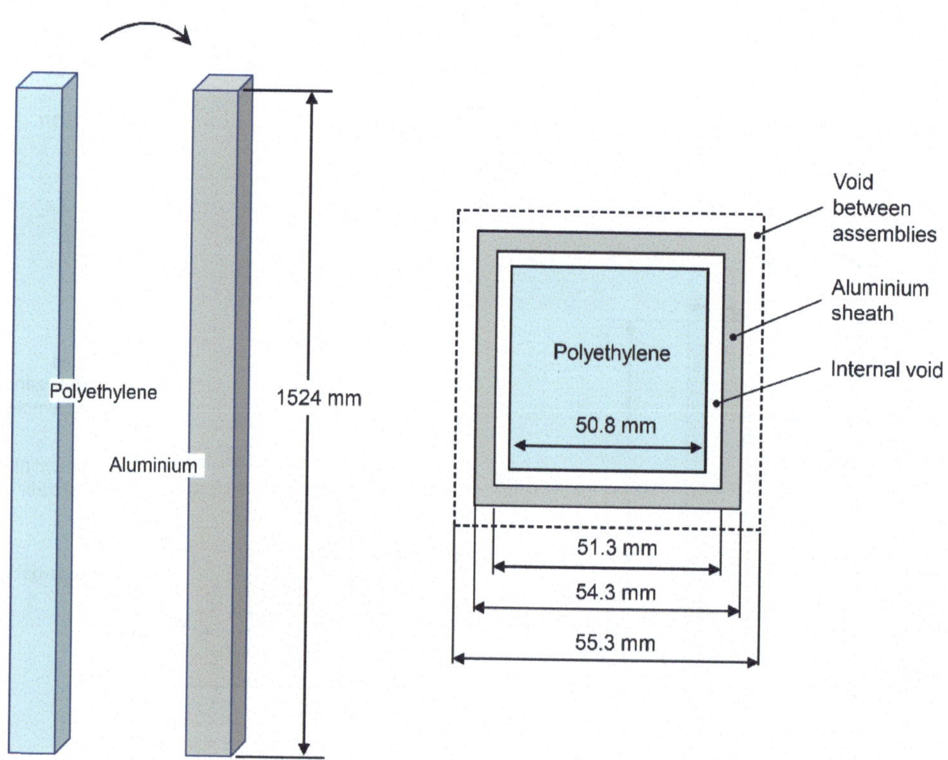

FIG. 2.4. Schematic of KUCA polyethylene reflector (Adapted from Ref. [2.6] with permission courtesy of Kyoto University).

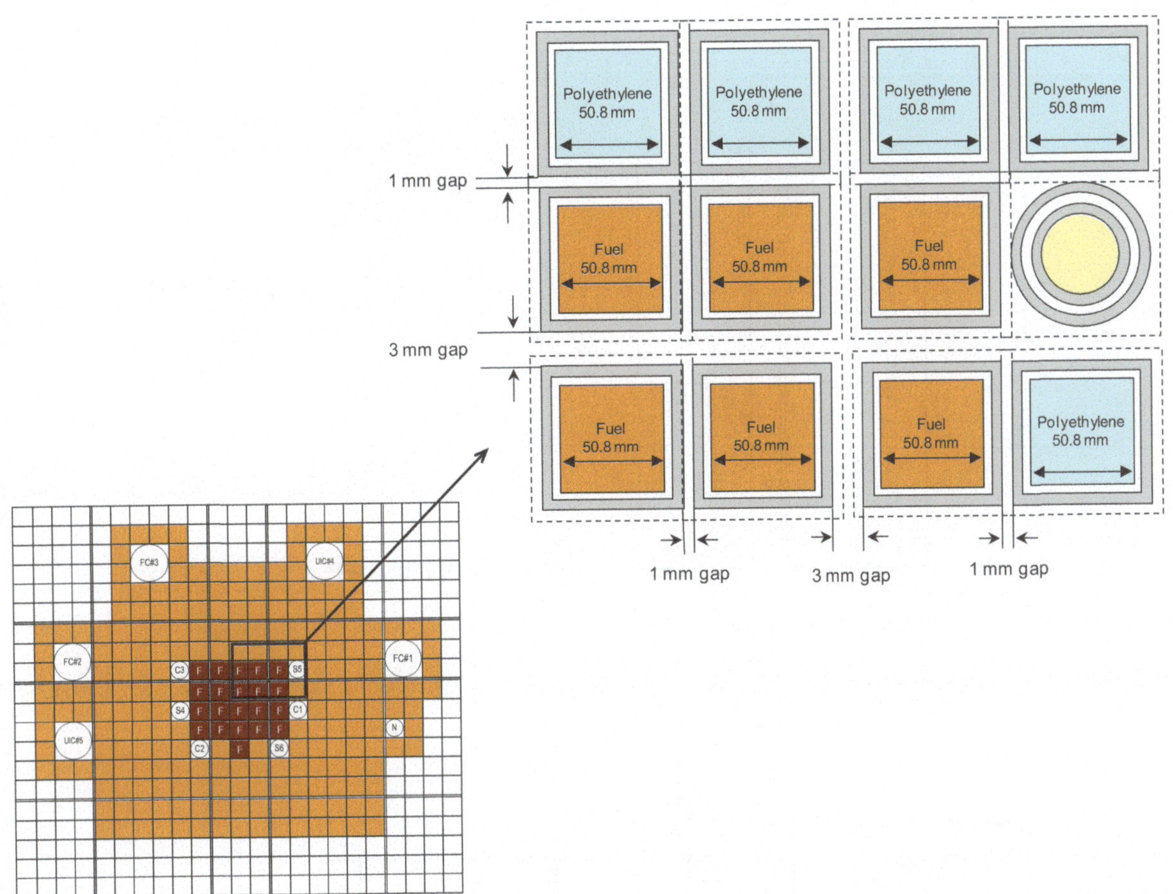

FIG. 2.5. Schematic of KUCA control (safety) rod (Reproduced from Ref. [2.6] with permission courtesy of Kyoto University).

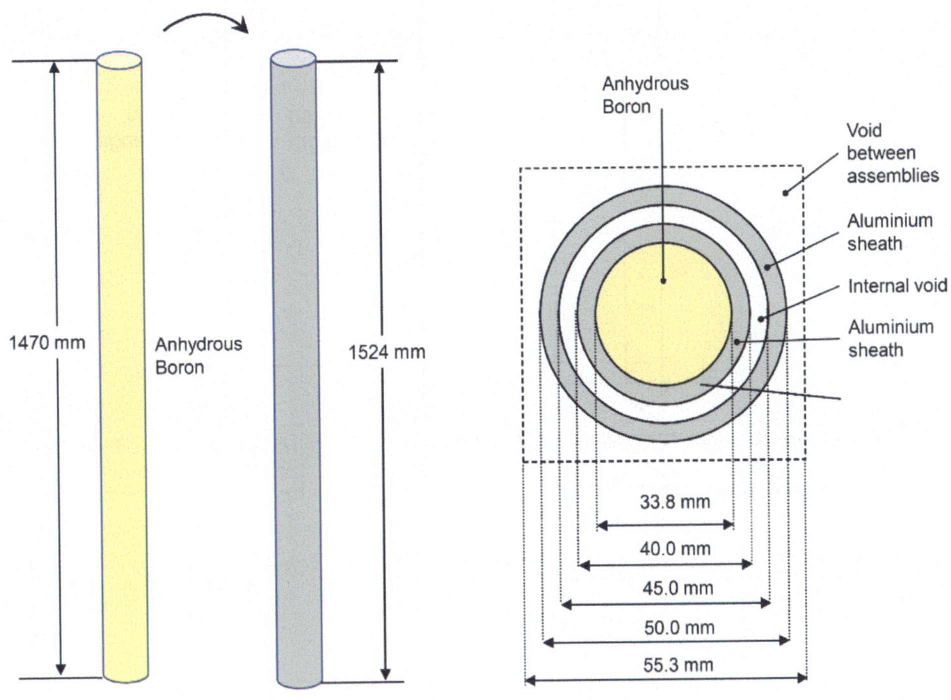

FIG. 2.6. Schematic of KUCA fuel assembly, polyethylene reflector and control rod (Reproduced from Ref. [2.6] with permission courtesy of Kyoto University).

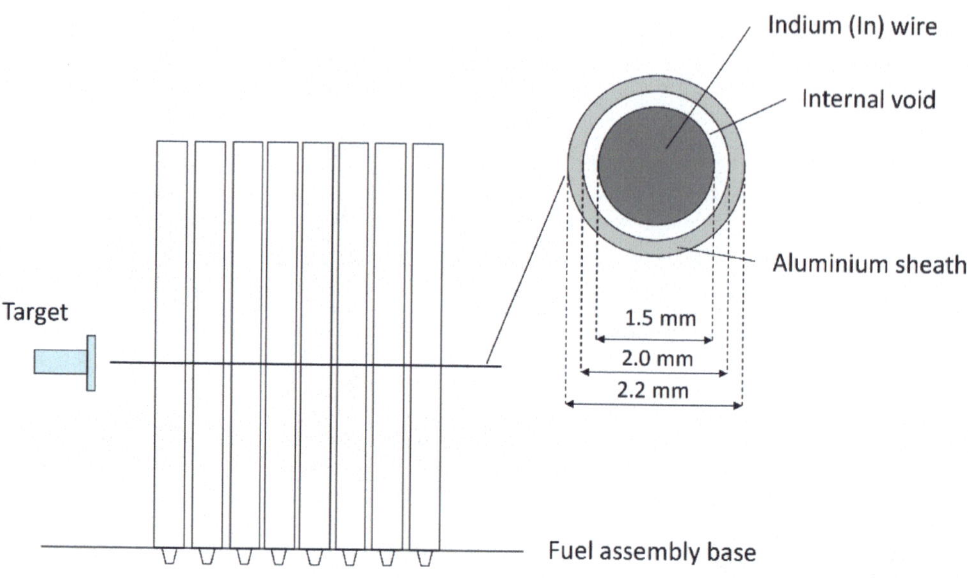

FIG. 2.7. Setting of indium wire (Reproduced from Ref. [2.6] with permission courtesy of Kyoto University).

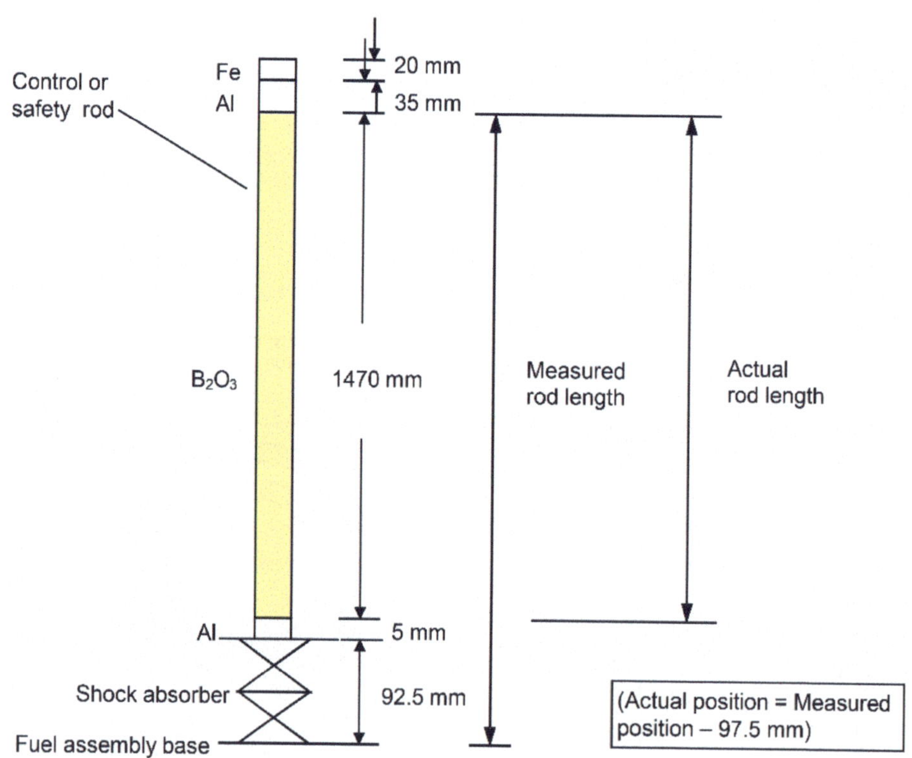

FIG. 2.8. Actual position of control (safety) rod (Reproduced from Ref. [2.6] with permission courtesy of Kyoto University).

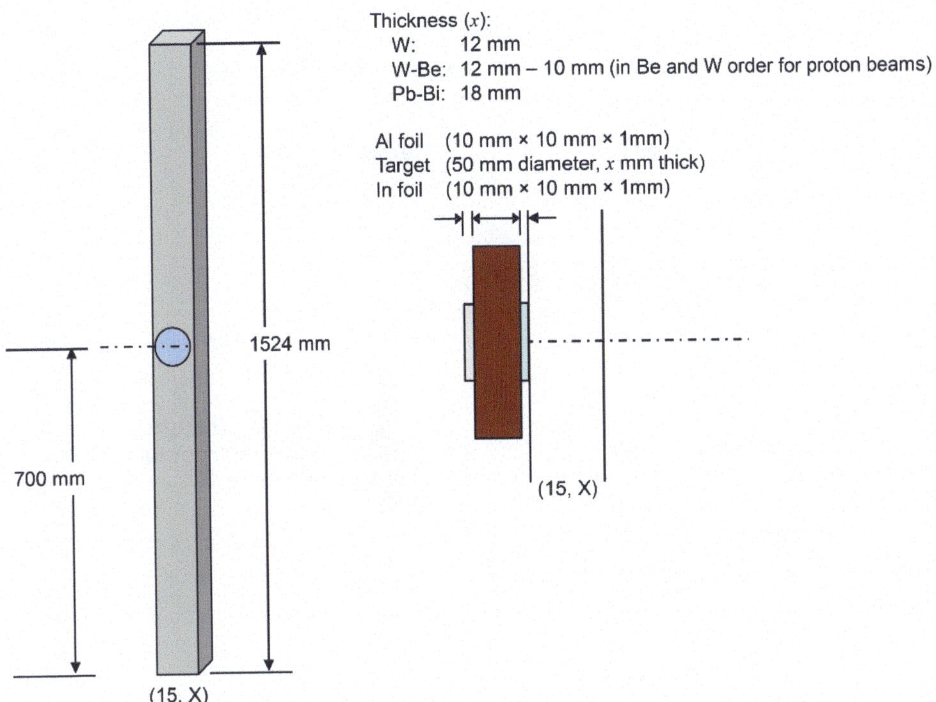

FIG. 2.9. Attachment of target, aluminium and indium foils at the core target location (Reproduced from Ref. [2.6] with permission courtesy of Kyoto University).

2.1.1.3. *The FFAG accelerator*

The FFAG accelerator proton beam had 100 MeV energy, 1 nA intensity, 20 Hz pulse frequency, 100 ns pulse width and a 40 mm diameter spot size at the target, producing a neutron yield at the target of around 1.0×10^8 s^{-1}. Figure 2.10 shows the side view of target and core configuration, the attachment of the gold foils is shown in Fig. 2.11, and the target configuration in Fig. 2.12.

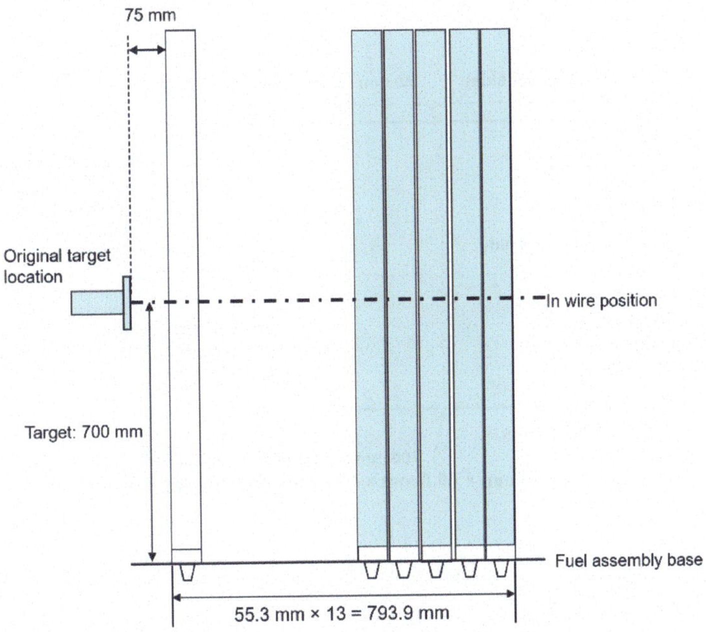

FIG. 2.10. Side view of target and core configuration with 100 MeV protons (Reproduced from Ref. [2.6] with permission courtesy of Kyoto University).

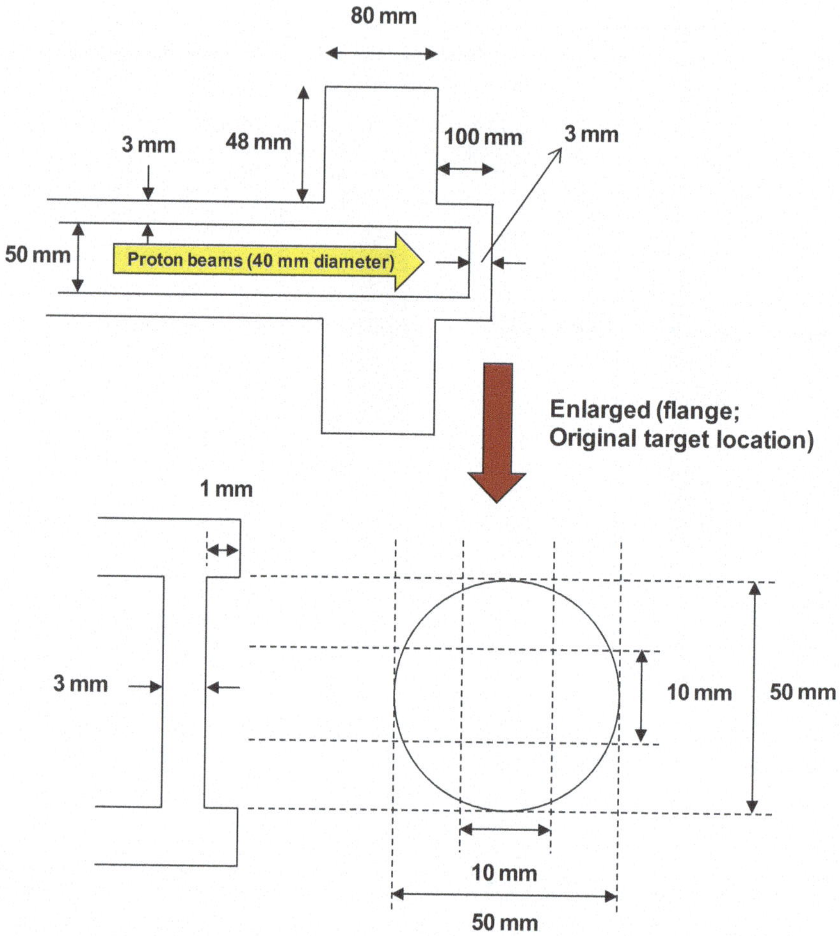

FIG. 2.11. Side view of UT fuel rod; (15, O) in Fig. 2.1 (Reproduced from Ref. [2.6] with permission courtesy Kyoto University).

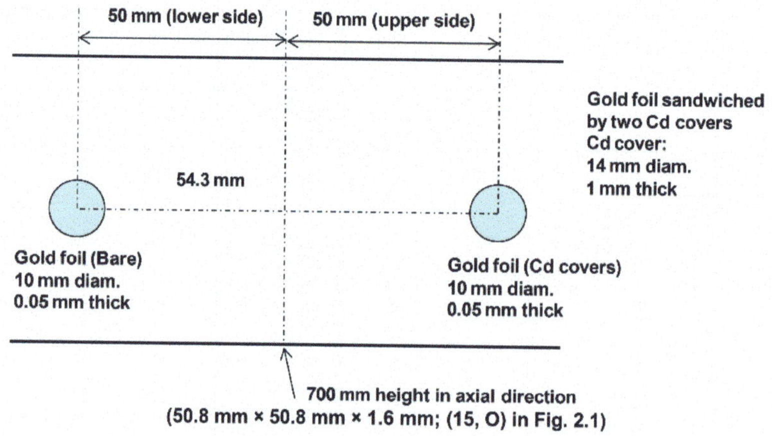

FIG. 2.12. Target configuration of the location of original target (Reproduced from Ref. [2.6] with permission courtesy of Kyoto University).

2.1.1.4. Experimental approach

This section summarises the detailed discussion in Ref [2.27]. The KUCA A-core was used to acquire neutron flux information from ^{115}In$(n, \gamma)^{116m}$In reactions, based on the assumption that the thermal neutron cross-sections of ^{235}U(n, f) are proportional to those of ^{115}In$(n, \gamma)^{116m}$In. A 1 mm diameter indium wire with a length of 800 mm was set vertically (14-13, P–A') as shown in Fig. 2.13. An aluminium foil and an indium foil, each with dimensions of 10 mm × 10 mm × 1 mm, were attached at the target location (15, H) to monitor protons and spallation neutrons through the ^{27}Al$(p, n+3p)^{24}$Na and ^{115}In$(n, n')^{115m}$In reactions, respectively. The normalized reaction rates methodology was used to compare the experimental and calculated results at the target and in the core. The reaction rates of ^{115}In $(n, n')^{115m}$In normalized by ^{27}Al$(p, n+3p)^{24}$Na and ^{115}In$(n, \gamma)^{116m}$In normalized by ^{115}In$(n, n')^{115m}$In were used to measure the neutron yield and neutron generation, respectively.

The proton beam had 100 MeV energy, 0.7 nA intensity, 20 Hz pulse rate, 100 ns pulse width, yielding 1.0×10^7 s^{-1} neutron generation. The irradiation time was approximately three hours.

The measured subcriticality of the core (2900 pcm) was obtained by fully inserting all control and safety rods, as shown in Fig 2.13, with their reactivity worths having been evaluated in advance by the rod drop and the positive period methods. The positive period method measures the reactivity change of a critical reactor in response to a reactivity addition from withdrawal of a control rod. The reactivity is plotted as a function of control rod position and is related to the stable reactor period though the inhour equation.

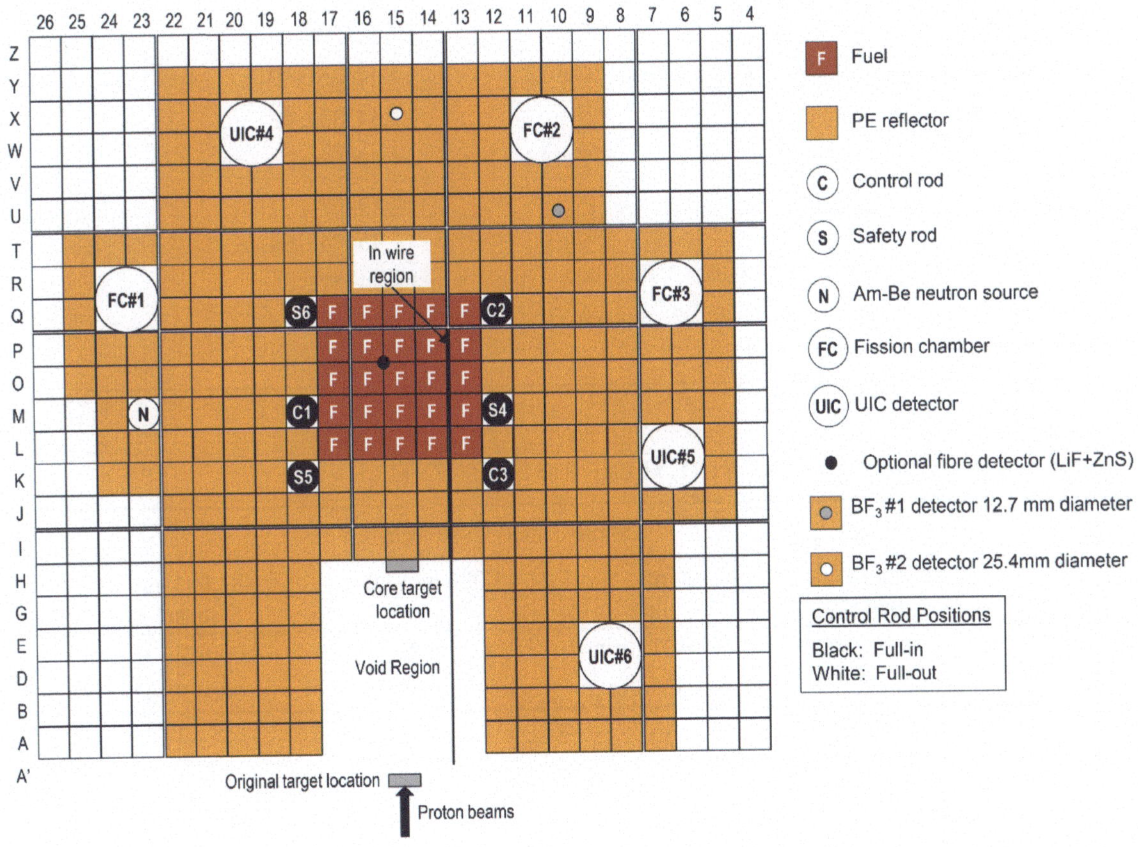

FIG. 2.13. Core configuration of ADS with 100 MeV protons for the indium wire reaction rate distribution measurements (Reproduced from Ref. [2.6] with permission courtesy of Kyoto University).

2.1.1.5. Description of KUCA components

Tables 2.1 to 2.4 show the atomic density data for KUCA reactor components that were needed for the neutronic calculations.

TABLE 2.1. ATOMIC DENSITIES OF CORE COMPONENTS [2.6]

Item	Isotope	Atomic density ($\times 10^{24}$/cm^3)
Control and safety rods	^{10}B	3.87448E-03
	^{11}B	1.68447E-02
	^{16}O	3.10787E-02
Aluminium sheath for the core element	^{27}Al	6.00385E-02
HEU fuel plate	^{234}U	1.13659E-05
	^{235}U	1.50682E-03
	^{236}U	4.82971E-06
	^{238}U	9.25879E-05
	^{27}Al	5.56436E-02
Beam tube component	^{54}Fe	3.55712E-03
	^{56}Fe	5.58391E-02
	^{57}Fe	1.28957E-03
	^{58}Fe	1.71618E-04
	^{50}Cr	7.51530E-04
	^{52}Cr	1.44925E-02
	^{53}Cr	1.64333E-03
	^{54}Cr	4.09060E-04
	^{58}Ni	5.10587E-03
	^{60}Ni	1.96674E-03
	^{61}Ni	8.54932E-05
	^{62}Ni	2.72597E-04
	^{64}Ni	6.94130E-05

TABLE 2.2. ATOMIC DENSITIES OF POLYETHYLENE REFLECTORS [2.6]

Isotope	Atomic density ($\times 10^{24}$/cm^3)			
	12.005 mm thick plate	6.300 mm thick plate	3.086 mm thick plate	483.085 mm long square rod
H	8.06560E-02	8.08711E-02	8.02167E-02	8.00083E-02
C	4.03280E-02	4.04356E-02	4.01084E-02	4.00042E-02

TABLE 2.3. ATOMIC DENSITIES OF FOILS [2.6]

Foil or wire	Isotope	Abundance (%)	Purity (%)	Atomic density ($\times 10^{24}$/cm^3)
Cd	^{106}Cd	1.25	99.99	5.39648E-04
	^{108}Cd	0.89	99.99	3.91477E-04
	^{110}Cd	12.51	99.99	5.59564E-03
	^{111}Cd	12.81	99.99	5.78677E-02
	^{112}Cd	24.13	99.99	1.10072E-02
	^{113}Cd	12.22	99.99	5.62419E-03
	^{114}Cd	28.72	99.99	1.33398E-02
	^{116}Cd	7.47	99.99	3.53884E-03
In	^{113}In	4.29	99.99	1.64406E-03
	^{115}In	95.71	99.99	3.66790E-02
Au	^{197}Au	100	99.95	5.90403E-02

TABLE 2.4. ATOMIC DENSITIES OF TARGET MATERIALS [2.6]

Target	Isotope	Abundance (%)	Atomic density ($\times 10^{24}$/cm^3)
W	^{180}W	0.12	8.04702E-05
	^{182}W	26.50	1.64613E-02
	^{183}W	14.31	9.00328E-03
	^{184}W	30.64	1.93829E-02
	^{186}W	28.43	1.81807E-02
Be	^{9}Be	100	1.23487E-01
Pb-Bi (44.5/55.5)	^{204}Pb	1.4	1.87461E-04
	^{206}Pb	24.1	3.25860E-03
	^{207}Pb	22.1	3.00266E-03
	^{208}Pb	52.4	7.15378E-03
	^{209}Bi	100	1.67670E-02

2.1.1.6. Dimension of solid targets

This section summarises the detailed discussion in Ref [2.27]. Two-layer targets with heavy (W, Pb-Bi) and light (Be) nuclide layers were used to study the neutron spectrum and neutron yield at the surface of the target. The targets were modelled using the MCNPX [2.28] and SRIM codes [2.29] with the JENDL/HE-2007 [2.30] library to investigate the high-energy neutron spectrum and determine the target thickness needed to fully stop incident protons within the heavy nuclide layer after passing through the Be layer. Table 2.5 gives the dimensions of the solid targets. The proton beam spot of 40 mm diameter adequately covered the targets.

TABLE 2.5. DIMENSION OF TARGETS [2.6]

Target[a]	Diameter (mm)	Thickness (mm)
W	50.0	12.0
W–Be	50.0	W: 12.0 Be: 10.0
Pb-Bi	50.0	18.0

[a] See also Fig. 2.9

2.1.1.7. Experimental configurations

The KUCA A-core was used, with fuel rods containing 60 fuel cells. The subcriticality was attained by full insertion and full withdrawal of control and safety rods, and substitution of fuel rods for polyethylene reflector rods, as shown in Fig. 2.14.

2.1.1.8. Experimental results

Table 2.6 shows the control rod positions at the critical state as a reference. This corresponds to configuration I-3 in Fig. 2.14, with control rod worths as shown in Table 2.7. Table 2.8 provides the detail of the detectors used to measure the reaction rate distribution. The results of reaction rate measurements are given in Table 2.9, and a graphical comparison of the normalized reaction rates is shown in Fig. 2.15.

TABLE 2.6. CONTROL ROD POSITIONS AT THE CRITICAL STATE [2.6]

Rod	Rod position (mm)
C1	676.92
C2	1201.16
C3	1201.43
S4	1200.00
S5	1200.00
S6	1200.00
Excess reactivity (pcm)	180

TABLE 2.7. CONTROL ROD WORTH [2.6]

Rod	Rod worth (pcm)
C1 (S4)	805
C2 (S6)	596
C3 (S5)	139

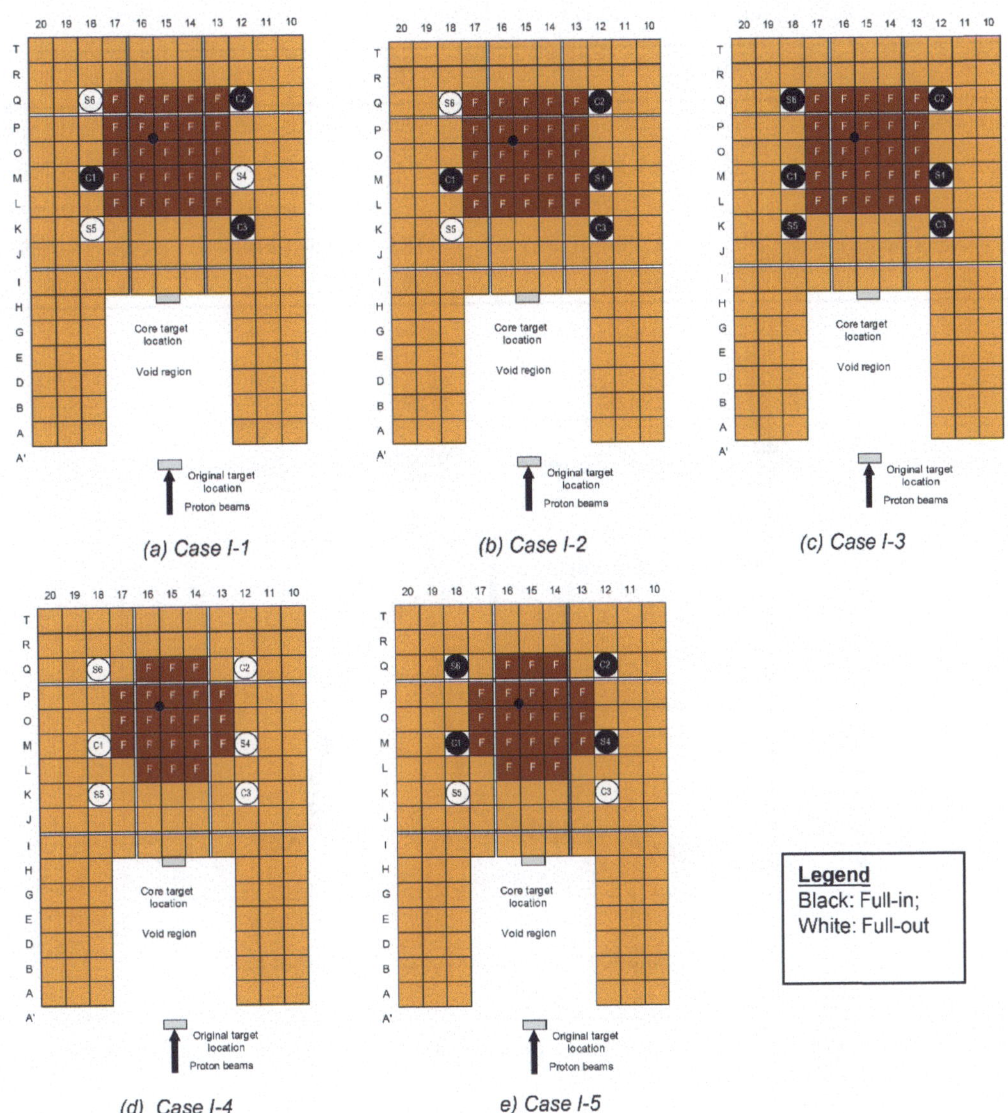

FIG. 2.14. Core configurations of ADS with 100 MeV protons used for subcriticality measurement by the PNS and the noise methods (Reproduced from Ref. [2.6] with permission courtesy of Kyoto University).

TABLE 2.8. DETECTORS FOR THE REACTION RATE MEASUREMENTS [2.6]

Reaction	Location	Dimension
^{115}In$(n, n')^{115m}$In	Target (Refer to Fig. 2.9)	Foil, 10 mm × 10 mm × 1 mm
^{115}In$(n, \gamma)^{116}$In	Core (Refer to Fig. 2.10)	Wire, 1 mm diameter, 800 mm long

TABLE 2.9. MEASURED REACTION RATES AND Cd RATIO [2.6]

	W target	W–Be target	Pb-Bi target
Au foil (Bare) ($s^{-1} \cdot cm^{-3}$)	(3.017 ± 0.734)E+06	(3.330 ± 0.026)E+06	(2.070 ± 0.158)E+06
Au foil (Cd covered) ($s^{-1} \cdot cm^{-3}$)	(2.240 ± 0.546)E+06	(2.406 ± 0.020)E+06	(1.497 ± 0.116)E+06
$^{115}In(n, n')^{115m}In$ ($s^{-1} \cdot cm^{-3}$)	(2.176 ± 0.078)E+05	(1.777 ± 0.015)E+05	(1.578 ± 0.051)E+05
$^{27}Al(n, n+3p)^{24}Na$ ($s^{-1} \cdot cm^{-3}$)	(1.036 ± 0.062)E+06	(1.185 ± 0.080)E+06	(0.928 ± 0.062)E+06
Cd ratio	1.35 ± 0.27	1.39 ± 0.02	1.38 ± 0.28

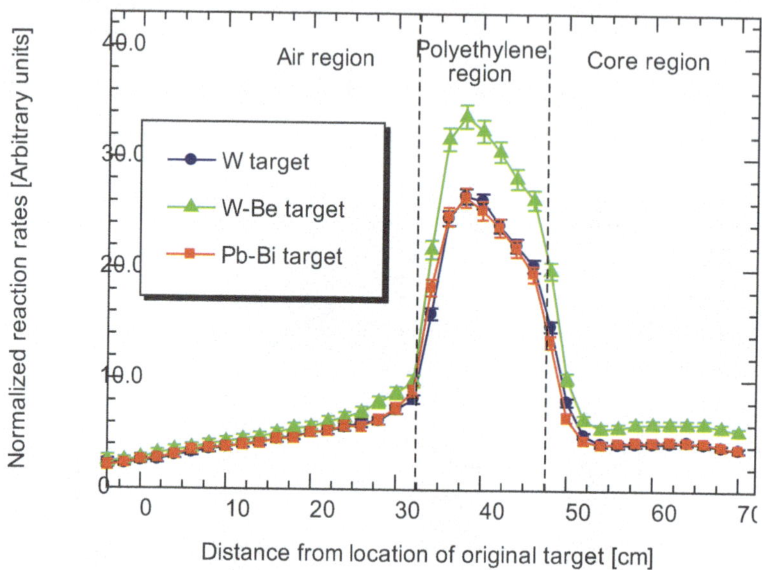

FIG. 2.15. Comparison of normalized reaction rates in ADS experiments (Reproduced from Ref. [2.27] with permission courtesy Taylor & Francis).

Either the pulsed neutron source (PNS) method or the noise method can be used to measure the kinetic parameters, as shown in Table 2.10.

TABLE 2.10. CORE CONDITIONS IN ALL THE CORES SHOWN IN FIG. 2.14 [2.6]

Case	Number of fuel rods	Rod insertion	100 MeV protons	
			PNS method	Noise method [a]
I-1 [b]	25	C1, C2, C3	Available	Available
I-2 [b]	25	C1, C2, C3, S4	Available	Available
I-3 [b]	25	C1, C2, C3, S4, S5, S6	Available	Available
I-4 [c]	21	All six rods withdrawn	Available	Available
I-5 [c]	21	C1, C2, S4, S6	Available	Available

[a] Noise method: Feynman-α method; [b] Subcriticality determined from the excess reactivity and the control rod worth; [c] Subcriticality determined from calculations.

2.1.2. Experiments: Phase II – Study on subcriticality measurements

2.1.2.1. Introduction

This section summarises the detailed information presented in Ref. [2.31]. The kinetic parameters of the ^{235}U-fuelled and Pb-Bi-zoned ADS core were studied experimentally using spallation neutrons from a solid Pb-Bi target irradiated with 100 MeV protons. The experiments used the KUCA A-core, see Section 2.1.1., with subcriticality between 1160 and 11 556 pcm. The variables examined included the detector position, neutron spectrum and subcriticality measurement methods. The effective delayed neutron fraction and the neutron generation time were calculated from the subcriticality and the prompt neutron decay constant using the PNS method. MCNP was used to model the neutron spectra produced from 100 MeV protons injected into the Pb-Bi target.

The nuclear data for Pb was validated using Monte Carlo analyses of sample reactivity experiments [2.32]. Earlier ADS experiments with 14 MeV neutrons used the PNS method [2.33] and the Feynman-α method [2.34], [2.35] to measure subcriticality at the MASURCA [2.3], YALINA [2.36] and VENUS-F [2.37], [2.38] facilities.

2.1.2.2. Description of the KUCA core

The KUCA A-core was configured as shown in Fig. 2.16 with three types of aluminium-sheathed fuel assemblies. The type "F" fuel assemblies contained 36 unit fuel cells with upper and lower polyethylene reflectors, as shown in Fig. 2.17. Table 2.11 shows the atomic densities for the moderator. The type "f" special fuel assemblies shown in Fig. 2.18 contained Pb-Bi reflector plates as described in Fig. 2.19 and Table 2.12. Type "16" partial fuel rods with 16 unit fuel cells as shown in Fig. 2.20 were used to reach a critical mass.

Subcriticality values ranging from 1160 to 2483 pcm were achieved by full insertion of control and safety rods in cases II-1, II-2 and II-3, and subcriticality values from 4812 to 11 556 pcm were achieved by replacing fuel assemblies with polyethylene assemblies in cases II-4, II-5 and II-6, see Table 2.13 and Fig. 2.21. In cases II-1, II-2 and II-3, the subcriticality was deduced experimentally by the rod drop method using the control and safety rod worths shown in Tables 2.14 and 2.15, and their calibration curves by the positive period method shown in Fig. 2.22. For cases II-4, II-5 and II-6, the subcriticality was calculated with MCNP6.1 [2.39] and the JENDL-4.0 library [2.40], because the substitution of the fuel rods changed the reactivities of the control and safety rods.

The 50 mm diameter × 18 mm thick Pb-Bi target was located inside the core at position (15, L) as shown in Fig. 2.16. The proton beam characteristics were 100 MeV energy, 0.7 nA intensity, 40 mm beam spot diameter, 20 Hz pulse repetition, 100 ns pulse width with a neutron yield of 1.0×10^7 s^{-1}.

2.1.2.3. Description of KUCA components

Tables 2.11 and 2.12 show the atomic density data used for neutronic calculations for the moderator and for the coatings of the Pb-Bi plate.

TABLE 2.11. ATOMIC DENSITY OF POLYETHYLENE MODERATOR "P" (FIGS 2.17–2.20) [2.6]

Element	Atomic density ($\times 10^{24}$/cm^3)	
	3.158 mm plate	254 mm square rod
H	7.77938E-02	7.97990E-02
C	3.95860E-02	4.08960E-02

TABLE 2.12. ATOMIC DENSITY OF COATING MATERIALS (FACE AND BASE) OVER LEAD-BISMUTH PLATE. [2.6]

Element	Isotope	Atomic density ($\times 10^{24}$/cm^3) Face; 1st layer	Atomic density ($\times 10^{24}$/cm^3) Base; 2nd layer
H	^{1}H	2.83301E-03	3.78991E-03
	^{2}H	4.25015E-07	5.64072E-07
C	–	2.27058E-03	4.03671E-03
O	^{16}O	2.06885E-03	4.58782E-04
	^{17}O	7.88039E-07	1.74753E-07
	^{18}O	4.14757E-06	9.19753E-07
Ti	^{46}Ti	2.50941E-05	5.36896E-05
	^{47}Ti	2.28983E-05	4.89918E-05
	^{48}Ti	2.31493E-04	4.95287E-04
	^{49}Ti	1.72522E-05	3.69116E-05
	^{50}Ti	1.69385E-05	3.62405E-05
Si	^{28}Si	–	1.86243E-05
	^{29}Si	–	9.43026E-07
	^{30}Si	–	6.25992E-07
S	^{32}S	4.16818E-04	–
	^{33}S	3.28997E-06	–
	^{34}S	1.84677E-05	–
	^{35}S	8.77326E-08	–
Ba	^{132}Ba	4.43050E-07	–
	^{134}Ba	1.06025E-05	–
	^{135}Ba	2.89167E-05	–
	^{136}Ba	3.44526E-05	–
	^{137}Ba	4.92619E-05	–
	^{138}Ba	3.14521E-04	–

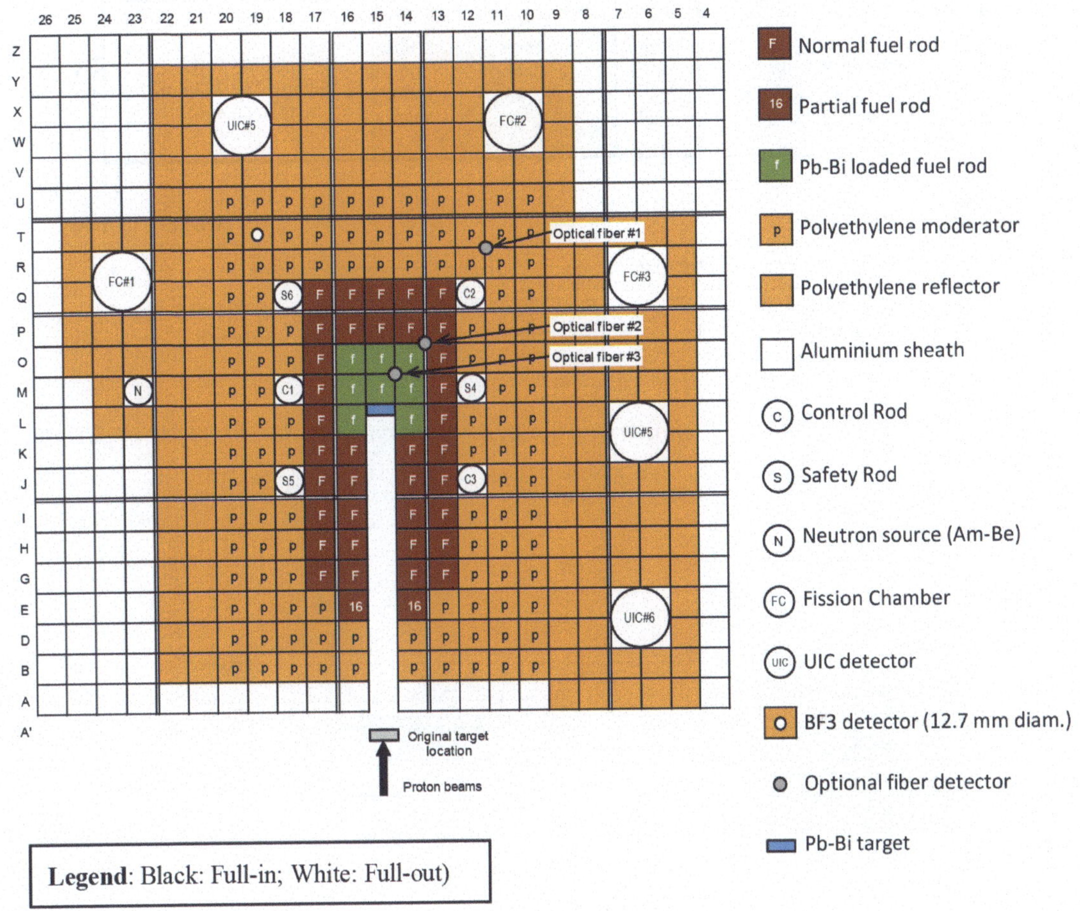

FIG. 2.16. Core configuration of ADS with 100 MeV protons used for the subcriticality measurements) (Reproduced from Ref. [2.31] with permission courtesy of Elsevier B.V.).

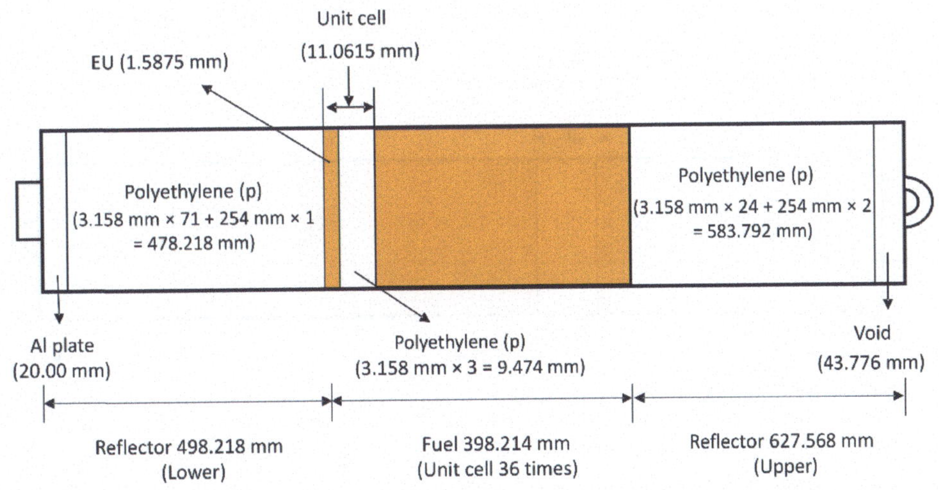

FIG. 2.17. Side view of fuel assembly "F" shown in Fig. 2.16 (Reproduced from Ref. [2.31] with permission courtesy of Elsevier B.V.).

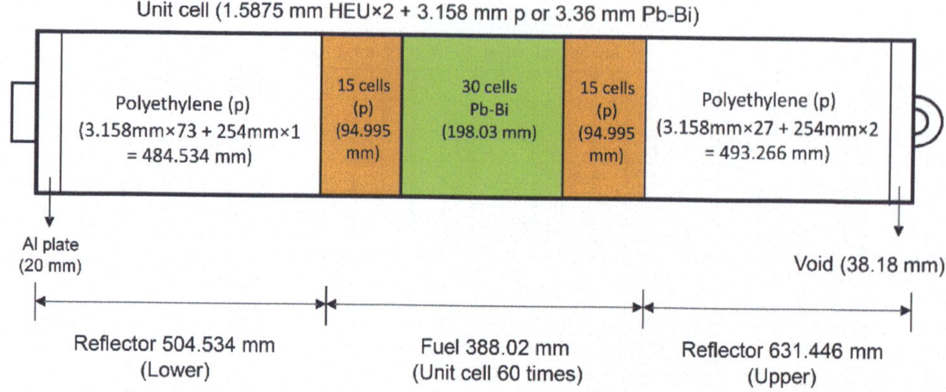

FIG. 2.18. Side view of fuel assembly "f" shown in Fig. 2.16 (Adapted from Ref. [2.31] with permission courtesy of Elsevier B.V).

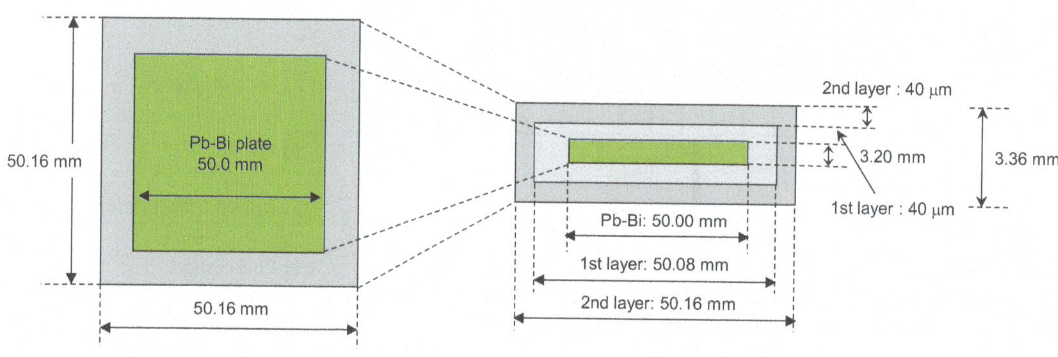

FIG. 2.19. Description of Pb-Bi plate covering over coating materials shown in Fig. 2.18 (Reproduced from Ref. [2.6] with permission courtesy of Kyoto University).

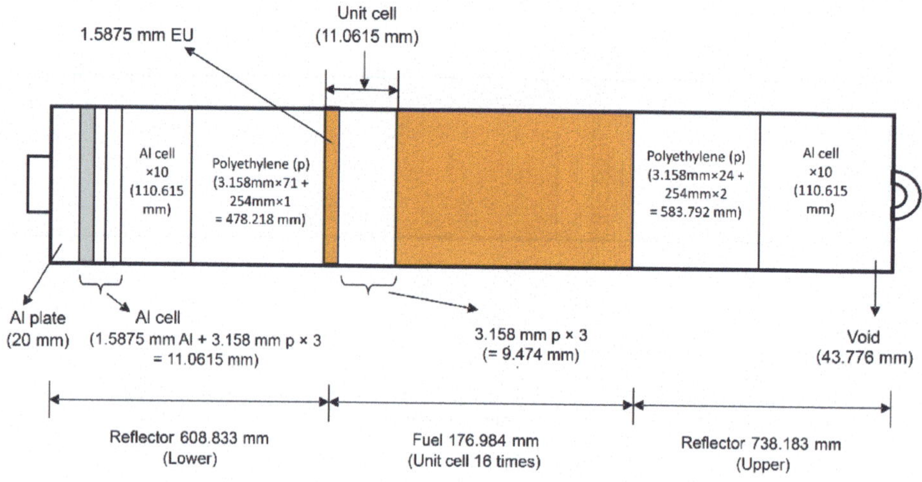

FIG. 2.20. Side view of fuel assembly "16" shown in Fig. 2.16 (Adapted from Ref. [2.31] with permission courtesy of Elsevier B.V.).

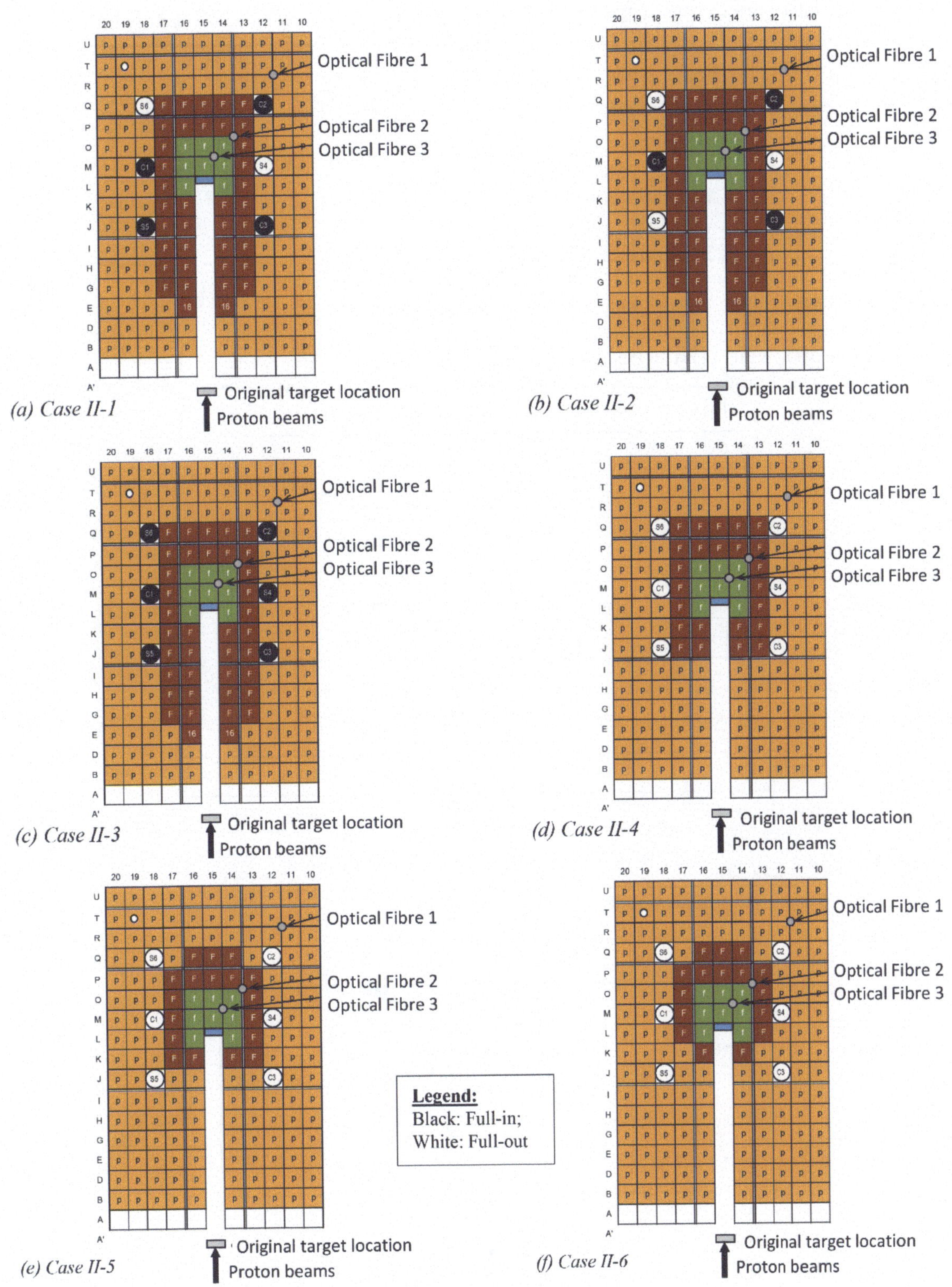

FIG. 2.21. Core configurations of ADS with 100 MeV protons used for subcriticality measurement by the PNS and the noise methods (Reproduced from Ref. [2.31] with permission courtesy of Elsevier B.V.).

TABLE 2.13. LIST OF CORE CONFIGURATIONS IN ALL THE CORES SHOWN IN FIG. 2.21 [2.6]

Case	Number of fuel rods	Rod insertion	100 MeV protons	
			PNS method [a]	Noise method [b]
II-1 [c]	46	C1, C2, C3	Available	Available
II-2 [c]	46	C1, C2, C3, S5	Available	Available
II-3 [c]	46	C1, C2, C3, S4, S5, S6	Available	Available
II-4 [d]	32	All six rods withdrawn	Available	Available
II-5 [d]	26	All six rods withdrawn	Available	Available
II-6 [d]	24	All six rods withdrawn	Available	Available

[a] PNS method: Pulsed neutron source method; [b] Noise method: Feynman-α method; [c] Subcriticality determined from the excess reactivity and the control rod worth; [d] Subcriticality determined from calculations.

TABLE 2.14. CONTROL ROD POSITIONS AT THE CRITICAL STATE [2.6]

Rod	Rod position (mm)
C1	1200.00
C2	525.39
C3	1200.00
S4	1200.00
S5	1200.00
S6	1200.00
Excess reactivity (pcm)	149±3

TABLE 2.15. CONTROL ROD WORTH [2.6]

Rod	Rod worth (pcm)
C1 (S4)	549±3
C2 (S6)	194±1
C3 (S5)	483±2

β_{eff} = 781±4 (p.m.)

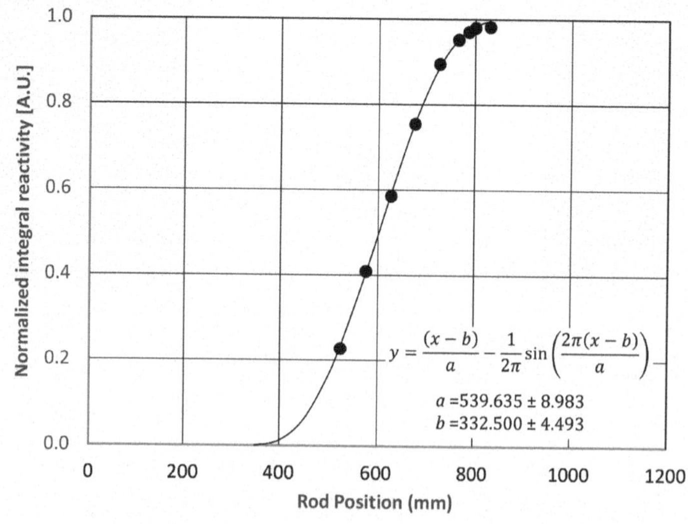

FIG. 2.22. Calibration curve of the C2 (also C1 and C3) control rod (Reproduced from Ref. [2.6] with permission courtesy of Kyoto University).

2.1.2.4. Measurements of kinetic parameters

This section is a summary of the more detailed information in Ref [2.31]. The experimental configuration for measurement of the kinetic parameters is shown in Fig. 2.16. 100 MeV protons were injected into the Pb-Bi target at location (15, L). Three optical fibre detectors [2.22] were used to record the time evolution of prompt and delayed neutrons, as shown in Table 2.16. The optical fibres were shaped with a mixture of ^{6}Li-enriched LiF and ZnS (Ag) scintillator pasted at the 1 mm diameter tip.

TABLE 2.16. POSITIONS OF OPTICAL FIBRES

Optical Fibre	Location	Description
#1	(12-11, T–R)	between the polyethylene moderator rods
#2	(14-13, P–O)	outside the Pb-Bi-zoned core
#3	(15-14, O–M)	inside the Pb-Bi-zoned core

The prompt neutron decay constant, α, was determined by fitting of the prompt neutron PNS measurements with the exponential function given in Eq. (2.1), as shown in Fig. 2.23.

$$N = C_{PNS} \cdot \exp(-\alpha t) + B_{PNS} \tag{2.1}$$

where N is the neutron counting rate, and C_{PNS} and B_{PNS} are constants obtained by least-squares fitting. The subcriticality $\rho_\$$ in dollar units was determined by the PNS method based on the area ratio method [2.33], using Eq. (2.2).

$$\rho_s = \frac{\rho}{\beta_{eff}} = -\frac{A_p}{A_d} \tag{2.2}$$

where ρ is the subcriticality in pcm units, β_{eff} the effective delayed neutron fraction, A_p the area of the decay curve by prompt neutrons and A_d the area of delayed neutrons.

To reduce the spatial higher-mode components of neutron flux, the extrapolated area ratio method according to Eq. (2.3) was used [2.41].

$$\rho_s = \frac{\rho}{\beta_{eff}} = -\exp(\alpha t_w) \frac{\int_{t_w}^{T} A_p(t)dt}{\int_{0}^{T} A_d(t)dt} \tag{2.3}$$

where t is the measurement time, and t_w the waiting time.

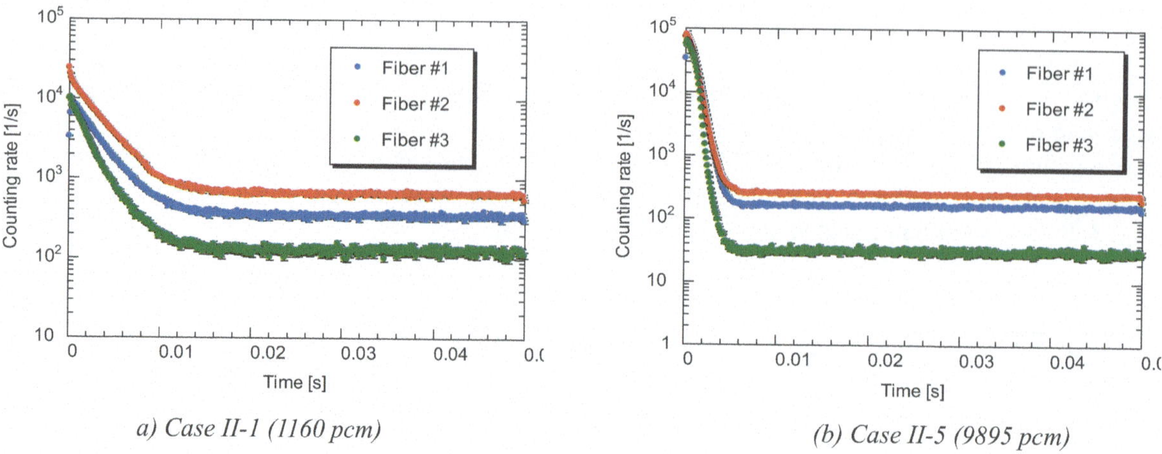

a) Case II-1 (1160 pcm)   (b) Case II-5 (9895 pcm)

FIG. 2.23. Time evolution of the counting rate by the PNS method (Reproduced from Ref. [2.31] with permission courtesy of Elsevier B.V.).

The Feynman-α method was also used to estimate the α value [2.34] with the use of neutron noise data such as those shown in Fig. 2.24. The α value was determined by least-squares fitting the Y value as a function of gate width t_g, taking delayed neutron effects into account [2.35], see Eq. 2.4.

$$Y = \frac{C_{\text{Noise}}}{\alpha^2}\left\{1 - \frac{1 - \exp(-\alpha t_s)}{\alpha t_s}\right\} + \frac{B_{\text{Noise}}}{t_g}\sum_{n=1}^{\infty}\left\{\frac{1}{n^4\alpha^2 T_0^2 + 4n^6\pi^2}\sin^2\left(\frac{n\pi}{T_0}W\right)\sin^2\left(\frac{n\pi}{T_0}t_s\right)\right\} \tag{2.4}$$

where C_{Noise} and B_{Noise} are the constants obtained from the noise data, and W is the pulse width as a fitting parameter and T_0 the proton beam pulse frequency (20 Hz). The method is described in more detail in Section 5.8.

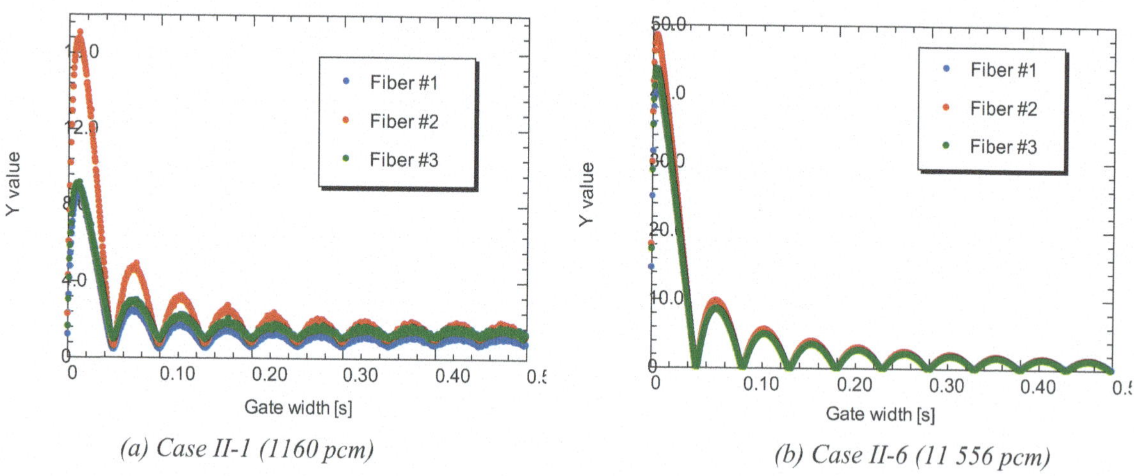

(a) Case II-1 (1160 pcm)   (b) Case II-6 (11 556 pcm)

FIG. 2.24. Neutron noise data by the Feynman-α method (Reproduced from Ref. [2.31] with permission courtesy of Elsevier B.V.).

26

2.1.3. Experiments: Phase III – Study on reaction rates

2.1.3.1. Introduction

This section is a summary of the detailed information presented in Ref. [2.42]. The Phase III studies examined the accuracy of reaction rates determined by the foil activation method in ADS driven with spallation neutrons, at varying subcriticalities near to the actual ADS operating mode k_{eff} = 0.95.

Uncertainties in the cross-section data and subcriticality levels were revealed by discrepancies between experiment and calculation in high-energy-threshold reactions in the MUSE experiments at MASURCA using 14 MeV neutrons [2.3]. The YALINA-booster experiments [2.36] revealed problems with precise reaction rate calculations of the high-energy neutrons in subcritical systems using MCNP5 [2.28] with common nuclear data libraries. At KUCA, ADS experiments were carried out with a 14 MeV pulsed neutron source [2.4], [2.43]–[2.45]. Good agreement between the experimental and MCNP reaction rate results confirmed the accuracy of the measured reaction rates for subcriticality between 50 and 1050 pcm [2.45].

This section describes the investigation of subcritical characteristics by measuring the reactor physics parameters of spallation neutrons and in fast neutron spectrum cores. A series of reaction rate experiments at various subcriticalities were undertaken at the KUCA facility [2.31], [2.46]–[2.49] with a HEU-fuelled and Pb-Bi-zoned core. The neutron spectrum information was acquired by the foil activation method in the Pb-Bi-zoned fuel region of the core, modelling the Pb-Bi coolant core locally around the central region. MCNP6.1 was used for the numerical analyses of reaction rates in high-energy thresholds [2.39] with JENDL/HE-2007, JENDL-4.0 [2.40], and JENDL/D-99 libraries [2.50]. The ratio of two different indium reaction rate distributions was used to analyse the neutron spectrum data from the activation foils.

2.1.3.2. Core configurations

The KUCA A-core shown in Fig. 2.16 was used, with subcriticality ranging between 2483 and 11 556 pcm, see Figs 2.21(c)–(f). The cores included normal fuel assemblies "F", partial fuel assemblies "16", Pb-Bi-zoned fuel assemblies "f", polyethylene moderator rods "p", and reflector rods. The "F" fuel assembly is shown in Fig. 2.17, and comprised 36 unit cells, each with an 1.579 mm HEU fuel plate and three polyethylene (p) plates totalling 9.474 mm, and upper and lower polyethylene reflectors. The partial fuel assemblies are shown in Fig. 2.20. These were similar to the F assemblies, but with 16 unit cells and longer polythene reflectors. The "f"-type Pb-Bi-zoned fuel assembly shown in Fig. 2.18 comprised 30 unit cells, each with two HEU plates and 3.36 mm Pb-Bi plate, and 30 unit cells with two HEU plates and a 3.158 mm polyethylene plate.

The subcriticality in pcm units in case II-3 (Fig. 2.21(c)) was determined from the control rod worth and its calibration curve measured by the rod drop method and the positive period method. The effective delayed neutron fraction β_{eff} (783 pcm) and the neutron generation time Λ (4.64×10^{-5}s), were calculated using MCNP6.1 with the JENDL-4.0 library. In cases II-3 to II-6 (Figs 2.21 (c)–(f)), the reference subcriticality in pcm units was obtained from MCNP6.1 eigenvalue calculations with the JENDL-4.0 library. The worth of the control and safety rods was estimated by substituting fuel assemblies with polyethylene rods.

2.1.3.3. Experimental settings

The experiments used spallation neutrons from a Pb-Bi target irradiated with 100 MeV protons from the FFAG accelerator. The proton beam had 100 MeV energy, 1.0 nA intensity, 20 Hz pulse frequency, 50 ns pulse width and 40 mm diameter spot size at the target.

A 10 mm × 10 mm × 1 mm indium foil was attached at the (15, L-M) boundary, as shown in Figs 2.21(c)–(f) to provide information on spallation neutrons for the reaction rate experiments, using the ^{115}In$(n, n')^{115m}$In reaction with a threshold energy of 0.4 MeV. Six activation foils were attached within the Pb-Bi-zoned core between the fuel assemblies at (15, M) and (15, O) to obtain neutron spectrum information over a wide range of energies, see Table 2.17. The foils were bare gold (Au-bare), gold sandwiched between two cadmium plates (Au-Cd), iron, aluminium, indium and nickel, arranged as shown in Fig. 2.25. The Au-bare foil was used as a reactor power normalization for the iron, aluminium, indium, and nickel foils. The Cd ratio was experimentally obtained using the Au-bare and Au-Cd foils. The neutron spectrum was modelled using MCNP at several positions around the Pb-Bi-zoned core as shown in Fig. 2.21 (c): the target location, fuel assembly (15, M), and two positions at the boundary between Pb-Bi-zoned and normal fuel regions, (14–13, P–O) and (14–13, L–K).

TABLE 2.17. MAIN CHARACTERISTICS AND DESCRIPTION OF ACTIVATION FOILS. [2.42]

Reaction	Dimension (mm)	Threshold (MeV)	Half-life	γ-ray energy (keV)	Emission rate (%)
^{197}Au$(n, \gamma)^{198}$Au (Bare and Cd^a covered)	8 mm diameter 0.05 mm thick	—	2.697 days	411.9	95.51
Cd covering plate	10 mm diameter 1 mm thick	—	—	—	—
^{115}In$(n, \gamma)^{116m}$In (wire)	1 mm diameter 680 mm long	—	54.12 months	1097.3 1293.54	55.7 85
^{115}In$(n, n')^{115m}$In (foil)	10×10×1	0.4	4.486 hours	336.2	45.08
^{58}Ni$(n, p)^{58}$Co	10×10×1	0.9	70.82 days	810.8	99.4
^{56}Fe$(n, p)^{56}$Mn	10×10×1	5.0	2.578 hours	846.8 1810.7	98.9 27.2
^{27}Al$(n, \alpha)^{24}$Na	10×10×1	5.6	14.96 hours	1368.6	100

^{a} Au foil (Cd covered) was sandwiched between two Cd plates (10 mm diameter × 1 mm thick).

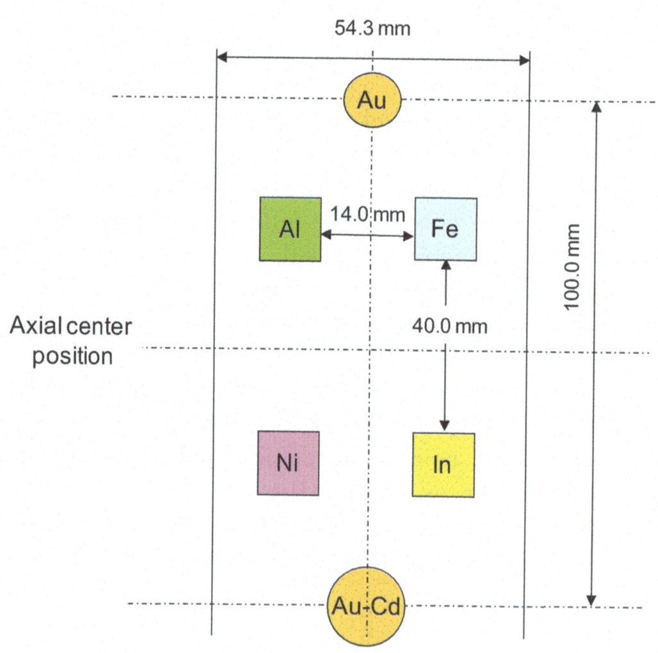

FIG. 2.25. Foil arrangement at the boundary between (15,M) and (15,O) shown in Figs 2.21(c)–(f) (Reproduced from Ref. [2.42] with permission courtesy of Taylor & Francis).

The neutron flux distribution was determined from the ^{115}In$(n, \gamma)^{116m}$In reaction using a 1mm diameter indium wire set in the gap between the Pb-Bi-zoned and normal fuel regions, (14-13, P-A), 700 mm above the bottom of the core. The normalization factor for spallation neutrons was obtained from the ^{115}In$(n, n')^{115m}$In reaction rates in the foil at the Pb-Bi target location on the (15, L)-(15, M) boundary.

2.1.4. Experiments: Phase IV – Study on effective delayed neutron fraction

2.1.4.1. Introduction

This section is a summary of the detailed discussion in Ref. [2.46] which describes the development of the methodology for measuring β_{eff} in an ADS driven by pulsed spallation neutrons.

The effective delayed neutron fraction, β_{eff}, can be estimated using the neutron yield and the fission spectrum of delayed neutrons [2.51]. Verification measurements of β_{eff} were made with the external neutron source set at the centre of the subcritical core to minimize higher-mode components of neutron flux [2.52]. The measurement methodology had been validated by work in the Fast Critical Assembly [2.53]–[2.60] and the Westinghouse Idaho Nuclear Company slab tank assembly [2.60]. Before operation of the ADS, the subcriticality in dollar units measured using a pulsed neutron source was compared with the subcriticality in pcm units calculated from β_{eff}. The β_{eff} measurements provided an additional verification of the methodology and of the reliability of the underlying nuclear data.

To simplify the β_{eff} measurement, the Nelson number method [2.61] based on the Rossi-α method [2.62] was used. β_{eff} is obtained from the source intensity, measured subcriticality and two calculated parameters to correct the location of the external neutron source. To apply the Nelson number method, a stable neutron source, such as ^{252}Cf, was assumed to be set at the centre of the near-critical core. In previous studies [2.63], β_{eff} was determined using an Am-Be neutron source external to the core at a subcriticality level of $k_{eff}= 0.97$.

This section describes the application of the Rossi-α method [2.35] to determine the value of β_{eff} value using both calculation and experiments. The KUCA ADS was configured with a pulsed spallation neutron source and $k_{eff}= 0.95$. Subcriticality was determined by the area ratio method and neutron noise data [2.41], and Monte Carlo calculations were used to determine the correction parameters.

2.1.4.2. Measurement methodologies

The Nelson number method based on the Rossi-α method [2.61] simplifies the measurement of β_{eff} by reducing the number of experimentally challenging parameters that are needed, such as detection efficiency, fission rate at the core centre, and the number of neutrons. However, β_{eff} needs to be measured near to the critical state with the external neutron source at the centre of the core. This approach can also be applied to PNS experiments using a modification of the Rossi-α method [2.35], [2.64].

2.1.4.3. ADS experiments

The KUCA A-core shown in Fig. 2.26 was used to measure β_{eff}. This core comprised 25 "F"-type fuel rods surrounded by polyethylene reflectors. The fuel rods had with unit cells consisting of a 1.589 mm HEU plate and a 3.186 mm polyethylene moderator plate, with upper and lower polythene reflectors, as shown in Fig. 2.3. The core spectrum was relatively hard with a hydrogen/uranium ratio of approximately 50.

Spallation neutrons were obtained by injecting 100 MeV protons from the FFAG accelerator into a 12 mm thick tungsten target located at position (15, H) in Fig. 2.26. The proton beam had an energy of 100 MeV, pulse frequency of 20 Hz, and 50 mm diameter beam spot size. The beam current was 50 pA for cases IV-1–IV-3 which are shown in Figs 2.27(a)–(c) and 75 pA for cases IV-4–IV-7 as shown in Figs 2.27(e)–(g).

The time evolution of the neutron flux following the injection of spallation neutrons was measured using a BF$_3$ detector installed at position (10, U) and an optical fibre detector installed at position (15–16, M–N), see Fig. 2.26, The fibre was coated with a mixture of 95% enriched ^{6}LiF and ZnS(Ag) [2.22] to enable scintillation detection of ^{6}Li(n,t)^{4}He reactions. For each case, the PNS experiments acquired neutron signals for 10 minutes for analyses by the area ratio method and the Rossi-α method. The neutron source pulse widths shown in Table 2.18 were calculated by fitting the detector signals with Eq. (2.5). The pulse width measured by the BF$_3$ detector was greater than that measured by the optical fibre because pulse width is dependent on the distance from the neutron source.

$$P_v(t_1, t_2)dt_1 dt_2 = B dt_1 dt_2 + D \sum_{n=1}^{1000} \frac{1}{\alpha^2 + \left(\frac{2n\pi}{T_0}\right)^2} e^{-\left(\frac{2n\pi}{T_0}\right)^2 \sigma^2} \cos\left\{\left(\frac{2n\pi}{T_0}\right)(t_2 - t_1)\right\} d\pi_1 dt_2 \qquad (2.5)$$

The derivation of this equation and the meaning of the individual terms are given in Chapter 5.

(a) Subcriticality

Subcriticality was obtained by full insertion of control and safety rods, and by substituting fuel assemblies with polyethylene rods, see Table 2.18. The excess reactivity was measured by the positive period method and control rod worth was measured by the rod drop method. In cases IV-1 to IV-3, the subcriticality was measured using the control rod worths and their calibration curve as determined by the positive period method. In cases IV-4 to IV-7, the area ratio method was used [2.41] because the substitution of fuel assemblies with polyethylene rods as shown in Fig. 2.27 changed the control rod worths.

TABLE 2.18. PULSE WIDTH AND MEASURED SUBCRITICALITY IN DOLLAR UNITS BY FUEL ROD SUBSTITUTION AND CONTROL ROD INSERTION [2.6]

Case	№ of fuel rods	Rod insertion pattern	Pulse width (s):		Subcriticality ($)	Methodology
			in optical fibre	in BF3 detectors		
IV-1	25	C1, C2, C3	(2.10 ± 0.05)E-04	(4.30 ± 0.20)E-04	1.20 ± 0.03	control rod worth
IV-2	25	C1, C2, C3, S4	(1.25 ± 0.05)E-04	(3.10 ± 0.20)E-04	2.01 ± 0.06	control rod worth
IV-3	25	C1, C2, C3 S4, S5, S6	(1.00 ± 0.05)E-04	(2.90 ± 0.20)E-04	2.66 ± 0.07	control rod worth
IV-4	21	—	(1.00 ± 0.05)E-04	(2.70 ± 0.10)E-04	3.32 ± 0.02	area ratio method
IV-5	21	C1, C2, S4, S6	(5.12 ± 0.25)E-05	(2.20 ± 0.05)E-04	6.13 ± 0.05	area ratio method
IV-6	19	—	(5.00 ± 0.42)E-05	(2.75 ± 0.02)E-04	6.80 ± 0.05	area ratio method
IV-7	19	C1, C2, S4, S6	(5.00 ± 0.42)E-05	(2.60 ± 0.02)E-04	10.16 ± 0.09	area ratio method

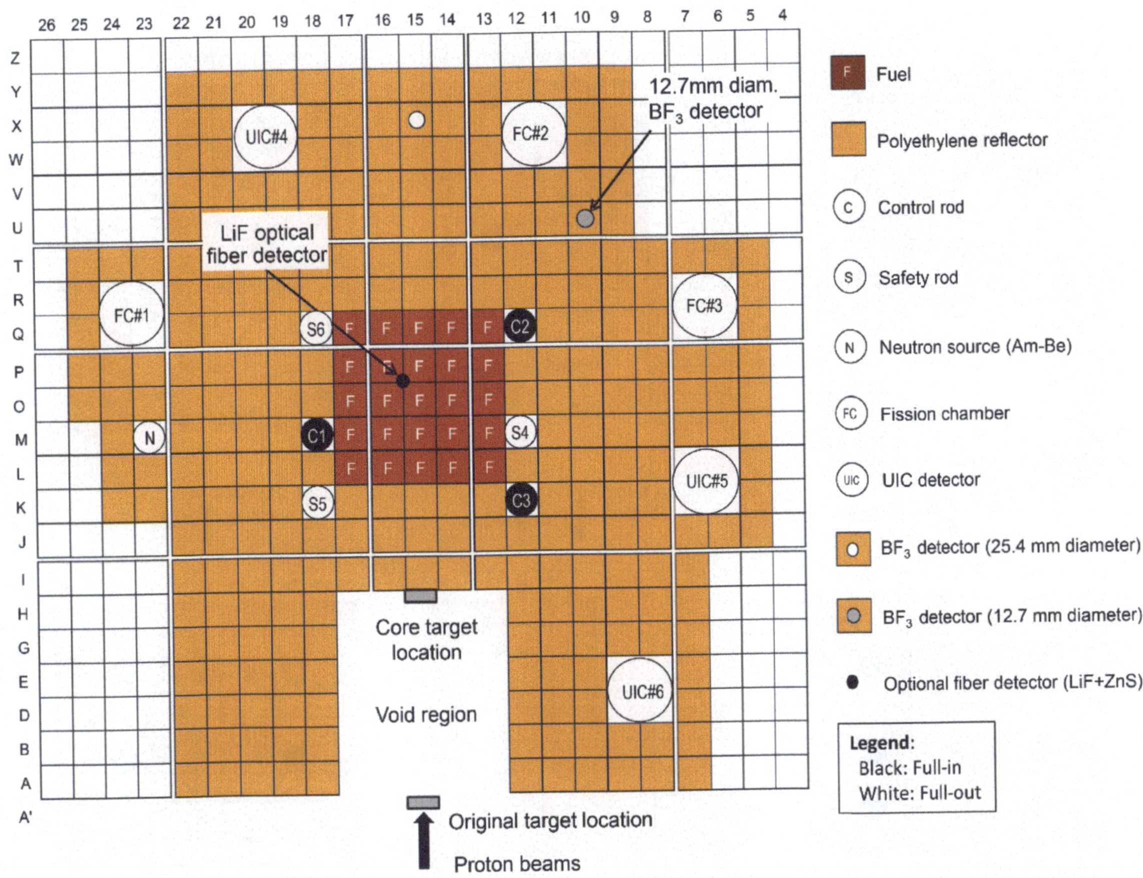

FIG. 2.26. KUCA core configuration for measurement of the effective delayed neutron fraction (Adapted from Ref. [2.6] with permission, courtesy of Kyoto University).

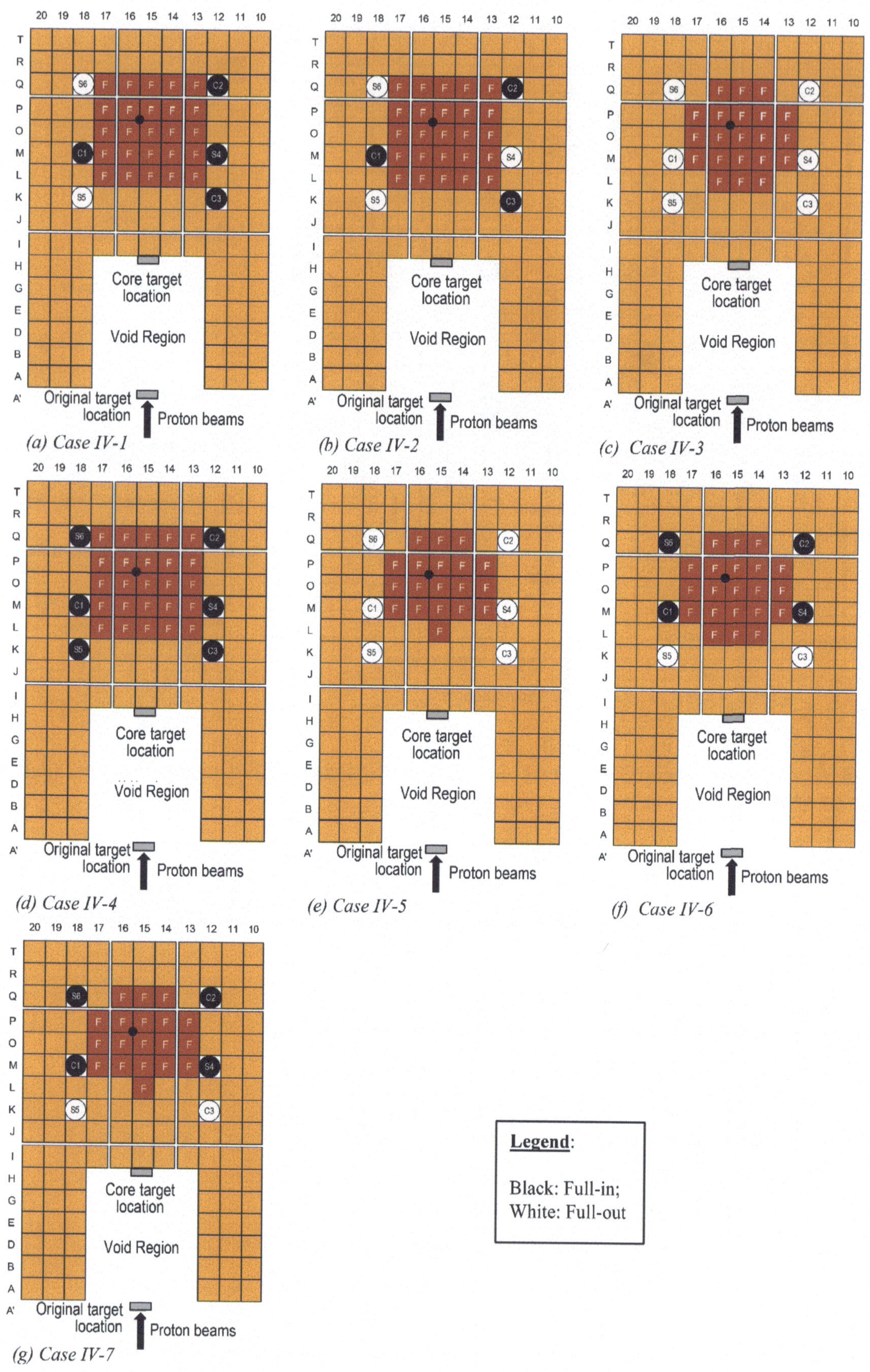

FIG. 2.27. KUCA core configurations for subcriticality measurements by the PNS and the noise methods (Reproduced from Ref. [2.6] with permission courtesy of Kyoto University).

2.1.5. Analyses: Phase I – Study of W, W–Be and Pb-Bi targets

2.1.5.1. Modelling objectives, codes, and data libraries

(b) KUCA (Japan)

This subsection is summarised from Ref. [2.27]. MCNPX simulations used JENDL/HE-2007 for high-energy protons and high-energy neutrons, JENDL-4.0 [2.40] for transport, and JENDL/D-99 [2.50] for reaction rates. The reaction rates were calculated from the tallies of the activation foils. The foils had a non-negligible reactivity and so were included in the simulated geometry and Monte Carlo transport calculations. The eigenvalue calculations used 1000 active cycles of 10^5 histories.

(a) SNU (Republic of Korea)

The effective multiplication factor, the control rod worth, and the excess reactivity were calculated by McCARD [2.65] for the base core configuration of Fig. 2.13 with control rod positions at the critical state, as shown in Table 2.6. McCARD k-eigenvalue calculations were performed with 100 inactive and 500 active cycles on 10^5 histories per cycle using JENDL-4.0 cross-section libraries. The prompt neutron decay constants of the five subcritical configurations in Fig. 2.14 were calculated by the McCARD α-static calculations using the α-iteration method [2.66]. The McCARD α-static calculations used 1000 iterations with 10^4 histories per iteration using the JENDL-4.0 cross-section libraries.

(b) Karlsruhe Institute of Technology (KIT) (Germany) [2.67]

Static analyses were performed by means of the European Reactor ANalysis Optimized Calculation System (ERANOS) [2.68] and the PARallel, TIme-dependent SN (PARTISN) v5.97 [2.69]. Transient analyses were performed using KIN3D/VARIANT [2.70] and KIN3D/PARTISN [2.71], [2.72] as extended by KIT. KIN3D is the kinetics and perturbation extension of the VARIANT/TGV nodal transport code [2.72], [2.73] included in the ERANOS code system.

The criticality level and the kinetic parameters from the 3-D (XYZ) ERANOS and PARTISN models of case I-4 were benchmarked using MCNPX2.7 [2.74] and the reference value from KUCA (MCNP6). Transport calculations used 40-group effective macroscopic cross-sections, with 10 groups below 1 eV. These cross-sections were generated from an ultra-fine-group European Cell Code (ECCO) [2.75] data library with 1968 energy groups and the JEFF3.1 reference data library. The VARIANT transport solver of ERANOS was used. The MCNPX2.7 calculations were performed using the JEFF3.1, JEFF3.2, and JENDL4.0 point-wise cross-section libraries.

(c) Politecnico di Torino (Italy)

The pulsed experiments carried out during Phase I were analysed using the MAρTA approach described in Section 5.2. The results were compared with those obtained using the area ratio method, and with the results provided by other CRP participants.

(d) ANL (USA)

Starting from the reference subcritical KUCA A-core case I-4 shown in Fig. 2.14(d), the following studies were carried out by changing the positions of detectors to those as shown in Fig. 2.28:

- Advances in the computation of the Sjöstrand, Rossi, and Feynman distributions [2.76],
- Calculation of the cross and auto power spectral densities for low neutron count rates from pulse mode detectors [2.77],
- Calculation of the prompt neutron decay constant for the KUCA facility driven by a stationary or pulsed external neutron source [2.79],
- Dead-time and spatial corrections for the KUCA subcritical assembly experiments [2.79].

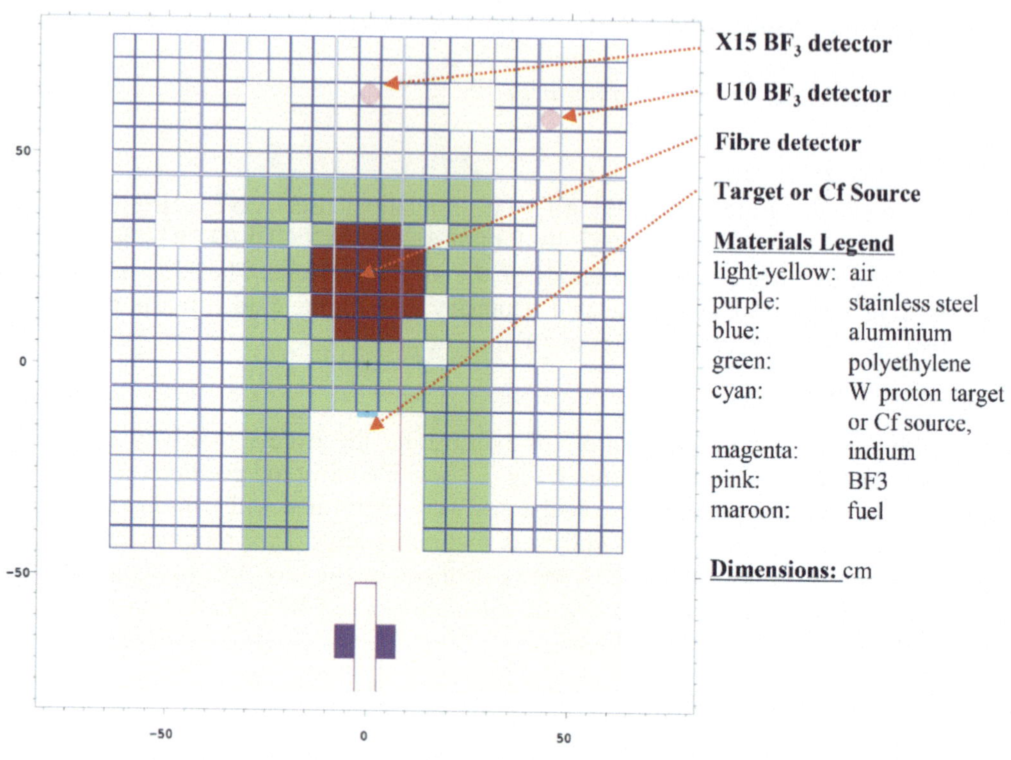

FIG. 2.28. Horizontal section of the KUCA facility; (Reproduced from Ref [2.78] with permission courtesy of Taylor & Francis).

i.   Advances in the computation of the Sjöstrand, Rossi, and Feynman distributions

This section is summarised from the detailed discussion in Ref. [2.76]. The proton beam driving the KUCA had 100 MeV energy, 100 μs pulse duration, 20 Hz frequency, and 1 nA current. The beam passed through a 3 mm thick stainless-steel window and then impinged on a 12 mm thick tungsten target. The 93.3% enriched HEU fuel had 0.0573 atoms/(b·cm) nominal density. In the numerical model, the nominal density of the fuel was reduced by 5% to 0.0544 atoms/(b·cm)) to improve the match with experimental data. BF₃ detectors were located at (10, U) and (15, X).

Previous studies modelled KUCA using MCNP6 (version 6.1.1beta) [2.80] and MCNPX (version 2.7.0) [2.74] with the JENDL-4.0 [2.40], JEFF-3.1.0 [2.81], and JEFF-3.2.0 [2.82] nuclear data libraries. The PTRAC and F8 cards were used for time dependent MCNP6 calculations. The volume of the (10, U) detector in the MCNP6 model was increased to reduce the statistical error. The ACE format of the JENDL-4.0 nuclear data library was prepared by the NJOY nuclear data processing system (version 2012) [2.83], which differed slightly from the earlier NJOY (version 99) [2.84].

KUCA was also modelled with the deterministic codes ERANOS (version 2.2) [2.85] and PARTISN (version 5.97) [2.81], [2.85]. The same spatial mesh with a maximum size of 5cm was used with both codes. The variational node method and the SP3 simplified transport approximation of the VARIANT module were used in the ERANOS calculation while the PARTISN calculations used the S4 discrete ordinates approximation. The air gaps between the aluminium frames, of either 1mm or 3mm, were not explicitly modelled in ERANOS and PARTISN, but were homogenized with aluminium when setting up the cross-sections and the spatial mesh values. This avoided the use of near-zero cross-sections in regions close to the core, which would have challenged the VARIANT calculations. As indicated in Figs 2.29 and 2.30, the zones for the fuel assemblies 3, 3B, and 3C, the polyethylene blocks 4, 4B, and 4C, the polyethylene plates 5 and 5B, and the control rods 2 and 2B were modelled. The suffixes 'B' and 'C' indicate a boundary or corner relative to the 3 mm air gap, see Fig. 2.31. The BF₃ detectors, indium wire, and target were not modelled. The control rods zones 2 and 2B were treated as a homogeneous aluminium/air mixture with a 10.3% aluminium volume fraction rather than just air. This corresponds to the volume fraction of the aluminium frame in a fuel assembly. The PARTISN source code was modified by KIT to include delayed neutron transport. The deterministic time dependent calculations modelled a 14.1 MeV neutron source instead of a 100 MeV proton source.

The macroscopic cross-sections with 40 energy groups, including 10 groups with energy less than 1 eV, were generated by the ERANOS ECCO module [2.75] using an ultra-fine, 1968 energy groups nuclear data library based on JEFF3.1.0 and the probability table method [2.86]. Figure 2.32(a) shows the ECCO model used to generate the macroscopic cross-sections of fuel zone 3. The model had a 2-D geometry with reflected boundary conditions, and a 0.47 cm homogenised aluminium-air region around the fuel and polyethylene plates with an aluminium volume fraction of 32.7%. The width of the homogenized zone takes into account the 15.6% air and aluminium volume fraction of the fuel assembly.

The calculations of the infinite multiplication factors were benchmarked using an equivalent MCNPX model, as shown in Table 2.19. The deterministic and Monte Carlo calculations showed good agreement within 60 pcm, and the variation between the ECCO results for 1968, 172, and 40 energy groups was less 30 pcm.

The same approach was used to generate the macroscopic cross-sections for the fuel zones 3B and 3C with aluminium and air volume fractions adjusted to account for differences in the air gaps between the fuel assemblies, see Fig. 2.32 (b) and (c). For example, the air gap between assemblies at positions (15, O) and (16, O) was 1 mm, the air gap between positions (15, P) and (15, Q) was 3 mm, and the fuel assembly at (16, Q) had a 3 mm air gap on two sides.

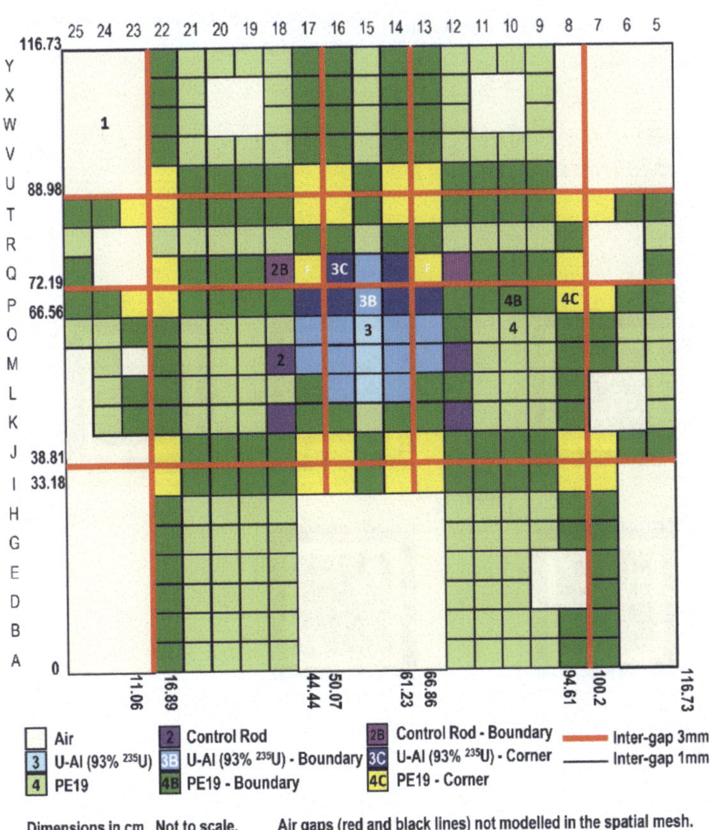

FIG. 2.29. Vertical section of the VARIANT and PARTISN model of the KUCA facility; (Reproduced from Ref. [2.76] with permission courtesy of Elsevier B.V.).

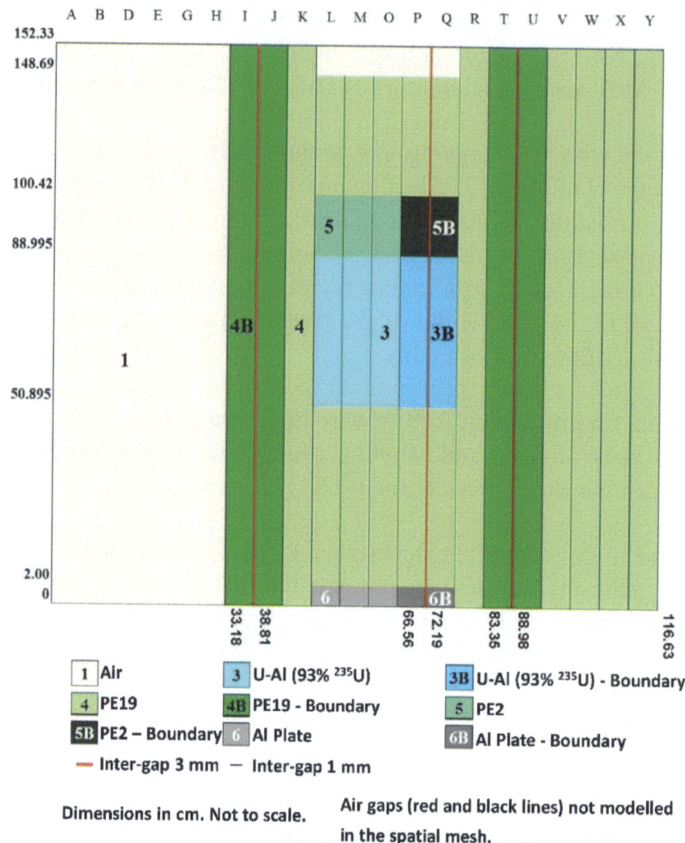

FIG. 2.30. Horizontal section of the VARIANT and PARTISN model of the KUCA facility; (Reproduced from Ref. [2.76] with permission courtesy of Elsevier B.V.).

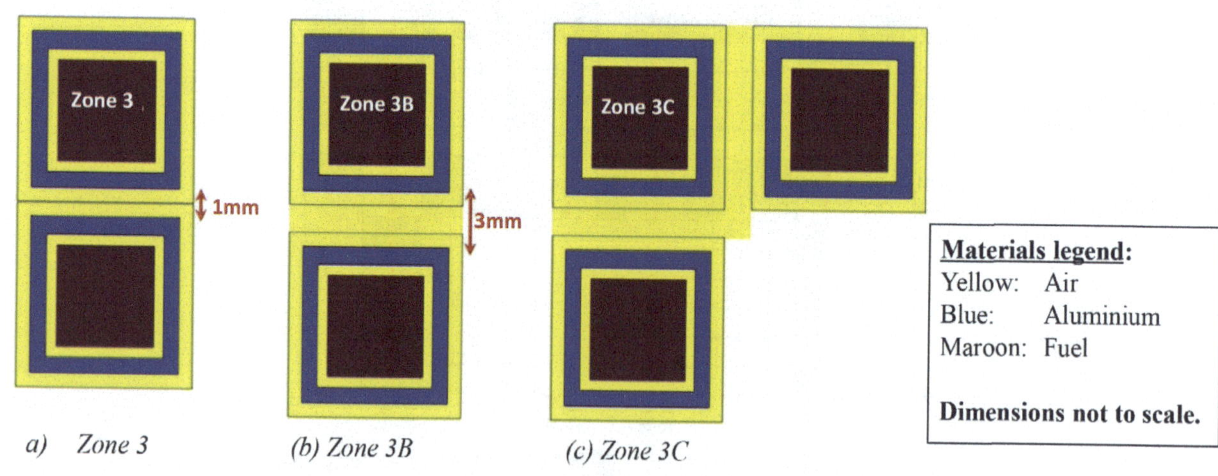

FIG. 2.31. Vertical cross-section of the actual fuel assemblies showing the air gaps between the aluminium frames (Reproduced from Ref. [2.76] with permission courtesy of Elsevier B.V.).

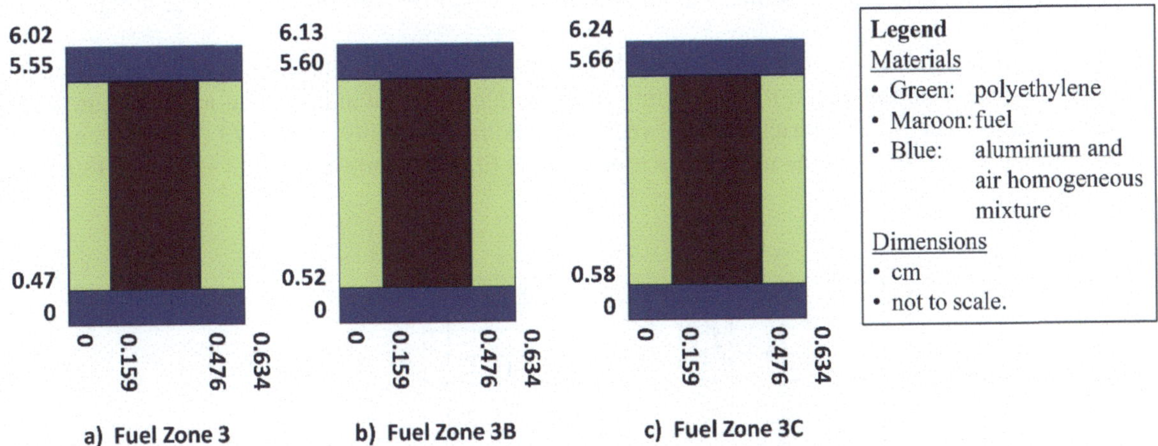

FIG. 2.32. Cross-section of the 2-D ECCO models of the fuel assembly zones (Reproduced from Ref. [2.76] with permission courtesy of Elsevier B.V.).

TABLE 2.19. ECCO AND MCNPX INFINITE MULTIPLICATION FACTORS FOR THE FUEL ASSEMBLY ZONE 3 (LEFT PLOT OF FIGS 2.32 AND 2.29) [2.76]

| | MCNPX | ECCO | | |
	Continuous Energy Heterogeneous 2D Model	1968 Energy Groups Heterogeneous 2D Model	172 Energy Groups Heterogeneous 2D Model	40 Energy Groups Homogeneous 2D Model
Infinite Multiplication Factors	1.83496 ± 0.00040	1.83523	1.83547	1.83520
Difference from MCNPX (pcm)	0	27	51	24

An ECCO model with 1-D cylindrical geometry and buckling was used to collapse the macroscopic cross-sections of reflector zones 4 and 5 from 1968 energy groups to 40. A fixed-source calculation and the neutron spectrum from the fuel assembly were used to generate the cross-sections. In zone 4 the materials and radii were: polyethylene up to 1 cm, polyethylene up to 2 cm, polyethylene up to 2.87 cm, air up to 2.89 cm, aluminium up to 3.06 cm, and air up to 3.12 cm. For the zones 4B and 5B, the 3 mm air gap was modelled by increasing the outer radius from 3.12 cm to 3.148 cm, and for zone 4C which had 3mm air gaps on two sides, to 3.176 cm. Table 2.20 shows the aluminium and air volume fractions used for various zones in the ECCO models.

TABLE 2.20. VOLUME FRACTIONS (%) USED IN THE INFINITE MEDIUM ECCO CALCULATIONS [2.76]

Zone	Air 1	Control rod 2	Control rod zone 2B	Aluminium 6	Aluminium 6B	Aluminium 6C
Air	100	89.64	89.83	3.6	5.3	7
Aluminium	0	10.36	10.17	96.4	94.7	93

ii.   Cross and auto power spectral densities calculation for low neutron counting from pulse mode detectors

The proton beam from the accelerator had 100 MeV energy, 100 μs pulse duration, 20 Hz pulse frequency, and 1 nA current. The beam passed through a 3 mm thick stainless-steel window, then impinged on a 12 mm thick tungsten target. The 93.3% enriched HEU fuel had a nominal density of 0.0573 atoms/(b·cm). A BF_3 detector was in location (10, U) in the moderator zone and the fibre detector in the fuel zone, see Fig. 2.33.

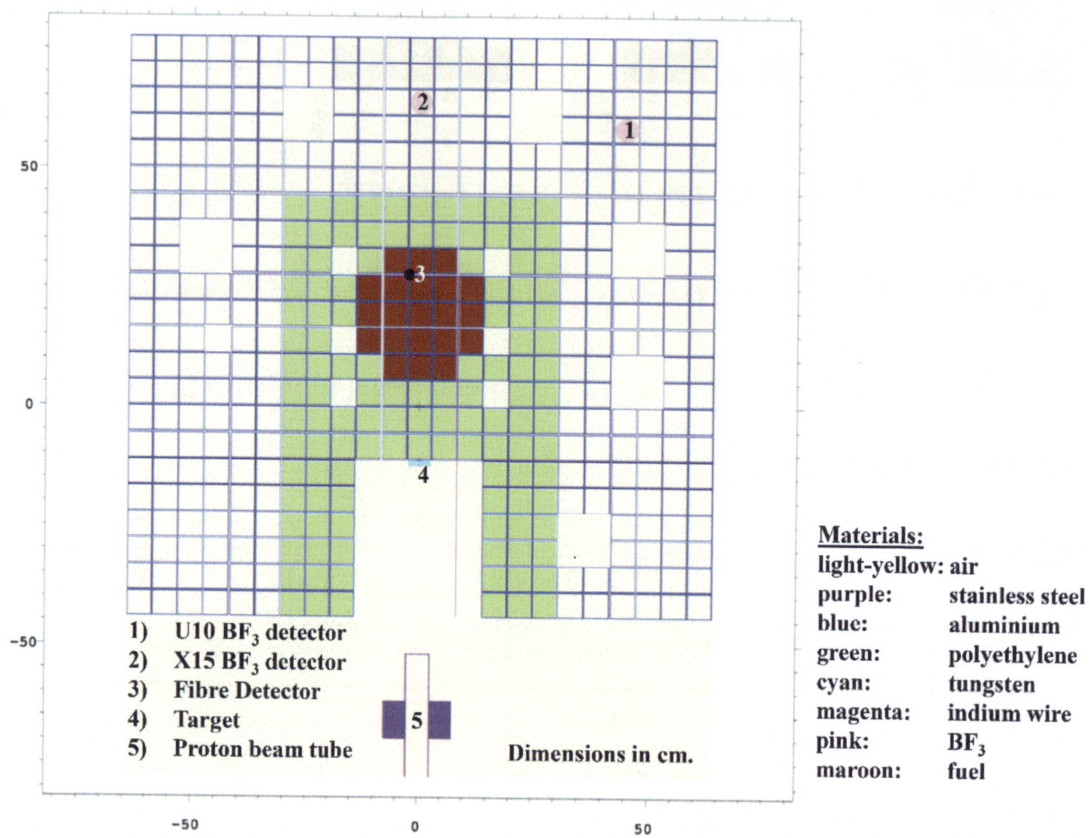

FIG. 2.33. Vertical section of the KUCA facility. Reproduced from Ref. [2.77] with permission courtesy of Elsevier B.V.).

iii. Calculation of the prompt neutron decay constant for the KUCA facility driven by a stationary or pulsed external neutron source

The prompt neutron decay constant of the KUCA A-core was calculated by changing the positions of detectors starting from the sub-critical reference core I-4 shown in Fig. 2.14 (d).

iv.   Dead-time and spatial corrections for the KUCA subcritical assembly experiments

The dead-time and spatial correction studies were carried out in the KUCA A-core by changing the positions of detectors starting compared to the sub-critical reference core I-4 shown in Fig. 2.14 (d).

(e) BARC (India)

The KUCA core configurations were modelled and analysed using the 1-D transport theory lattice code WIMSD4 [2.87], the 2-D transport theory lattice code TWOTRAN [2.88], [2.89] and the 3-D multigroup diffusion theory code KINFIN [2.90]. 3-D core calculations were performed using neutron cross-sections in three energy groups. The WIMSD4 code and the ENDF/B-VI cross-section data library (69 neutron energy groups) were used to generate three-group cross-sections (macroscopic) of all core materials at lattice level, which were subsequently used as input to TWOTRAN and KINFIN for assembly and core level calculations, respectively.

The objective of the KUCA analysis was to investigate the validity of KINFIN to analyse static and kinetic experiments in the determination of subcriticality. Using KINFIN, prompt fundamental α eigenvalue and K eigenvalue were evaluated.

KINFIN is a 3-D space time kinetics code which solves the few-group time dependent diffusion equation with delayed neutrons. KINFIN solves the multigroup neutron diffusion equation by finite difference methods in Cartesian geometry, calculates core reactivity, core fission power distribution, and multigroup fluxes in and around the core region. It has several modules for different evaluations. KINFIN was used to calculate the prompt and delayed α-modes, the λ-modes, as well as fixed source and time dependent problems.

One of the modules in KINFIN was related with the evaluation of the prompt α mode. The time eigenvalue equation was used to describe the fast evolution of the system as well as to characterize the reactivity of the system. The fundamental mode or time eigenvalues was defined by assuming exponential time behaviour of the neutron flux in the transport/diffusion equation. The solutions of the time dependent neutron diffusion/transport equation are known as α modes, with the eigenvalues α. The theoretical model used in KINFIN to evaluate the fundamental α mode is described in Ref. [2.89].

KINFIN was used for the analysis of decay of a single pulse of neutron using its space time kinetics module. In this module, the time dependent neutron diffusion equation was solved by an integration method, based on a final volume approach in space and an implicit discretization in time.

2.1.5.2. Experimental and numerical results and discussion: Criticality

This section is a summary of the detailed discussion in Ref. [2.27] describing the numerical analyses of the high-energy neutron spectrum. As discussed in Section 2.1.1.6, two-layer targets were used with a combination of heavy (W, Pb and Bi) and light (Be) nuclides, with thickness sufficient to fully stop the proton beam within the target.

The MCNPX modelling showed that the high-energy neutron spectrum resulting from the injection of 100 MeV into solid targets with W, W-Be or Pb-Bi was similar regardless of the target material, as shown in Figs 2.34 and 2.35, with the exception of the energy range from 1–10 MeV. Of particular interest was the effect this had on neutron multiplication in the core. The difference in the neutron spectrum from the W-Be two-layer target compared with single targets as shown in Fig. 2.35 was attributed to the high-energy neutron scattering in the Be layer. The proton beams penetrated the Be layer and were stopped inside the W layer.

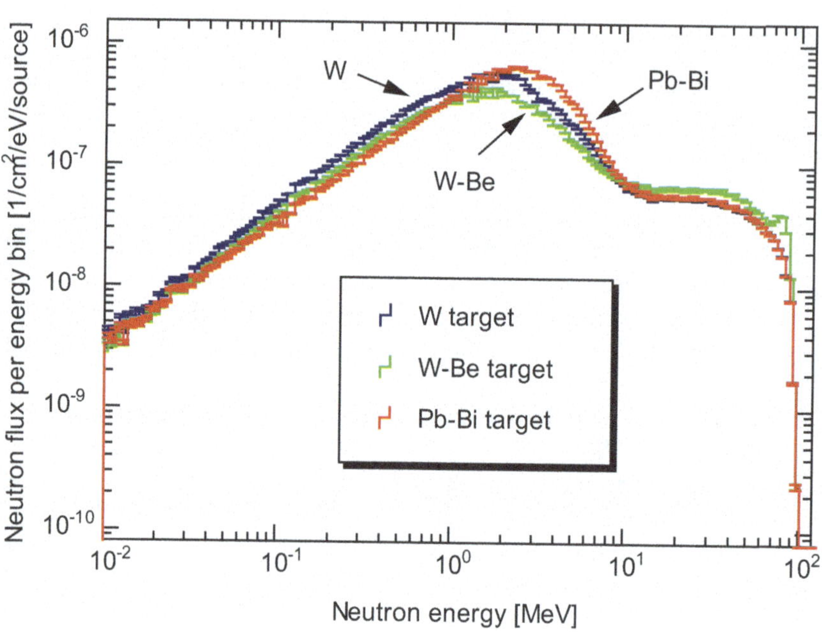

FIG. 2.34. Comparison between neutron spectra with the use of W, W-Be and Pb-Bi targets (Reproduced from Ref. [2.27] with permission courtesy of Taylor & Francis).

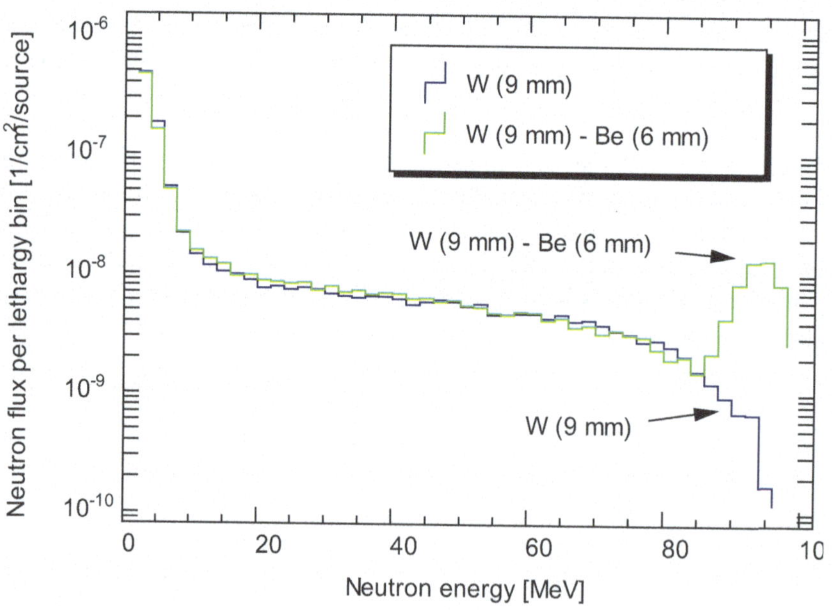

FIG. 2.35. Comparison between neutron spectra with the use of W and W-Be targets (Reproduced from Ref. [2.27] with permission courtesy of Taylor & Francis).

The statistical errors on the subcriticalities in the MCNP eigenvalue calculations were within 0.01% $\Delta k/k$ (10 pcm), within the relative difference of 5% between the experimental and the calculated results. The reaction rates in the fixed source calculations were within 3%, as determined by 10^8 histories.

Table 2.21 compares the measured and calculated neutron yields according to Eq. (2.7). Table 2.22 shows the neutron multiplication M and the subcritical multiplication factor k_s obtained from ^{115}In$(n, \gamma)^{116m}$In reaction rates with subcriticality 2900 pcm.

TABLE 2.21. COMPARISON OF MEASURED AND CALCULATED NEUTRON YIELD IN EQ. (2.7) [2.27]

Target	Calculation	Experiment	C/E [a]
W	0.57 ± 0.01	0.21 ± 0.01	2.66 ± 0.01
W–Be	0.32 ± 0.01	0.15 ± 0.01	2.13 ± 0.04
Pb-Bi	0.38 ± 0.01	0.17 ± 0.01	2.27 ± 0.04

[a] the ratio of calculated to experimental results

TABLE 2.22. NEUTRON MULTIPLICATION AND SUBCRITICAL MULTIPLICATION FACTORS DERIVED FROM ^{115}In$(n, \gamma)^{116m}$In REACTION RATES IN SUBCRITICALITY 2900 PCM [2.27]

Target	Neutron multiplication, M			Subcritical multiplication factor k_s		
	Calculation	Experiment	C/E	Calculation	Experiment	C/E
W	1.73 ± 0.01	1.85 ± 0.02	0.93 ± 0.01	0.42033 ± 0.00100	0.45874 ± 0.01003	0.92 ± 0.02
W–Be	2.29 ± 0.01	2.36 ± 0.03	0.97 ± 0.01	0.56355 ± 0.00100	0.57662 ± 0.01510	0.98 ± 0.03
Pb-Bi	1.95 ± 0.01	1.94 ± 0.02	1.01 ± 0.01	0.48830 ± 0.00100	0.48488 ± 0.01335	1.01 ± 0.03

Table 2.23 compares the k static parameters calculated by McCARD with experimental results. Equation (2.6) was used to calculate the excess reactivity. It can be seen that McCARD overestimated the criticality by 185 pcm, and that calculated and experimental values for excess reactivity and control rod worth agree within the 99% confidence intervals.

$$\Delta\rho_{ex} = 1/k_{critical} - 1/k_{withdrawn} \tag{2.6}$$

where $k_{critical}$ is the effective neutron multiplication factor at the critical state and $k_{withdrawn}$ the effective neutron multiplication factor with all control rods withdrawn.

TABLE 2.23. REACTIVITY COMPARISONS BETWEEN MCCARD AND EXPERIMENTAL RESULTS OF PHASE I IN THE PB-BI BENCHMARK [2.89]

	McCARD (SD)	Experiment
k at the critical state	1.00185 (0.00012)	1.00000
Excess reactivity (pcm)	211 (18)	180
Rod worth CR1 (pcm)	851 (18)	805
Rod worth CR2 (pcm)	574 (18)	595
Rod worth CR3 (pcm)	172 (18)	139

Table 2.24 compares the McCARD and MCNP6.1 calculations of the effective multiplication factor for the Pb-Bi benchmark Phase I experimental configurations.

TABLE 2.24. EFFECTIVE MULTIPLICATION FACTORS OF LEAD-BISMUTH BENCHMARK PHASE I [2.89]

Case	McCARD[a] (SD) [SNU]	MCNP6.1 (SD) [from KUCA]
I-1	0.98871 (0.00013)	0.99093 (0.00009)
I-2	0.98017 (0.00013)	0.98274 (0.00009)
I-3	0.97373 (0.00013)	0.97627 (0.00009)
I-4	0.97450 (0.00012)	0.97662 (0.00009)
I-5	0.95411 (0.00013)	0.95680 (0.00009)

[a] JENDL-4.0

Table 2.25 presents the criticality calculations for case I-4 [2.67] and shows the good agreement between the deterministic and the Monte Carlo results.

TABLE 2.25. CASE I-4: RESULTS OF CRITICALITY CALCULATIONS [a]

Data library	ERANOS/VARIANT	PARTISN	MCNPX2.7	KINFIN-WIMSD4-TWOTRAN
JEFF3.1	0.97309	0.97324	0.97409±0.0002	
JEFF3.2			0.97418±0.0002	
JENDL4.0			0.97251±0.0002	
ENDF/B-VI				0.9645

[a] Reference criticality level: 0.97302

Table 2.26 presents the kinetic parameters calculated by ERANOS and KIN3D/PARTISN [2.67]. A code extension by KIT was used for the 3-D ERANOS results. There was good agreement between ERANOS and KIN3D/PARTISN, and the mean neutron generation time was in good agreement with the KUCA reference. For β_{eff} the codes showed a discrepancy of about 20 pcm.

TABLE 2.26. KINETIC PARAMETERS OF CASE I-4

Parameter	ERANOS/VARIANT	KIN3D/PARTISN	KUCA Reference
β_{eff} (pcm)	829	827	807
Λ (s)	3.027E-05	3.032E-5	3.050E-05

Table 2.27 compares the calculated prompt neutron decay constant α for the various configurations with the experimental results. BARC also estimated α for case I-4 tungsten target using the static alpha-eigenvalue and time-dependent calculations of KINFIN, with results of 1262 s^{-1} and 1270 s^{-1}, respectively.

TABLE 2.27 MEASURED PROMPT NEUTRON DECAY CONSTANT DEDUCED BY LEAST-SQUARED FITTING IN THE PNS METHOD [2.6]

Case	Target	Neutron decay constant α (s^{-1})			McCARD α-iteration (From SNU[a])
		BF$_3$ #1 in (10, U)	BF$_3$ #2 in (15, X)	Optical fibre	
Case I-1	W	737.1 ± 56.0	739.6 ± 9.5	706.4 ± 83.3	–
	W–Be	732.5 ± 5.2	747.8 ± 5.5	782.0 ± 5.1	–
	Pb-Bi	737.7 ± 10.9	731.9 ± 10.6	756.5 ± 12.5	738.9 ± 6.8
Case I-2	W	1059.9 ± 10.0	1075.0 ± 5.8	1215.2 ± 12.5	–
	W–Be	1062.7 ± 10.0	1085.6 ± 4.0	1101.9 ± 5.3	–
	Pb-Bi	1070.1 ± 21.5	1072.0 ± 20.0	1080.8 ± 8.3	1065.5 ± 8.4
Case I-3	W	1364.2 ± 8.4	1372.5 ± 9.9	1379.6 ± 68.4	–
	W–Be	1377.2 ± 6.5	1383.2 ± 6.8	1402.3 ± 10.3	–
	Pb-Bi	1338.0 ± 3.7	1358.3 ± 4.0	1381.5 ± 7.1	1369.9 ± 9.7
Case I-4	W	1052.0 ± 7.5	1033.3 ± 8.8	1204.5 ± 16.6	–
	W–Be	1033.2 ± 3.1	1027.7 ± 3.5	1073.8 ± 4.1	–
	Pb-Bi	1006.3 ± 5.6	1004.1 ± 6.1	1083.2 ± 5.3	1029.6 ± 7.0
Case I-5	W	1844.6 ± 21.3	1845.9 ± 17.7	1922.5 ± 33.3	-
	W–Be	1770.5 ± 3.7	1697.0 ± 4.6	1909.1 ± 3.9	-
	Pb-Bi	1785.2 ± 9.4	1742.1 ± 8.7	1895.5 ± 11.1	1883.7 ± 10.0

[a] Seoul National University

Figs 2.36 and 2.37 show the time evolution of prompt and delayed neutrons for cases I-3 and I-5 at the BF$_3$ detector #1 location (10, U).

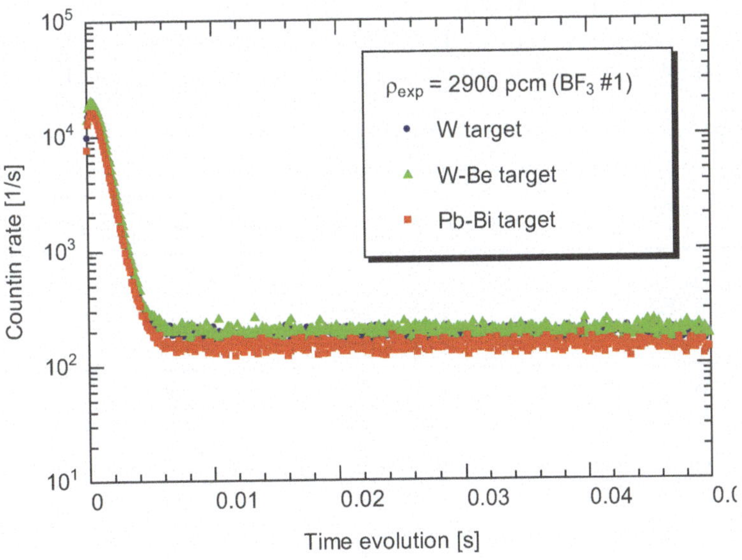

FIG. 2.36. Experimental time evolution of prompt and delayed neutrons at BF$_3$ detector position (10, U) in Case I-3 (Reproduced from Ref. [2.6] with permission courtesy of Kyoto University).

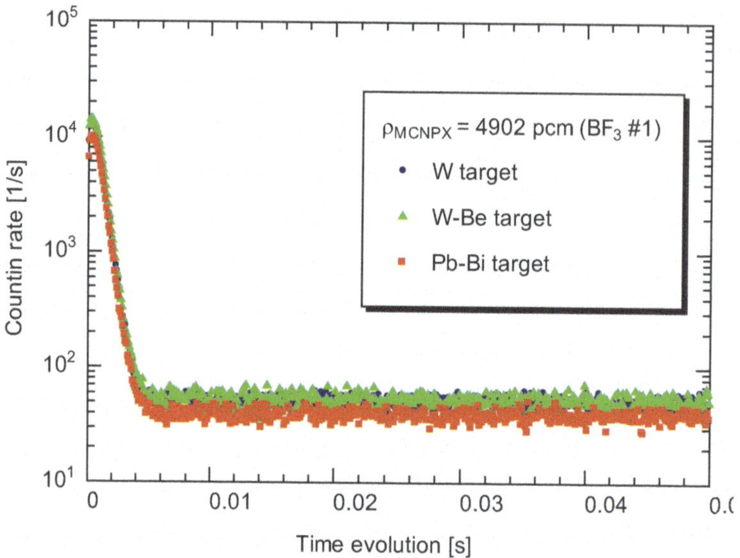

FIG. 2.37. Experimental time evolution of prompt and delayed neutrons at BF$_3$ detector position (10, U) in case I-5 (Reproduced from Ref. [2.6] with permission courtesy of Kyoto University).

The PNS methods analyse the response of a subcritical system to an external source pulse. The area method assesses the absolute reactivity level in $ units following a neutron pulse into a subcritical system as ratio of the areas under the prompt and delayed peaks [2.91]. The reactivity value depends upon the detector position if the system response cannot be approximated by point kinetics. However, Ref. [2.92] shows that the spatial effects can be accommodated using a homogeneous (no external source) time-independent transport calculation and an inhomogeneous (with source) prompt simulation. The reactivity at a given detector position can be calculated as the ratio of the prompt and delayed reaction rates, derived from the total and prompt reaction rates. Table 2.28 presents the reactivity results using this method for cases I-1 to I-5 [2.6]. Tables 2.29 and 2.30 show the reactivities calculated using the neutron decay constant and the noise method [2.6].

TABLE 2.28. MEASURED SUBCRITICALITY DEDUCED BY THE EXTRAPOLATED AREA RATIO METHOD

Case	Target	Subcriticality ρ (pcm) (by area ratio method) [a]		
		BF$_3$ #1 in (10, U)	BF$_3$ #2 in (15, X)	Optical fibre
Case I-1 (reference: ρexp = 1360 pcm)	W	1132 ± 17	1393 ± 20	1888 ± 27
	W–Be	1145 ± 18	1458 ± 21	1473 ± 21
		1166 ± 18	1387 ± 20	1390 ± 20
Case I-2 (reference: ρexp = 2165 pcm)	W	1877 ± 27	2363 ± 35	3002 ± 44
	W–Be	1879 ± 28	2463 ± 36	2229 ± 33
	Pb-Bi	1926 ± 29	2356 ± 35	2252 ± 33
Case I-3 (reference: ρexp = 2900 pcm)	W	2398 ± 35	3156 ± 46	3822 ± 58
	W–Be	2502 ± 39	3405 ± 51	2939 ± 45
		2624 ± 39	3238 ± 48	2977 ± 45
Pb-Bi Case I-4 (reference: ρMCNPX = 2773 pcm)	W	2682 ± 39	3291 ± 48	3672 ± 55
	W–Be	2688 ± 41	3483 ± 52	2910 ± 43
	Pb-Bi	2738 ± 41	3302 ± 49	2922 ± 44
Case I-5 (reference: ρMCNPX = 4902 pcm)	W	4944 ± 78	6386 ± 101	6228 ± 99
	W–Be	5004 ± 83	7050 ± 114	4929 ± 77
	Pb-Bi	4903 ± 77	6356 ± 92	4912 ± 74

[a] (β_{eff} = 807 pcm; Λ = 3.050×10^{-5} s)

TABLE 2.29. MEASURED PROMPT NEUTRON DECAY CONSTANTS DEDUCED BY LEAST-SQUARED FITTING IN THE NOISE METHOD

Case	Target	Neutron decay constants α (s^{-1})			McCARD α-iteration (From SNU)
		BF$_3$ #1 in (10, U)	BF$_3$ #2 in (15, X)	Optical fibre	
Case I-1	W	632.5 ± 7.1	665.4 ± 12.5	636.6 ± 10.8	–
	W–Be	641.8 ± 21.2	683.0 ± 13.5	705.6 ± 11.8	–
	Pb-Bi	616.2 ± 30.4	650.7 ± 26.8	714.2 ± 19.4	738.9 ± 6.8
Case I-2	W	878.7 ± 22.2	869.1 ± 28.2	1089.8 ± 60.2	–
	W–Be	874.7 ± 12.6	912.8 ± 23.7	1073.4 ± 60.4	–
	Pb-Bi	885.0 ± 25.3	898.5 ± 36.3	970.2 ± 62.6	1065.5 ± 8.4
Case I-3	W	1086.4 ± 21.6	1103.1 ± 26.8	1407.5 ± 57.1	–
	W–Be	1094.3 ± 11.8	1159.6 ± 16.4	1296.0 ± 18.6	–
	Pb-Bi	1124.8 ± 13.9	1132.8 ± 23.4	1308.5 ± 19.5	1369.9 ± 9.7
Case I-4	W	889.0 ± 11.6	941.8 ± 20.5	1072.8 ± 23.8	–
	W–Be	880.1 ± 12.8	975.5 ± 35.0	999.7 ± 11.4	–
	Pb-Bi	900.0 ± 17.5	931.1 ± 10.5	1021.5 ± 23.3	1029.6 ± 7.0
Case I-5	W	1487.9 ± 48.9	1451.6 ± 35.4	1965.8 ± 91.7	–
	W–Be	1456.5 ± 18.8	1575.3 ± 28.5	1746.6 ± 23.2	–
	Pb-Bi	1390.4 ± 95.4	1431.6 ± 96.1	1682.6 ± 142.9	1883.7 ± 10.0

TABLE 2.30. MEASURED SUBCRITICALITY DEDUCED BY THE NOISE METHOD

Case	Target	Subcriticality ρ (pcm) (by Feynman-α method) [a]		
		BF$_3$ #1 in (10, U)	BF$_3$ #2 in (15, X)	Optical fibre
Case I-1 (reference: ρ_{exp} = 1360 pcm)	W	1122 ± 25	1515 ± 40	1406 ± 35
	W–Be	1151 ± 66	1581 ± 43	1667 ± 38
	Pb-Bi	1073 ± 94	1459 ± 83	1699 ± 61
Case I-2 (reference: ρ_{exp} = 2165 pcm)	W	1873 ± 69	2285 ± 87	3119 ± 184
	W–Be	1861 ± 41	2450 ± 74	3057 ± 185
	Pb-Bi	1892 ± 78	2396 ± 112	2667 ± 191
Case I-3 (reference: ρ_{exp} = 2900 pcm)	W	2507 ± 67	3169 ± 83	4320 ± 175
	W–Be	2531 ± 39	3383 ± 52	3898 ± 59
	Pb-Bi	2624 ± 45	3282 ± 73	3946 ± 61
Case I-4 (reference: ρ_{MCNPX} = 2773 pcm)	W	1905 ± 38	2560 ± 64	3055 ± 74
	W–Be	1877 ± 41	2884 ± 108	2778 ± 37
	Pb-Bi	1938 ± 55	2519 ± 35	2861 ± 72
Case I-5 (reference: ρ_{MCNPX} = 4902 pcm)	W	3731 ± 150	4486 ± 109	6430 ± 280
	W–Be	3634 ± 60	4954 ± 86	5601 ± 73
	Pb-Bi	3434 ± 291	4411 ± 294	5360 ± 436

[a] β_{eff} = 807 pcm; Λ = 3.050×10^{-5} s

The reactivity (in pcm) at the detector positions for case I-4 was also calculated using ERANOS and PARTISN, assuming a 14 MeV external source located at the target position shown in Fig. 2.13. Table 2.31 shows that these results were in reasonable agreement with the average experimental values for the W, W–Be, and Pb-Bi targets.

TABLE 2.31. REACTIVITIES CALCULATED USING THE PNS AREA METHOD [2.67].

Detector	VARIANT (pcm)	PARTISN (pcm)	Experiment W target (pcm)	Experiment W–Be target (pcm)	Experiment Pb-Bi target (pcm)
(15, X)	3111	3279	3291 ± 48	3483 ± 52	3302 ± 49
(10, U)	2819	2866	2682 ± 39	2688 ± 41	2738 ± 41
(16-15, O-P)	3040	3050	3672 ± 55	2910 ± 43	2922 ± 44

The KIN3D/VARIANT and the KIN3D/PARTISN codes were used for the 3-D (XYZ) transient analyses, with a focus on the direct and improved quasistatic options. The former is based on a fully implicit time discretization scheme. If the neutron flux is linear as a function of time, the time-dependent equation can be converted into a sequence of quasi-steady-state problems with an additional source term which are solved using the VARIANT/TGV code with KIN3D. In the improved quasistatic scheme, the flux shape is assumed to vary linearly during each time step. The calculations of reactivity and other kinetic parameters use the variational nodal perturbation theory with the adjoint flux as a weighting function.

The KIN3D/PARTISN procedure used the PARTISN instead of VARIANT as the transport solver. The KIN3D/VARIANT and the KIN3D/PARTISN calculation schemes were similar, the difference being that VARIANT performed quasi-steady-state calculations with two extra source terms for the past flux and the delayed neutron source, while PARTISN restarted transient calculations with one extra source term related to delayed neutrons [2.71].

Previously analysis of ADS experiments in the MUSE [2.93], YALINA [2.94], and FREYA [2.95] zero-power facilities using KIN3D/VARIANT and KIN3D/PARTISN showed that the codes could simulate the experimental results and properly describe the spatial kinetics effects. For this reason, they were deemed suitable to analyse the Phase-I KUCA experiments and investigate the performance of the direct and improved-quasistatic kinetics models employed in the codes.

For the KUCA analysis, ECCO was used to generated 20-energy-group effective neutron cross-sections. Table 2.32 shows the criticality level and kinetic parameters computed with 20 energy groups, which agree well with the 40-energy-group values shown in Tables 2.25 and 2.26.

TABLE 2.32. ERANOS AND KIN3D/PARTISN CRITICALITY AND KINETIC PARAMETERS AT 20-ENERGY GROUPS [2.67]

Parameter	ERANOS	KIN3D/PARTISN
k_{eff}	0.97222	0.97201
β_{eff} (pcm)	825.1	820.5
Λ (s)	3.017E-05	3.022E-5

KIN3D/VARIANT was used to model the detector responses to a 100 ns neutron pulse in case I-4 using with the improved-quasistatic kinetics option and a 14 MeV external neutron source. Figs 2.38 and 2.39 compare the calculated and experimental results in the fuel region and the reflector with the detector responses normalized to the value at 5.0×10^{-5} seconds. As can be seen from the figures, KIN3D/VARIANT reproduced the experimental observations with particularly good agreement up to 6 msec for all the detectors. In the fuel region the agreement between calculated and experimental values extends to the area dominated by the delayed neutrons, up to 0.03 seconds. The codes slightly overestimated the delayed neutron reaction rate in the reflector region. Overall, it was shown that the deterministic codes could reproduce the static and time-dependent behaviour of the case I-4 core.

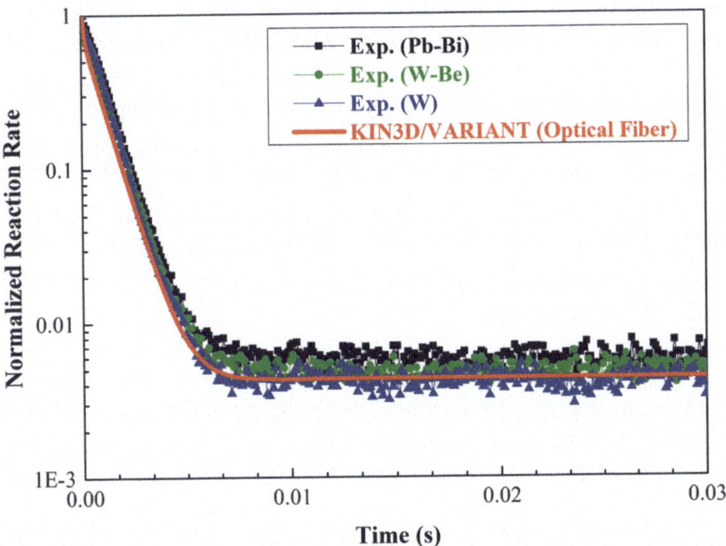

FIG. 2.38. Experimental [2.96] and KIN3D/VARIANT responses in the position (16-15, O-P) (Reproduced from Ref. [2.67] with permission courtesy of ANS).

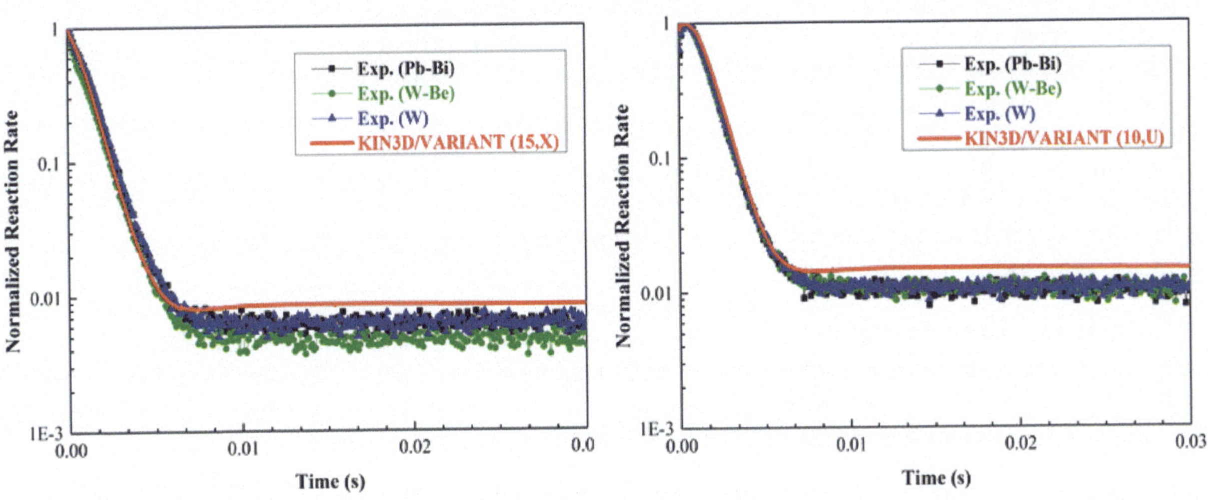

FIG. 2.39. Experimental [2.96] and KIN3D/VARIANT responses in the positions (15, X) and (10, U) (Reproduced from Ref. [2.67] with permission courtesy of ANS).

2.1.5.3. Target neutron yield

Sections 2.1.5.3 and 2.1.5.4 are summaries of the detailed discussion in Ref. [2.27]. The neutron yield is defined as the number of neutrons per proton, as shown in Eq. 2.7.

$$Y_{n,p} = \frac{S_n}{S_p} \tag{2.7}$$

where S_n is the number of spallation neutrons generated at the target and S_p the number of protons injected into the target. This equation was used with MCNPX and JENDL/HE-2007 to calculate the neutron yield.

The neutron yield was determined experimentally using Al and In activation foils, using the reactions ^{27}Al$(p, n+3p)^{24}$Na for high-energy protons and ^{115}In$(n, n')^{115m}$In for high-energy neutrons. The results are presented in Table 2.21 and show that the W target had a larger neutron yield than the other targets, and also that the calculated neutron yields were over a factor of two higher than the experimental values. This discrepancy was attributed to the defocussing and scattering of high-energy protons before they impacted the target. The protons travelled in a vacuum to the original target location at (15, A') in Fig. 2.13, but thereafter travelled in air to the new target location at (15, H). To increase the accuracy, the proton beam spot size and the proton energy spectrum at the target needs to be considered in more detail.

2.1.5.4. Reaction rate distribution in core region

The measured reaction rates were dependent upon the target material, as can be seen in Fig. 2.15. The reaction rates were highest for the two-layer target W-Be target with the peak thermal reaction rate in the polyethylene region. Neutron multiplication was measured using the ^{115}In$(n, \gamma)^{116m}$In reaction rates in the core region taking advantage of the proportionality of the ^{115}In$(n, \gamma)^{116m}$In and ^{235}U(n, f) cross-sections in the thermal region [2.17].

2.1.5.5. Application of the prompt neutron decay constant, Sjöstrand, Rossi, and Feynman methods for subcriticality evaluation

To study the time evolution of prompt and delayed neutrons at BF$_3$ detector position (10, U), see Figs 2.36 and 2.37, external neutrons were injected into the KUCA core with the control and safety rods fully inserted, as shown in Fig. 2.13, and with a subcriticality of 2900 pcm. The effect of detector position was investigated by setting BF$_3$ detectors at (15, X) and (10, U) and an optical fibre detector in the core at (16–15, O–P) [2.22]. Table 2.27 presents the experimental results for the prompt neutron decay constant α and compares them with McCARD estimates. It can be seen that α is dependent upon both the detector location and the choice of target. The McCARD estimates agreed within 1% with the average results from the three detectors, with the exception of the deepest subcritical core in Case I-5.

The measured subcriticality is presented in Table 2.28. The values of β_{eff} and Λ were estimated by diffusion-based calculations in x-y-z dimensions and 107 energy groups. As discussed in Ref. [2.23], the measured subcriticality depends on the external neutron spectrum which is affected by the target materials, and the neutron detector location.

In Case I-5 shown in Fig. 2.14 (e), the control rods C1 and C2 and the safety rods S4 and S6 were fully inserted, resulting in a subcriticality level of 4902 pcm as calculated by MCNPX. Table 2.28 shows that the optical fibre detector at position (16–15, O–P) accurately measured the subcriticality level, except for the W target. This influence of the target emphasized the importance of the experimental analyses.

The following discussion is a summary of the detailed information presented in Ref. [2.76]. The assembly effective multiplication factor of a subcritical assembly driven by a pulsed mode particle accelerator, can be obtained by either Eqs (2.8) or (2.9) [2.97]. Both of these equations are based on the point kinetics approximation [2.98]

$$k_{eff} \cong \frac{1}{1+\dfrac{A_p \beta_{eff}}{A_d}} \tag{2.8}$$

$$k_{eff} \cong \frac{1}{1-\beta_{eff}-\alpha\Lambda} \tag{2.9}$$

where A_p and A_d are the areas underneath the curve of the prompt and delayed neutron counts distributions during the pulse period, as shown in Fig. 2.40, and β_{eff} is the effective delayed neutron fraction.

Equation (2.8) was first introduced by Sjöstrand [2.33]. It does not use the prompt neutron generation time Λ, which is strongly dependent on the subcriticality level of the assembly. β_{eff} is not highly sensitive to the subcriticality level of the assembly [2.99], [2.100]. The prompt neutron decay constant α in Eq. (2.9) is derived from the slope of the neutron counts distribution as shown in Fig. 2.40, with the time interval for fitting α chosen such that the neutron counts distribution is linear on a logarithm scale. β_{eff} and Λ are usually derived from computer simulations in criticality mode, without modelling the external neutron source. The other parameters are obtained either experimentally or computationally including the external neutron source.

Equation (2.8) applies only to a subcritical assembly driven by a pulsed neutron source. Equation (2.9) can be applied to a subcritical assembly driven by either a pulsed neutron source or a neutron source such as ^{252}Cf. In the latter case, α can be obtained by fitting the Feynman or Rossi distributions [2.101]–[2.103]. Equation (2.9) cannot be used if the assembly is deeply subcritical because the decay of prompt neutrons has multiple

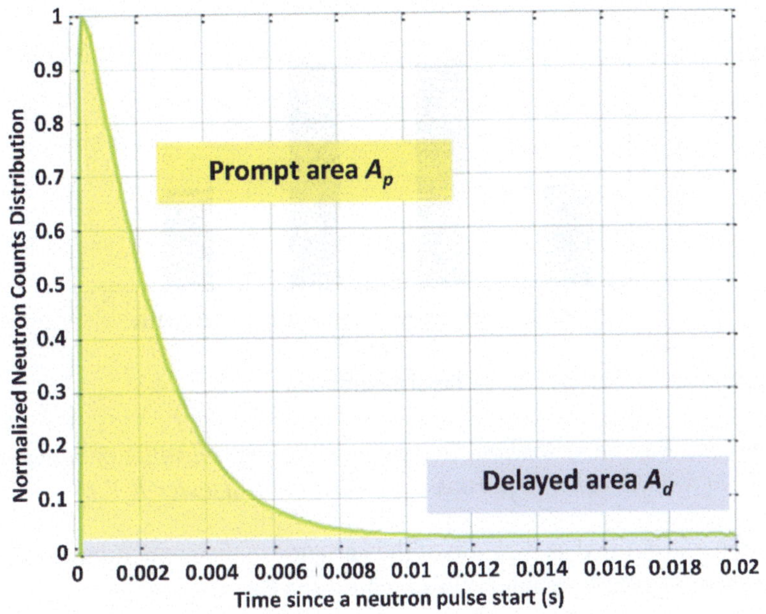

exponential functions with different decay constants.

FIG. 2.40. Neutron counts distribution of a subcritical assembly driven by an pulsed neutron source (Reproduced from Ref. [2.76] with permission courtesy of Elsevier B.V).

The Rossi distribution counts pairs of neutron captures as a function of the time interval Δt between the two capture events, see Fig. 2.41 [2.104]. This distribution has a random part due to captures of neutrons from different fission chains, and a correlated part due to captures of neutrons from the same fission chain. The correlated part exponentially decreases following the decay of prompt neutrons as shown in Eq. (2.10).

$$Rossi(\Delta t) = A + Be^{-\alpha \Delta t} \tag{2.10}$$

where Δt is the time interval between two neutron capture events, A is a constant set by neutron captures from different fission chains, B is a constant set by neutron captures from the same fission chain, and α is the prompt neutron decay constant.

Radioactive decay and particle scattering processes are described by the Poisson probability distribution. The probability of each event is independent of the previous event, and the distribution variance is equal to the mean. This does not apply to neutron capture events because these are correlated due to the fission chains.

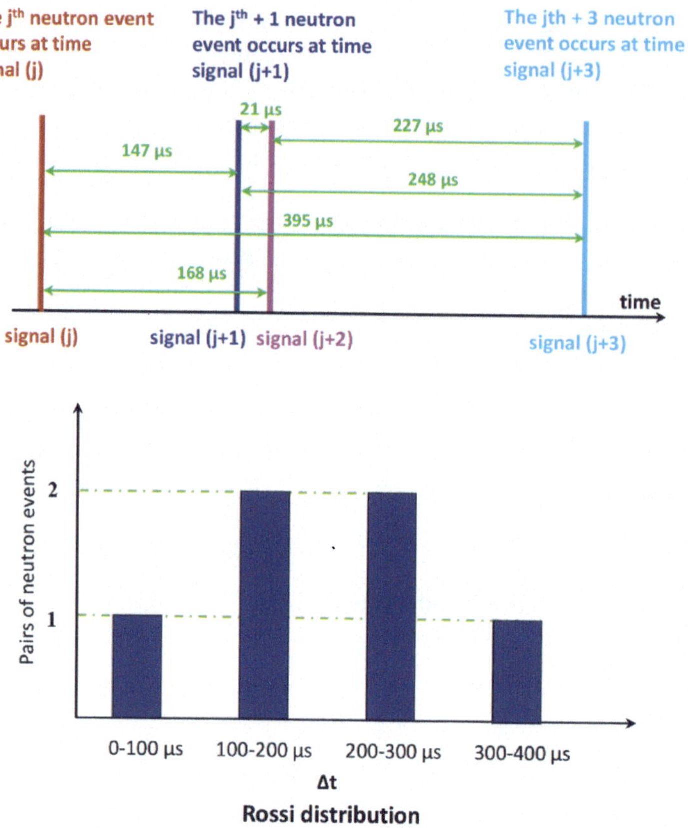

FIG. 2.41. Illustration of the Rossi distribution (bottom plot) for a quadruplet of neutron events (top plot) [2.104] (Reproduced from Ref. [2.76] with permission courtesy of Elsevier B.V.).

The Feynman distribution Y is defined as the ratio of the variance to the mean minus one, of the neutron counts array $c(\Delta t_g)$ as a function of a fixed time interval Δt_g (gate), see Eq. (2.11).

$$Feynman(\Delta t_g) = Y(\Delta t_g) \equiv \frac{var\left(c(\Delta t_g)\right)}{mean\left(c(\Delta t_g)\right)} - 1 \approx \frac{B}{\alpha^2}\left(1 - \frac{e^{-\alpha\Delta t_g}}{\alpha\Delta t_g}\right) \tag{2.11}$$

As an example, Fig. 2.42 shows 10 neutron events that are detected between 0.02 and 0.92 μs. The corresponding neutron counts array $c(\Delta t_g)$ for different gate values is given in Figs 2.43–2.45.

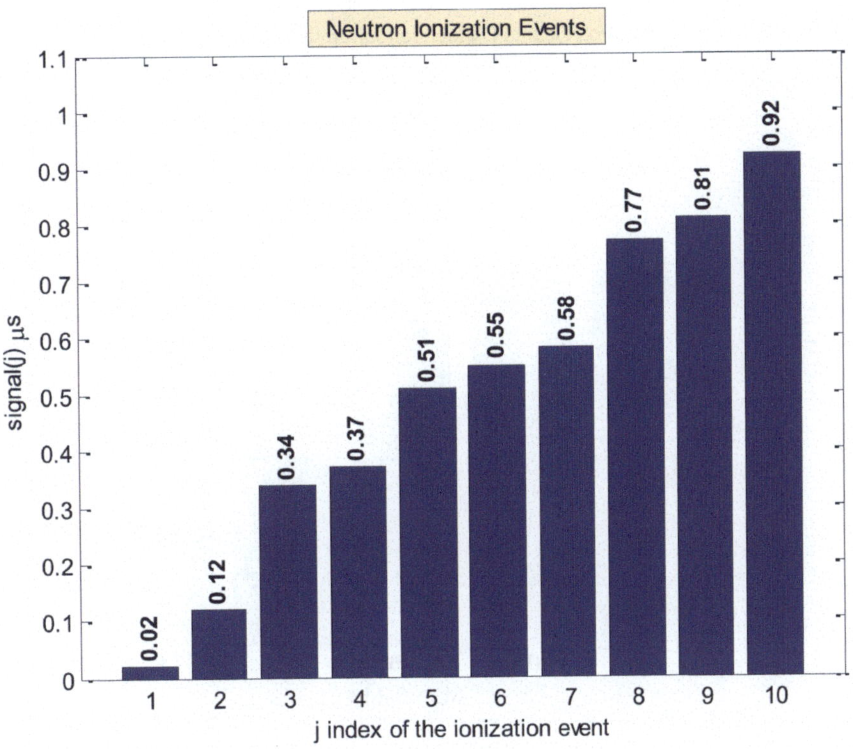

FIG. 2.42. Example of 10 detected neutron events contributing to a Feynman distribution (Reproduced from Ref. [2.76] with permission courtesy of Elsevier B.V.).

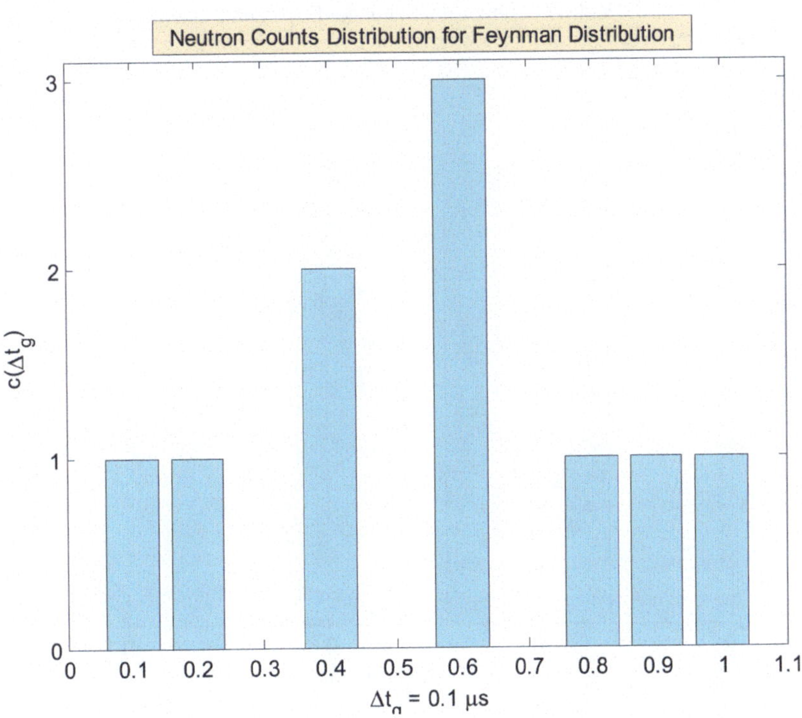

FIG. 2.43. The neutron counts array c(Δtg) for the neutron events from Fig. 2.42 with a 0.1 µs gate. The first column is the time bin 0–0.1 µs, the second 0.1–0.2 µs, etc. (Reproduced from Ref. [2.76] with permission courtesy of Elsevier B.V.).

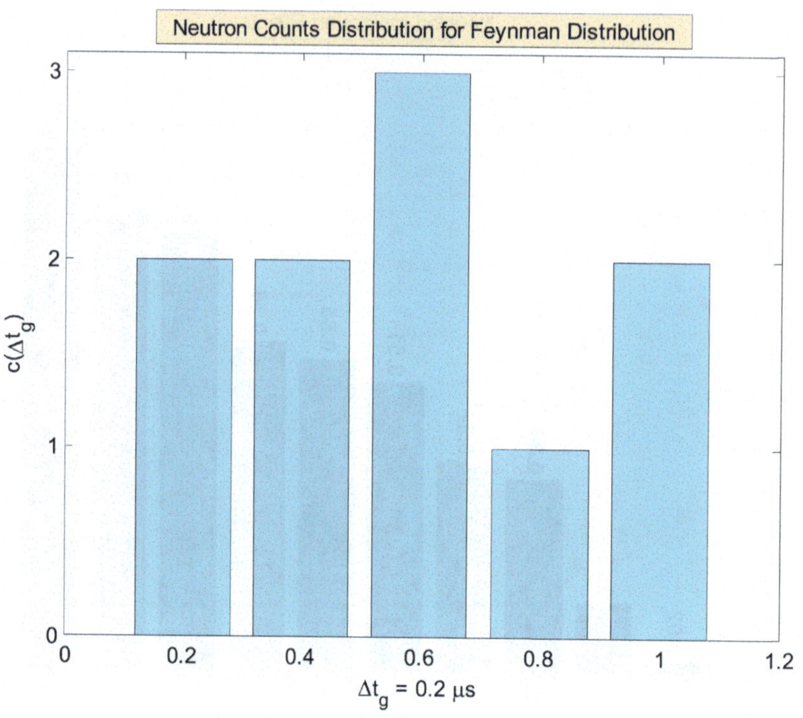

FIG. 2.44. The neutron counts array c(Δt$_g$) for the neutron events from Fig. 2.42 with a 0.2 μs gate. The first column is the time bin 0–0.2 μs; the second 0.2–0.4 μs, etc. (Reproduced from Ref. [2.76] with permission courtesy of Elsevier B.V.).

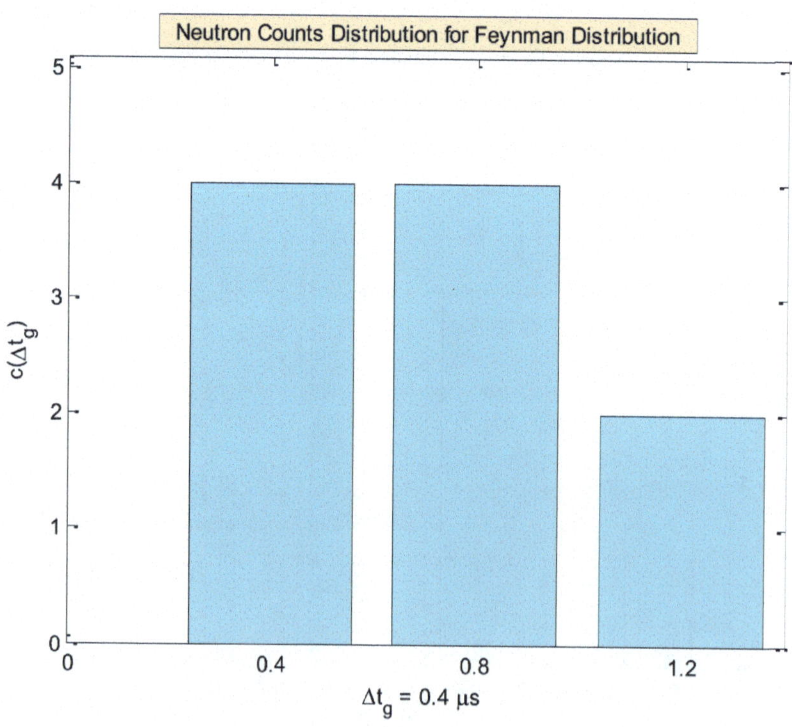

FIG. 2.45. Illustration of the neutron counts array c(Δt$_g$) for the neutron events from Fig. 2.42 with a 0.4 μs gate. The first column is the time bin 0–0.4 μs; the second 0.4–0.8 μs, etc. (Reproduced from Ref. [2.76] with permission courtesy of Elsevier B.V.).

To apply the Sjöstrand method, the following assumptions were made: the subcritical system was driven by a pulsed neutron source; the detector response to delayed neutrons was at equilibrium during the pulse; the pulse period was much shorter than the half-lives of the delayed neutron precursors and much longer than the prompt neutron half-life. With these assumptions, the total area under neutron counts distribution (see Fig. 2.40) was the integral of the detector reaction rate r during the pulse period T, as shown in Eq. (2.12).

$$A = A_p + A_d = \int_0^T r(t)dt = \int_0^\infty r_1(t)dt \qquad (2.12)$$

where t is the time, r is the reaction rate from a single accelerator pulse when delayed neutrons are at equilibrium, r_1 is the reaction rate from a single accelerator pulse when delayed neutrons are not at equilibrium. The latter term can be rewritten as shown in Eq. (2.13), with one integration over the pulse period and the second overs the period after the pulse.

$$\int_0^\infty r_1(t)dt = \int_0^T r_1(t)dt + \int_T^\infty r_1(t)dt \qquad (2.13)$$

Although prompt neutrons dominate the first term, it is not exactly equal to the prompt area A_p because some delayed neutrons are emitted during the pulse period. Analogously, the second term is dominated by delayed neutrons, but is not exactly equal to the delayed area A_d. The computer simulation of the neutron counts distribution needs to take into account the equilibrium status of the delayed neutrons. To do this, the second term can be decomposed into a series of integrals as shown in Eq. (2.14) [2.105], [2.106].

$$\int_T^\infty r_1(t)dt = \sum_{n=1}^\infty \int_{nT}^{(n+1)T} r_1(t + nT)dt \qquad (2.14)$$

This has the drawback that a large array of $\sim 5 \times 10^7$ elements needs to be processed. A more efficient alternative is to apply Eq. (2.15) in which the integrated contribution of delayed neutrons to the reaction rate after the pulse period is divided by the pulse period and added to the reaction rate from a single pulse during the period r_1. This yields the reaction rate with delayed neutrons at equilibrium r.

$$r(t) = r_1(t) + \frac{1}{T}\int_T^\infty r_1(t)dt \quad \forall\, t\, \varepsilon[0,T] \qquad (2.15)$$

Ref. [2.107] introduced this method, which requires a much less computing time and reduces the statistical uncertainty for the captures of delayed neutrons.

The following computational procedure was used to obtain the detector reaction rate r_1 using MCNP [2.97]:

1. Model one pulse of the particle accelerator; setting the particle emission duration to match the accelerator pulse width;

2. Include the delayed neutron contribution in the neutron capture reaction rate tally for the detector, e.g. set the cut card first entry to 500×108 so that the neutrons are transported up to 500 s;

3. Run the simulation in analogue mode without variance reduction techniques or multiprocessing option (because of the PTRAC card);

4. The PTRAC card coupled to the F8 tally writes the timestamps when a neutron is captured in the ^{3}He detector; the input has the following input card: PTRAC event = cap file = asc type = n write = all max = 2E9 coinc = lin = tally = 8;

5. Apply the special capture treatment to the F8 tally; the input for scoring neutron captures by 3He atoms contains: FT8 cap -1 -1 2003;

6. Combine the results of multiple simulations with different random number seeds.

For the application of noise methods, it was assumed that the subcritical system was driven by a spontaneous fission neutron source, with the delayed neutron count rate at equilibrium. The Rossi and Feynman distributions can be constructed from the timestamps data from Step 4 in the above procedure. The MCNP timestamps did not include the time of emission of the neutron from the external source and had a millisecond precision, which was insufficient for the noise methods. These issues were solved by patching the ptrak.F90 file (line 168) of MCNP 6.1.1 beta version, as shown in Fig. 2.46.

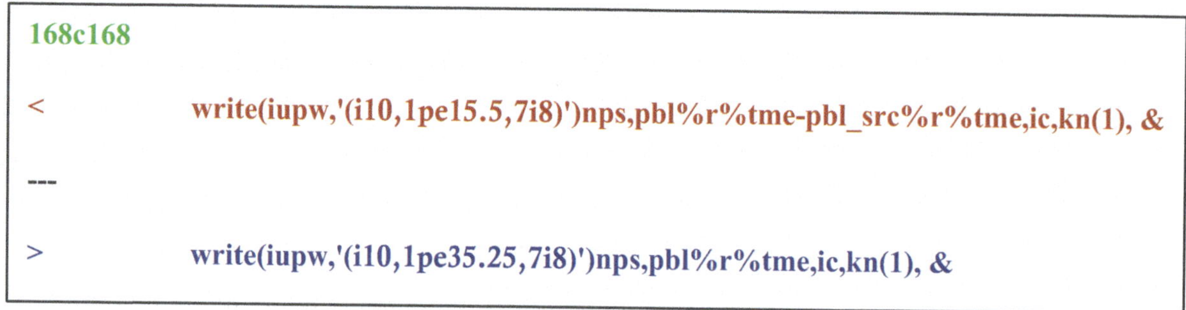

FIG. 2.46 patch to the ptrak.F90 file (line 168) of the MCNP 6.1.1 beta version.

The computational procedure above was adapted to the spontaneous fission neutron source by replacing steps 1 and 2 as follows [2.97]:

1. Model the neutron source for a long time, e.g., 1 h;

2. Include the delayed neutron contribution in the detector neutron capture reaction rate tally; e.g., set cut card first entry to 3600×108 so that the neutrons are transported up to 1 h;

Table 2.33 compares the Monte Carlo and deterministic calculations of the kinetic parameters in criticality mode (no external neutron source). All approaches determined the effective multiplication factor k_{eff} to be ~0.973. There was a 100 pcm difference in results from the JEFF-3.2.0 and JENDL-4.0 nuclear data libraries. The detectors at (15, X) and (10, U) were not explicitly modelled. MCNP6 calculations have shown that this does not have a significant effect on the kinetic parameters.

TABLE 2.33. KINETIC PARAMETERS OF THE KUCA SUBCRITICAL ASSEMBLY [2.76]

Code (Institution)[a]	Nuclear data library	k_{eff}	β_{eff} (pcm)	Λ (μs)
MCNP6 (ANL)	JENDL-4.0	0.97305 ± 4	820 ± 5	33.39 ± 0.04
MCNPX (KU)	JENDL-4.0	0.97302	807	30.5
MCNPX (KIT)	JENDL-4.0	0.97251	–	–
MCNPX (KIT)	JEFF-3.2.0	0.97398 ± 28	–	–
MCNPX (KIT)	JEFF-3.1.0	0.97349 ± 29	–	–
PARTISN (KIT)	JEFF-3.1.0	0.97324	–	–
VARIANT (KIT)	JEFF-3.1.0	0.97309	837	31.125

[a] Argonne National Laboratory (ANL), Kyoto University (KU), Karlsruhe Institute of Technology (KIT).

Figures 2.47 and 2.48 show the time evolution of the ^{10}B(n, α)^{7}Li reaction rates during the pulse for the detectors at (15, X) and (10, U). The MCNP6 calculations used Eq. (2.15) and assumed 100 MeV protons. The deterministic simulations assumed 14.1 MeV neutrons. All calculations were normalized to the experimentally determined average delayed neutron contribution to the reaction rate. There was good agreement between experiment and calculation, with PARTISN closer than VARIANT. Both codes used the same macroscopic cross-section library, mesh structure, and external neutron source. The reason for the spread in results at the beginning of the pulse for the (15, X) detector was not identified.

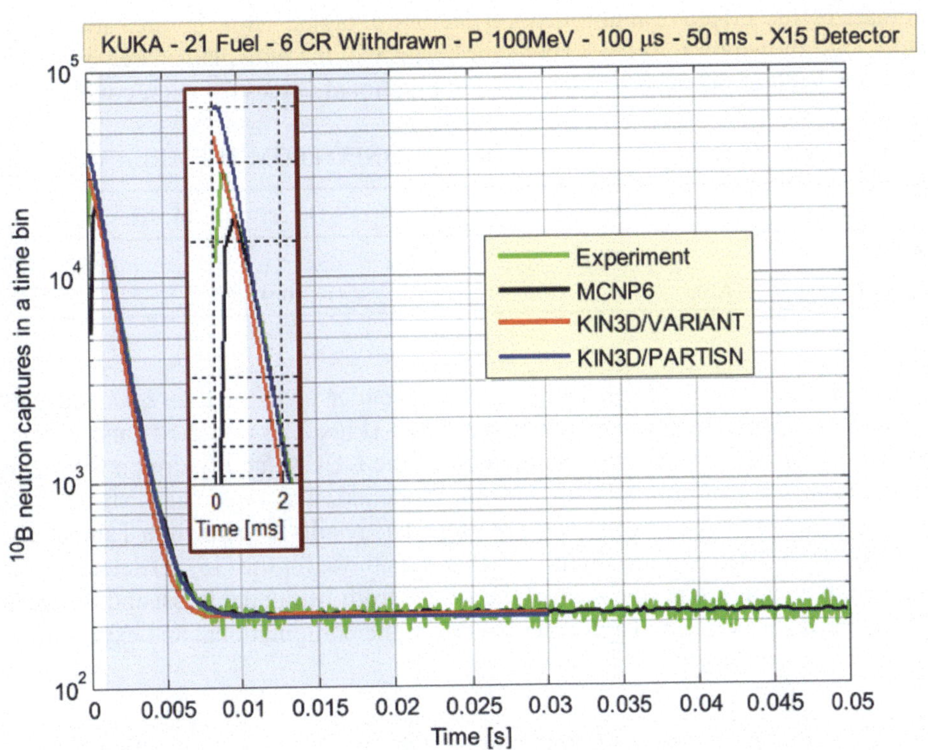

FIG. 2.47. $^{10}B(n,\alpha)^7Li$ reaction rate in the detector at position (15, X) during the pulse period (Reproduced from Ref. [2.76] with permission courtesy of Elsevier B.V.).

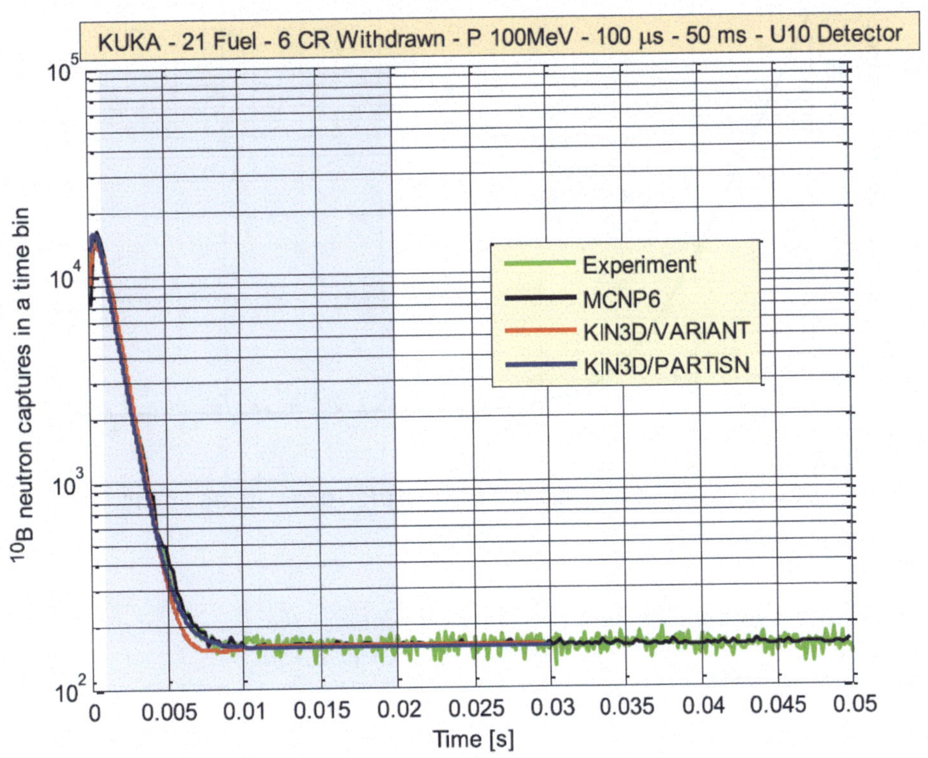

FIG. 2.48. $^{10}B(n,\alpha)^7Li$ reaction rate in the detector at position (10, U) during the pulse period (Reproduced from Ref. [2.76] with permission courtesy of Elsevier B.V.).

Table 2.34 summarizes the effective multiplication factors obtained with the area method (Eq. (2.8)). No experimental values are shown for the (15, X) detector because of the unexplained behaviour at the start of the pulse. For the (10, U) detector the experimental and computational values data agree within 40 pcm.

TABLE 2.34. EFFECTIVE MULTIPLICATION FACTOR OBTAINED FROM THE AREA METHOD [2.76]

Source of the Detector Data – Method	Detector in (15, X)	Detector in (10, U)
Experiment – Area	–	0.97370
MCNP6 with JENDL-4.0 data – Area	0.97487	0.97331

Figures 2.49–2.52 show the fitting of the data from the lavender areas of Figs 2.47 and 2.48. The prompt neutron decay constant α was obtained by fitting the Rossi (Figs 2.53 and 2.54) and Feynman (Fig. 2.55) distributions obtained from a single MCNP6 simulation of the KUCA facility driven for one hour by a ^{252}Cf source at the target location. The proton beam was not modelled. The Feynman distribution for the (10, U) detector had large statistical fluctuations and was not used. The results are shown in Table 2.35. It can be seen that the values for the (10, U) detector are slightly smaller than those for the (15, X) detector. The experimental values were obtained with a pulsed proton source, the values from fitting the Rossi and Feynman distributions assumed a ^{252}Cf source. There was good agreement between the computational and experimental values.

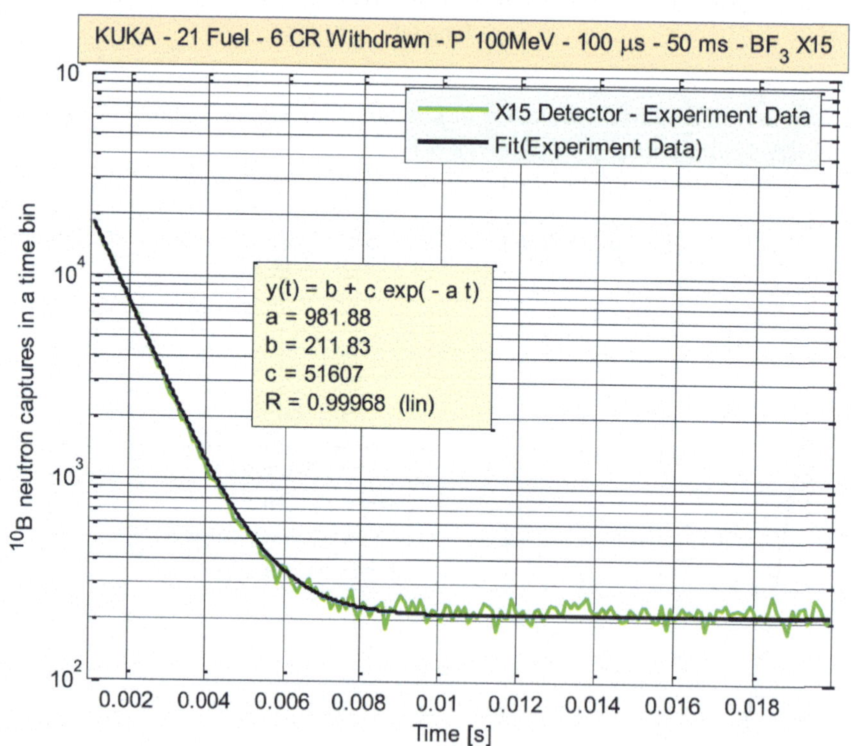

FIG. 2.49. Fitting the experimental $^{10}B(n,\alpha)^{7}Li$ reaction rate data for the detector at position (15, X) (Reproduced from Ref. [2.76] with permission courtesy of Elsevier B.V.).

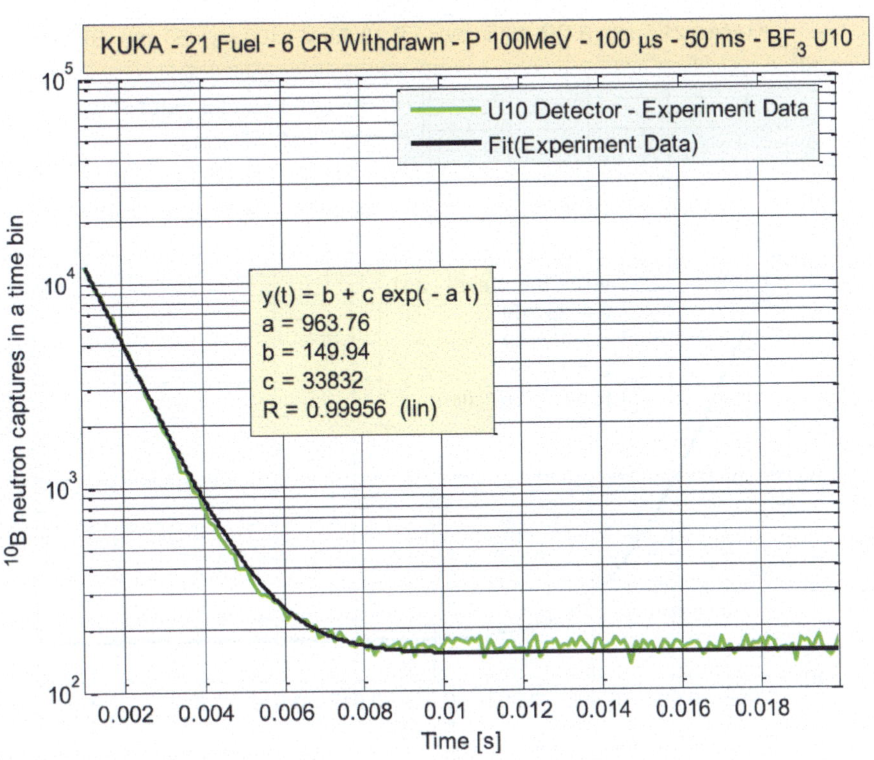

FIG. 2.50. Fitting the experimental $^{10}B(n,\alpha)^7Li$ reaction rate data for the detector at position (10, U) (Reproduced from Ref. [2.76] with permission courtesy of Elsevier B.V.).

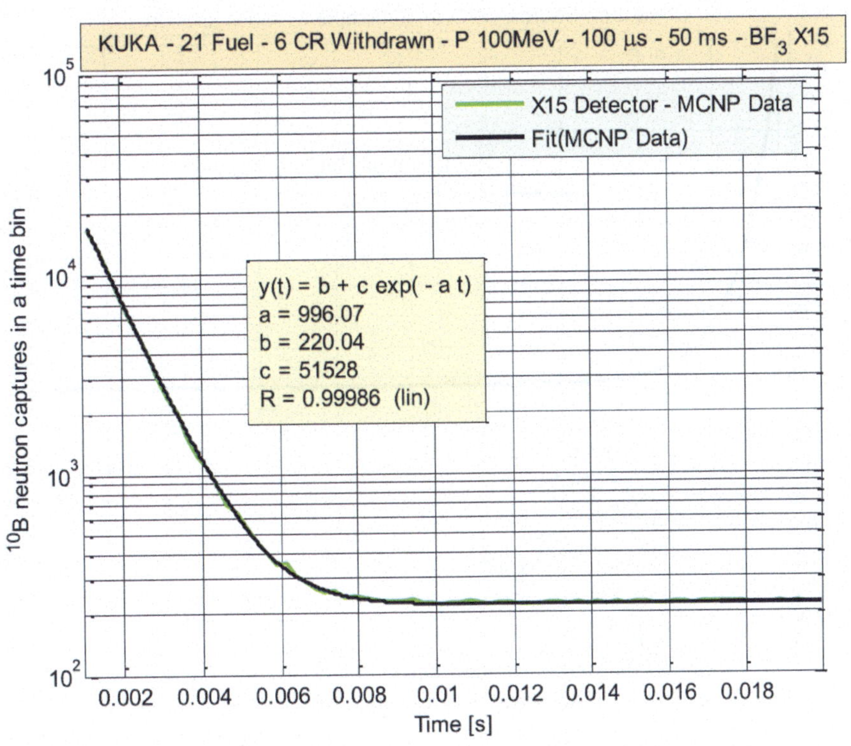

FIG. 2.51. Fitting the MCNP6 $^{10}B(n,\alpha)^7Li$ reaction rate data for the detector at position (15, X) (Reproduced from Ref. [2.76] with permission courtesy of Elsevier B.V.).

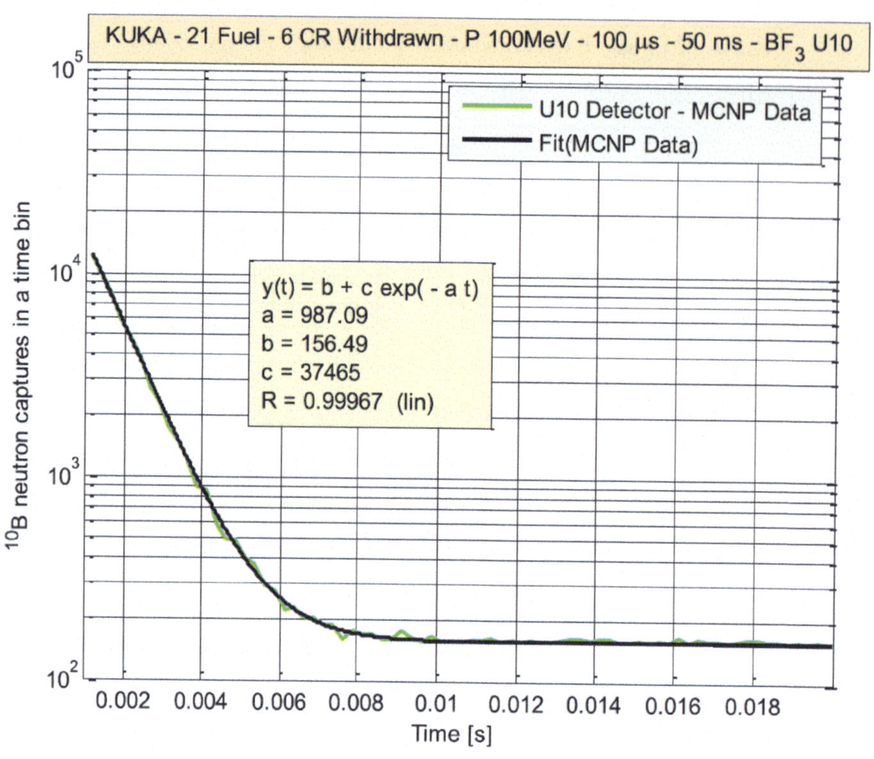

FIG. 2.52. Fitting the MCNP6 $^{10}B(n,\alpha)^7Li$ reaction rate data for the detector at position (10, U) (Reproduced from Ref. [2.76] with permission courtesy of Elsevier B.V.).

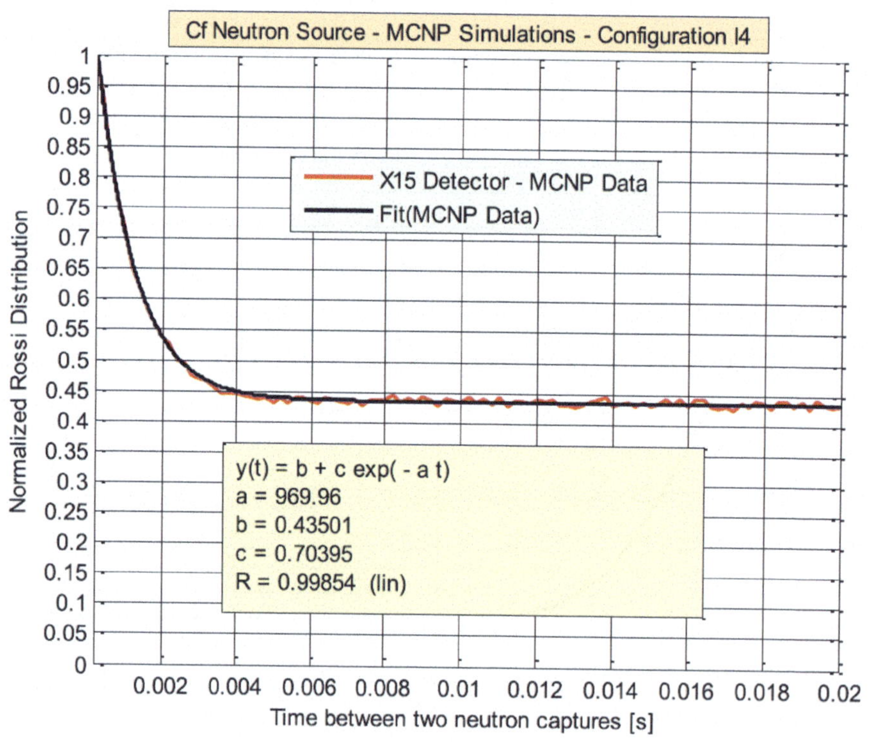

FIG. 2.53. Normalized Rossi distribution from MCNP6 simulations for the detector at position (15, X) (Reproduced from Ref. [2.76] with permission courtesy of Elsevier B.V.).

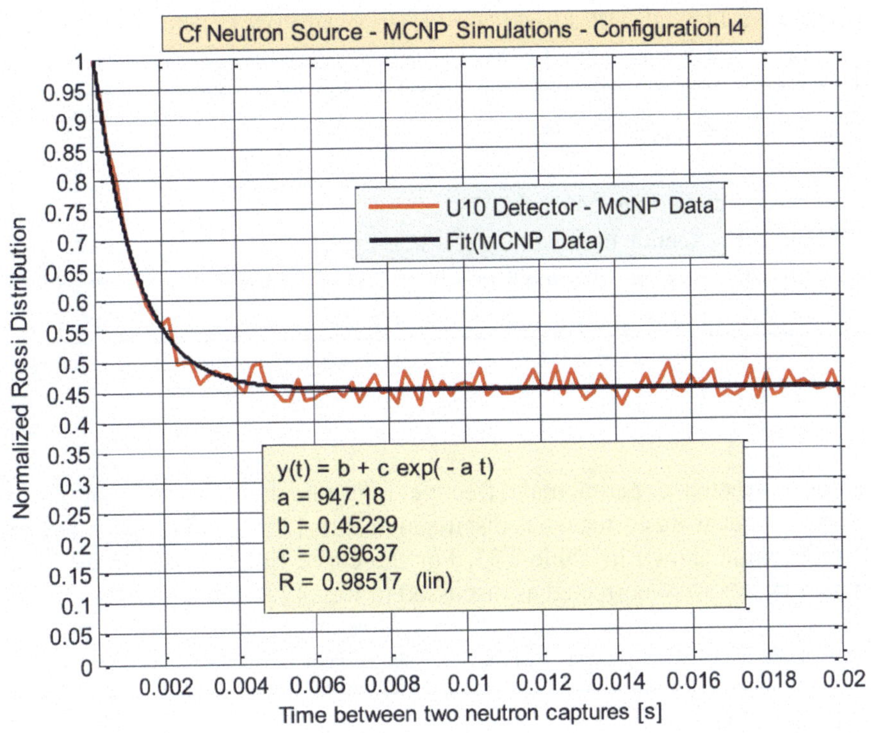

FIG. 2.54. Normalized Rossi distribution for the detector at position (10, U) from MCNP6 simulations (Reproduced from Ref. [2.76] with permission courtesy of Elsevier B.V.).

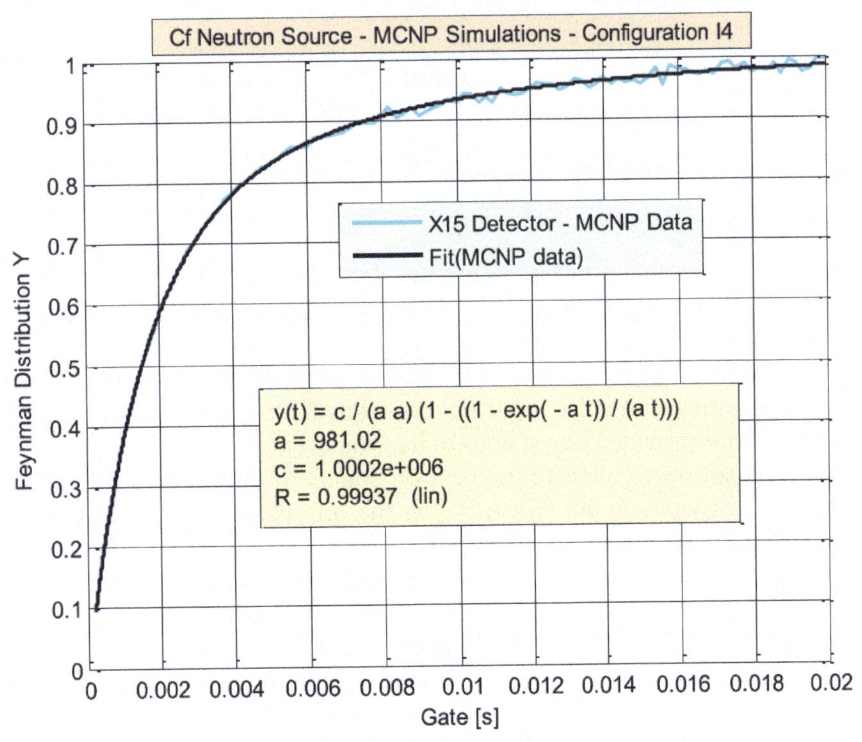

FIG. 2.55. Normalized Feynman distribution for the detector at position (15, X) from MCNP6 simulations (Reproduced from Ref. [2.76] with permission courtesy of Elsevier B.V.).

TABLE 2.35. ABSOLUTE VALUE OF THE PROMPT NEUTRON DECAY CONSTANT OBTAINED BY FITTING THE CURVES PLOTTED IN FIGS 2.49–2.55 [a] [2.76]

Source of the Detector Data – Method	Detector in (15, X) (s^{-1})	Detector in (10, U) (s^{-1})
Experiment – Neutron counts fitting (Figs 2.49 and 2.50)	981.88	963.76
MCNP6 with JENDL-4.0 data – Neutron counts fitting (Figs 2.51 and 2.52)	996.07	987.09
MCNP6 with JENDL-4.0 data – Rossi distribution fitting (Figs 2.53 and 2.54)	969.96	947.18
MCNP6 with JENDL-4.0 data – Feynman distribution fitting (Fig. 2.55)	981.02	–

[a] The prompt neutron decay constant obtained from MCNP6, with JENDL-4.0 data, in criticality mode (no external neutron source), and with the detectors modelling, is ~1076.96 s^{-1}.

Table 2.36 shows the corrected experimental effective multiplication factors. For the area method, the experimental reactivity values were corrected as discussed in Refs [2.97] and [2.108]. The results were within 90 pcm of the MCNP6 value shown in Table 2.33. For the curve fitting methods the experimental prompt neutron decay constant values were corrected as discussed in Refs [2.97] and [2.103].

TABLE 2.36. CORRECTED EXPERIMENTAL EFFECTIVE MULTIPLICATION FACTOR[a] [2.76]

Source of the Detector Data – Method	Detector in (15, X)	Detector in (10, U)
Experiment – Area [2.108])	–	0.97346
Experiment – Neutron counts fitting [2.103])	0.97355	0.97387
Experiment – Rossi distribution fitting [2.103])	0.97265	0.97248
Experiment – Feynman distribution fitting [2.103])	0.97304	–

[a] The effective multiplication factor obtained from MCNP6, with JENDL-4.0 data and in criticality mode (no external neutron source), is 0.97305 ± 4.

2.1.5.6. Calculation of the cross and auto power spectral densities from pulse mode detectors at low neutron count rates

The following is a summary of the detailed discussion provided in Ref. [2.79]. Neutron noise methods including time-domain and frequency-domain analysis are well known [2.51], [2.101], and [2.109 –2.112]. The cross and auto power spectral density method was used to obtain the prompt neutron decay constant α for a particle accelerator driven subcritical assembly. This section describes a new method to determine the cross and auto power spectral densities with pulse mode detectors and a low neutron count rate [2.112]. A detector system operating in pulse mode provided timestamps of neutron events with nanosecond precision which were processed using MATLAB scripts to allocate the neutron counts to time bins. The counting statistics were significantly improved by aggregating and redistributing the timestamps from several thousand accelerator pulses. The results are similar to those obtained from detectors operating in current mode which measure average neutron counts in the time bins [2.113], [2.20]. Operating detectors in pulse mode allows the gamma contribution to the detector signal to be discriminated [2.114], [2.115], although dead-time effects are possible if the neutron count rate is high [2.116]. Neutron detectors can be operated in either pulse or current modes. Fission chambers may, in addition, be operated in the Campbelling mode for a better gamma discrimination in a high gamma flux environment [2.117]. The pulse operation mode of the detector is unrelated to the particle accelerator pulsed operation mode.

The prompt neutron decay constant of an accelerator-driven subcritical assembly can be obtained by fitting the peaks of the cross and auto power spectral densities [2.113], [2.20], using the point kinetics approximation [2.98], and the assumption that detector neutron captures occur when delayed neutrons are at equilibrium [2.106], [2.76].

This study used experimental data from KUCA [2.76], processed by MATLAB without using particle transport simulation software. It assumed a subcritical assembly driven by an external pulsed neutron source with a short pulse duration relative to the pulse repetition period t_r, for example 5 to 100 μs compared to 10 to 50 ms. This configuration allowed prompt neutrons to completely decay within the accelerator pulse period while the delayed neutrons to remained approximately constant. Consequently, the counting statistics did not change with time when the acquisition time was long enough (e.g.,10 particle accelerator pulses).

The external accelerator-driven neutron source as a function of time is given by the Dirac com function shown in Eq. (2.16) [2.113].

$$s(t) = s_0 \sum_{m=-\infty}^{+\infty} \delta(t - mt_r) \tag{2.16}$$

where s_0 is the number of neutrons generated per pulse, m is an integer index, and δ is the Dirac delta function. The Fourier transform of Eq. (2.16) gives the source as a function of frequency, see Eq. (2.17).

$$S(j\omega) = \int_{-\infty}^{+\infty} dt\, s(t) \exp(-j\omega t) = \frac{2\pi s_0}{t_r} \sum_{m=-\infty}^{+\infty} \delta(\omega - \omega_m) \tag{2.17}$$

where ω_m is the angular repetition frequency defined in Eq. (2.18).

$$\omega_m = \frac{2\pi m}{t_r} \tag{2.18}$$

Equation (2.19) shows the neutron population as a function of frequency averaged over the whole core due to the point kinetics assumption.

$$N(j\omega) = G(j\omega)S(j\omega) = \frac{2\pi s_0}{t_r} \sum_{m=-\infty}^{+\infty} \frac{\delta(\omega - \omega_m)}{\alpha + j\omega} \tag{2.19}$$

where α is the prompt neutron decay constant and $G(j\omega)$ is the zero-power reactor transfer function on the frequency domain [2.101]. The neutron population on the time domain is given by the inverse Fourier transform of $N(j\omega)$, as shown in Eq. (2.20).

$$n(t) = \frac{1}{2\pi} \int_{-\infty}^{+\infty} d\omega\, N(j\omega) \exp(j\omega t) = \frac{s_0}{t_r} \sum_{m=-\infty}^{+\infty} \frac{\exp(j\omega_m t)}{\alpha + j\omega_m} \tag{2.20}$$

The cross-correlation function $\varphi(t_1, t_2)$ can be defined as the ensemble average of two separate detector counts, $c_a(t_1)$ and $c_b(t_2)$ [2.113], representing the average neutron counts at times t_1 and t_2 of detectors a and b, respectively. The detector counts are averaged over a small arbitrary time bin, dt. The cross-correlation function is given by Eq. (2.21), with variable τ, as defined in Eq. (2.22).

$$\varphi(\tau) = \lim_{t_1 \to \infty} \frac{1}{t_1} \int_{-\frac{t_1}{2}}^{+\frac{t_1}{2}} dt_a\, c_a(t_1)c_b(t_1 + \tau) \tag{2.21}$$

$$\tau = t_2 - t_1 \tag{2.22}$$

The detector counts c are proportional to the fission frequency λ_f, the neutron population n, and the detector efficiency ε, as shown in Eq. (2.23) . The suffix a denotes the detector.

$$c_a(t) \cong \lambda_f n(t)\varepsilon_a = \nu\, \Sigma_f n(t)\varepsilon_a \tag{2.23}$$

where ν is the number of neutrons produced per fission event and Σ_f the fission macroscopic cross-section. When detector a counts a neutron from a fission chain, detector b can eventually count ν-1 correlated neutrons from the same fission chain and ν uncorrelated neutrons from other fission chains. The cross power spectral density $\Phi_{ab}(\omega)$ is defined as the frequency domain cross-correlation function. Building on earlier work in reactor noise analyses [2.51], [2.101], [2.109]–[2.111], Yamamoto derived Eq. (2.24) for the cross power spectral density averaged over all fission events [2.118].

$$\Phi_{ab}(\omega) = \varepsilon_a\varepsilon_b\lambda_f \frac{s_0\langle\nu(\nu-1)\rangle}{\alpha t_r(\alpha^2+\omega^2)} - \varepsilon_a\varepsilon_b \frac{s_0}{t_r(\alpha^2+\omega^2)} + 2\pi\varepsilon_a\varepsilon_b \frac{s_0^2}{t_r^2}\sum_{m=-\infty}^{+\infty}\frac{\delta(\omega-\omega_m)}{(\alpha^2+\omega_m^2)} \tag{2.24}$$

on the right side of the equation, the first term describes correlated fission chains, second describes correlated neutrons from a spallation event, and the third term describes uncorrelated fission chains. The spallation process occurs only with very high energy protons (several hundred MeV) [2.119]. The prompt neutron decay constant can be found by fitting the peaks of the real part of the cross power spectral density.

The following discussion is a summary of the information presented in Ref. [2.77]. The calculation of the cross power spectral density is discussed in Refs [2.118] and [2.120]. If the detector counts are sampled over M time bins, the circular correlation function R_{ab} is given by Eq. (2.25).

$$R_{ab}(n) \equiv \sum_{k=0}^{M-1} c_a(t_k)c_b\big(t_{M-(n-k)}\big); \; n \in \{0,2,\dots,M-1\} \tag{2.25}$$

where $c_a(t_k)$ is the neutron count from detector a within a time bin of width Δt starting at time t_k. If the delayed neutrons are at equilibrium, the detector signals are periodic such that $c_a(t_k) = c_a(t_{k+M})$ and $c_b(t_k) = c_b(t_{k+M})$ [2.118]. Equations (2.26) and (2.27) give the real and imaginary parts of the cross power spectral density.

$$Re\{\Phi_{ab}(\omega_k)\} = 2\sum_{n=0}^{M-1}\big(R_{ab}(n) + R_{ba}(n)\big)\cos\left(\frac{2\pi kn}{M}\right) \tag{2.26}$$

$$Im\{\Phi_{ab}(\omega_k)\} = 2\sum_{n=0}^{M-1}\big(R_{ba}(n) - R_{ab}(n)\big)\sin\left(\frac{2\pi kn}{M}\right) \tag{2.27}$$

where k is an integer index extending from 0 to $M/2$ and ω_k is as defined in Eq. (2.28), assuming $\Delta t M \gg t_r$.

$$\omega_k = \frac{2\pi k}{\Delta t M}; \quad k \in \left\{0,2,\dots,\frac{M}{2}\right\} \tag{2.28}$$

If only the fundamental mode existed, the imaginary part of the cross power spectral density would be zero. The auto power spectral densities Φ_{aa} for detector a, and Φ_{bb}, for detector b, are calculated similarly to the cross power spectral density, except that for Φ_{aa} it is assumed that $c_a = c_b$ and it is only necessary to process the counts from detector a. Likewise, for Φ_{bb} only c_b is used.

Neutron detectors may operate in either current or pulse mode. In the former, they measure the current induced by many ionization processes, in the latter, they measure the voltage induced by a single ionization process. Equation (2.25) can be directly applied to current mode detectors because the counts are already averaged over the time bin Δt set by the digital acquisition system. When detectors operate in pulse mode, the output of the digital acquisition system consists of the timestamps for each ionization process. Detectors operating in pulse mode may experience dead-time effects, causing some neutron counts to be lost [2.116]. This can be overcome by using a low count rate (e.g., below 10 000 counts per second) such that the dead-time effects are negligible [2.97]. When the count rate is low, the "timestamps" can be converted to the "average" format to improve the counting statistics [2.112]. The application of Eqs. ("2.25")–("2.28") is then straightforward, enabling the prompt neutron decay constant to be determined from the fitting procedure, as discussed Refs [2.20], [2.113], and [2.118]. Kitamura et al. [2.121] pioneered the use of timestamps to determine the auto power spectral density and achieved this by programming the digital acquisition system. This section describes an alternative approach that does not require a modification of the digital acquisition system but instead uses a collapsing procedure on the raw data implemented through MATLAB scripts.

The experiment was carried out over 12 000 proton accelerator pulses with a pulse repetition period t_r 50 ms. The collected timestamps recorded the start of each accelerator pulse, and the neutron captures in the (10, U) detector in the reflector zone and in the fibre detector in the fuel zone, see Fig. 2.33. The arrays c_a for the detector at (10, U) and c_b for the fibre detector each contained 5000 time bins (M=5000) and thus represented a total time interval of 10 particle accelerator pulses. The MATLAB script redistributed the neutron counts from 12 000 accelerator pulses into these 10 accelerator pulses. This "collapsing procedure" provides flexibility because the time bin Δt and the number of pulses for the construction of R_{ab} and R_{ba} can be arbitrarily changed in the MATLAB script at any time.

Figure 2.56 shows the resulting arrays c_a and c_b and Figs 2.57 and 2.58 show the circular correlation functions R_{ab}, R_{ba}, R_{aa}, and R_{bb} derived therefrom (Eq. (2.25)). Figure 2.59 shows the arrays c_a and c_b from 10 pulses without the collapsing procedure, and Figs 2.60 and 2.61 give the corresponding correlation functions R_{ab}, R_{ba}, R_{aa}, and R_{bb}.

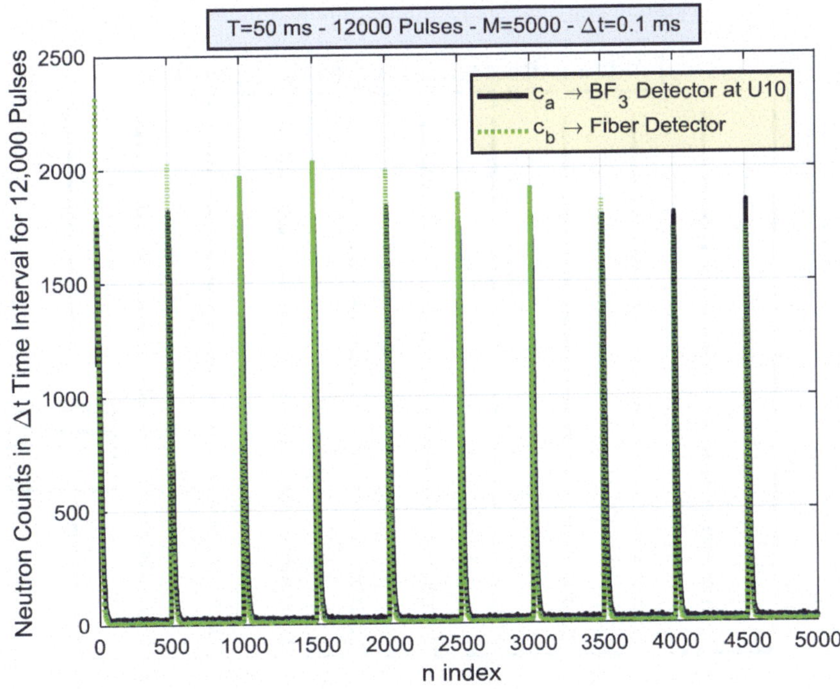

FIG. 2.56. Neutron counts arrays ca and cb from 12 000 pulses (Reproduced from Ref. [2.77] with permission courtesy of Elsevier B.V.).

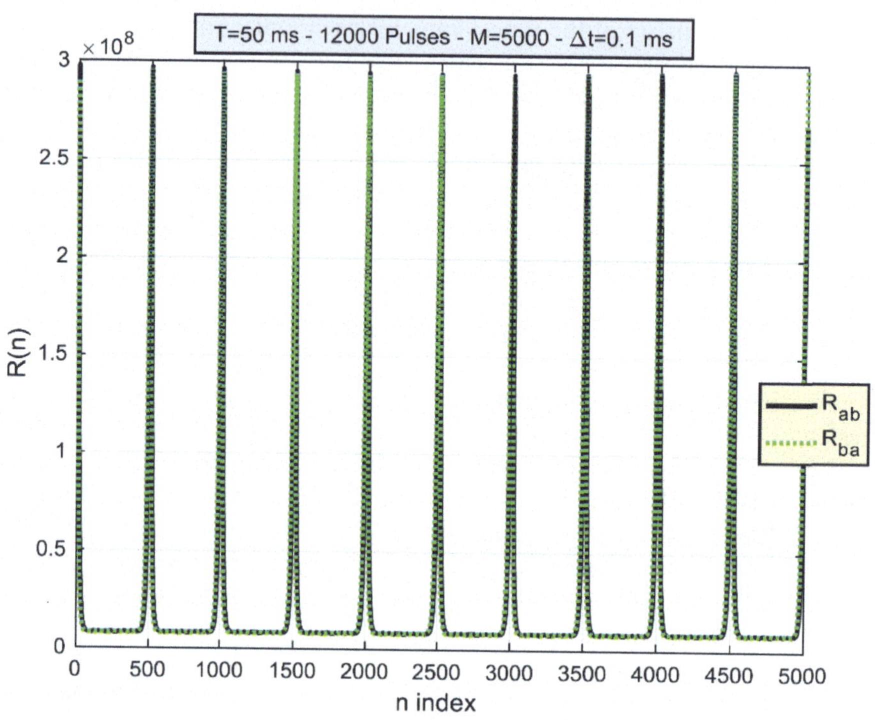

FIG. 2.57. Circular correlation functions Rab and Rba from 12 000 pulses (Reproduced from Ref.[2.77] with permission courtesy of Elsevier B.V.).

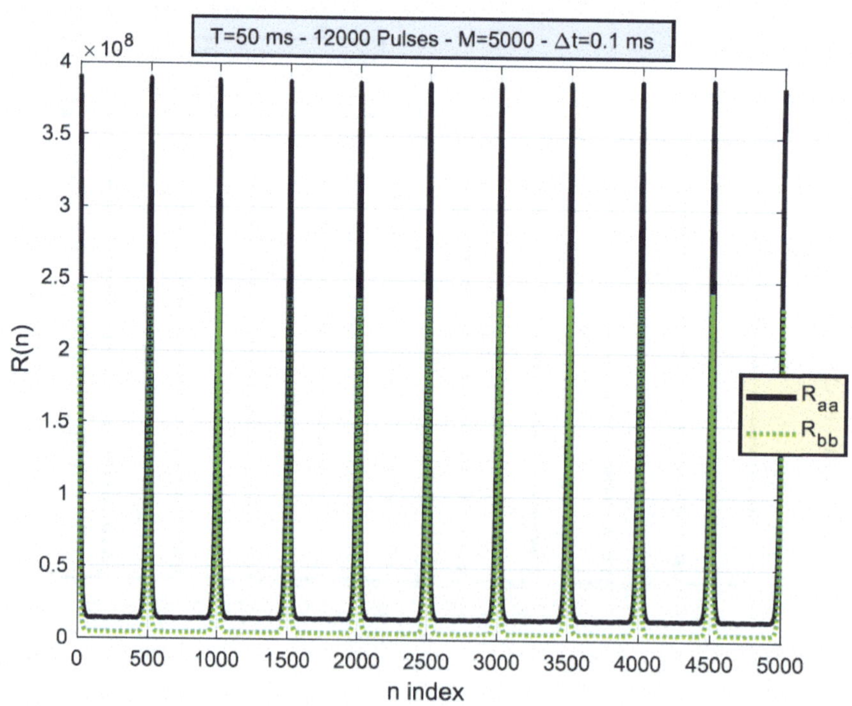

FIG. 2.58. Circular correlation functions Raa, and Rbb from 12 000 pulses (Reproduced from Ref. [2.77] with permission courtesy of Elsevier B.V.).

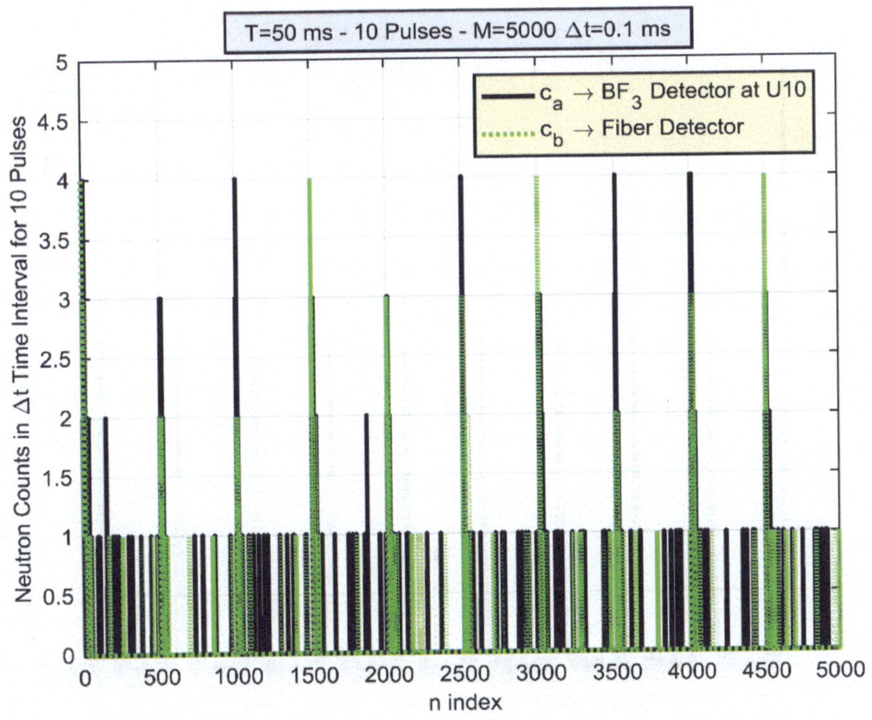

FIG. 2.59. Neutron counts arrays ca and cb from 10 pulses without the collapsing procedure (Reproduced from Ref. [2.77] with permission courtesy of Elsevier B.V.).

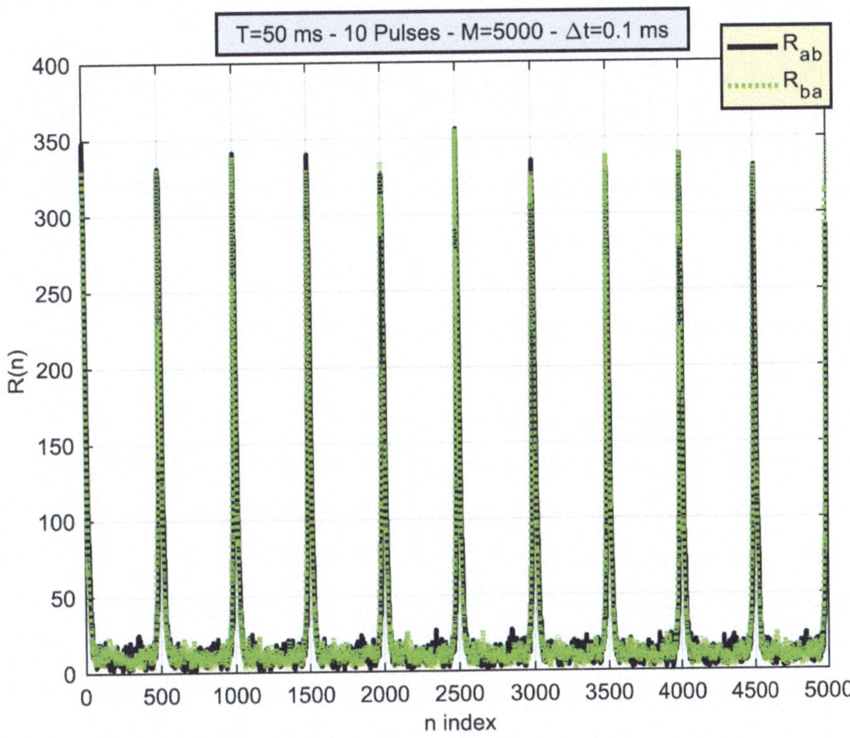

FIG. 2.60. Circular correlation functions Rab and Rba from 10 pulses without the collapsing procedure (Reproduced from Ref. [2.77] with permission courtesy of Elsevier B.V.).

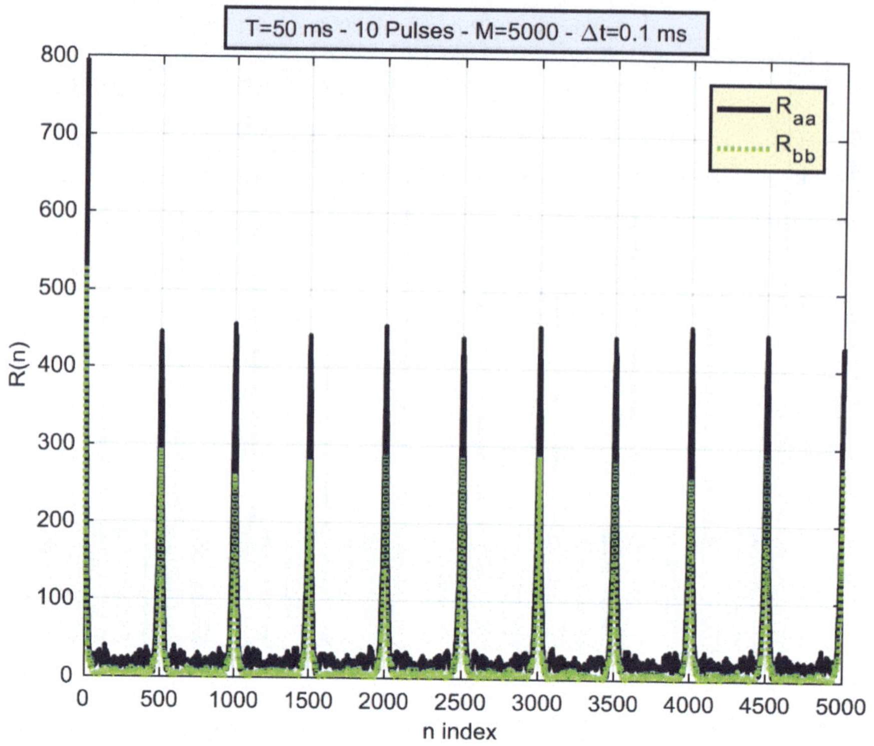

FIG. 2.61. Circular correlation functions Raa, and Rbb from 10 pulses (Reproduced from Ref. [2.77] with permission courtesy of Elsevier B.V.).

Figure 2.62 shows the real and imaginary parts of the cross power spectral density, as given by Eqs (2.26) and (2.27). Figs 2.63 and 2.64 show the real parts of the auto power spectral densities Eq. (2.26). There is no imaginary part of the auto power spectral densities, as the Fourier transform of an even function. Figures 2.65–2.67 show the corresponding data from 10 pulses without the collapsing procedure.

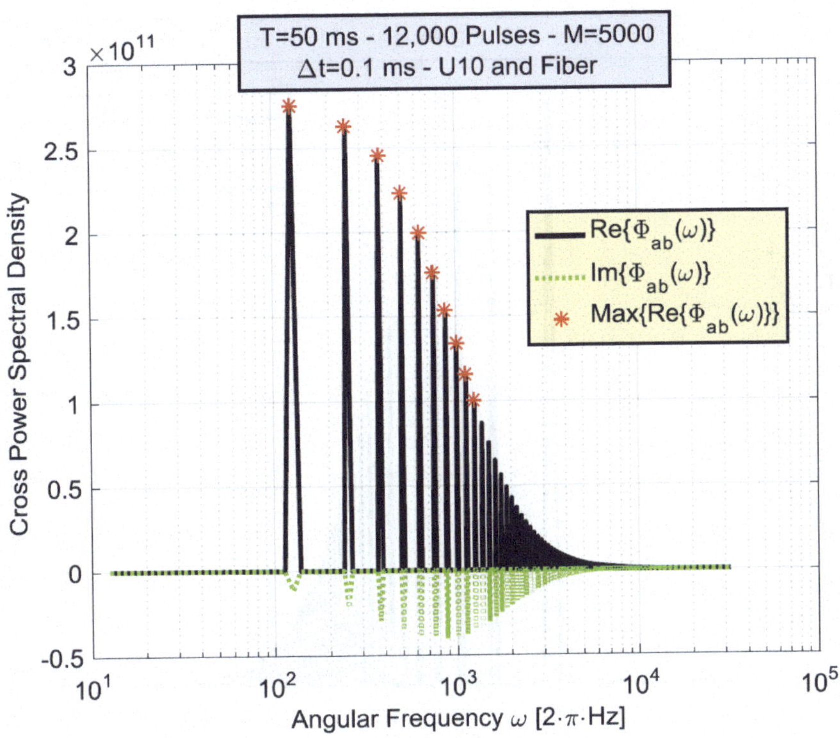

FIG. 2.62. Real and imaginary parts of the cross power spectral density Φab as a function of the angular frequency from 12 000 pulses (Reproduced from Ref. [2.77] with permission courtesy of Elsevier B.V.).

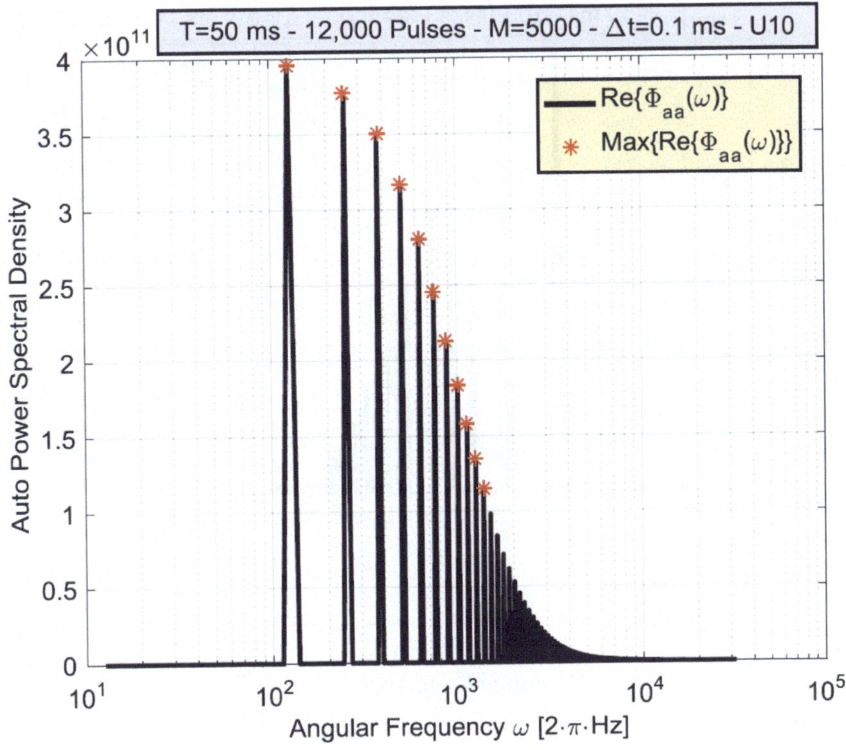

FIG. 2.63. The auto power spectral density Φaa as a function of the angular frequency from 12 000 pulses (Reproduced from Ref. [2.77] with permission courtesy of Elsevier B.V.).

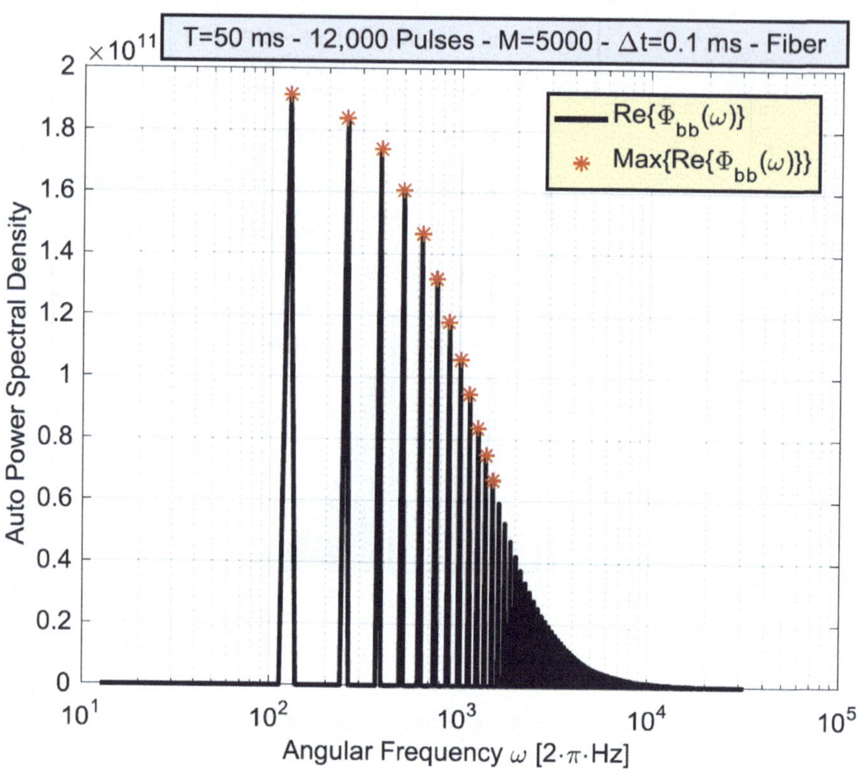

FIG. 2.64. The auto power spectral density Φbb as a function of the angular frequency from 12 000 pulses (Reproduced from Ref. [2.77] with permission courtesy of Elsevier B.V.).

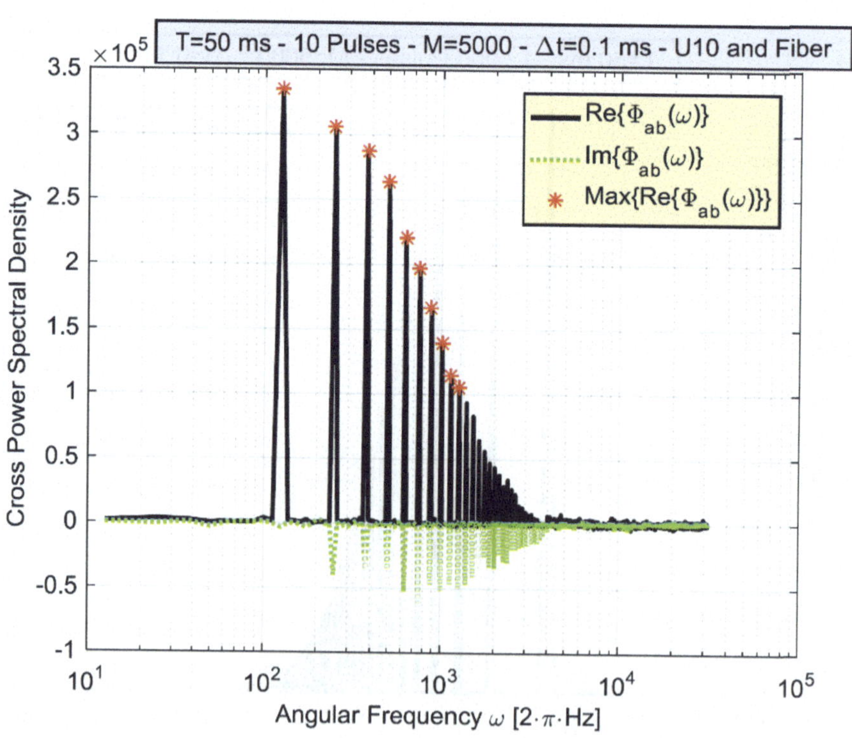

FIG. 2.65. Real and imaginary parts of the cross power spectral density Φab as a function of the angular frequency from 10 pulses without the collapsing procedure (Reproduced from Ref. [2.77] with permission courtesy of Elsevier B.V.).

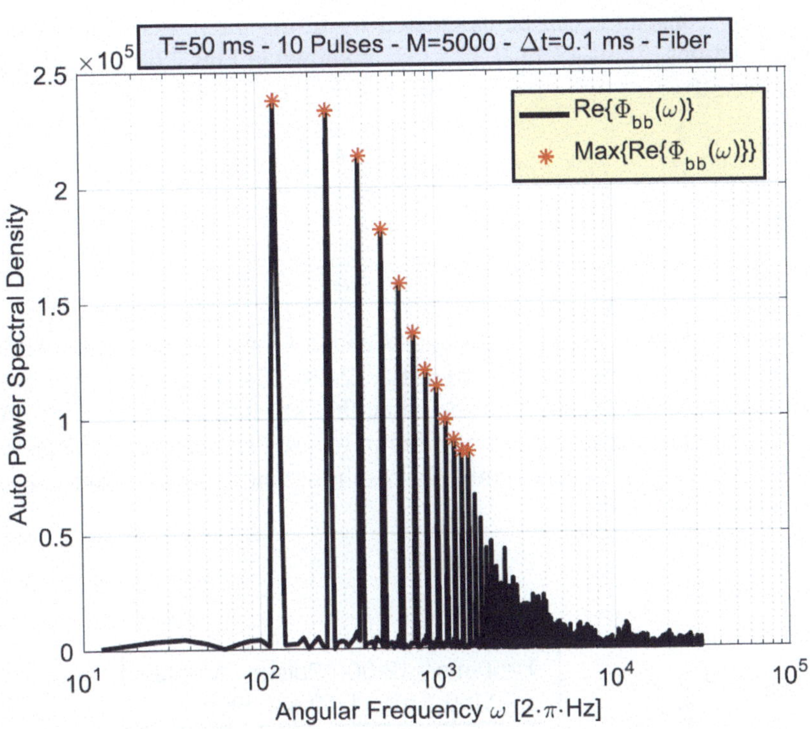

FIG. 2.66. The auto power spectral density Φaa as a function of the angular frequency from 10 pulses without the collapsing procedure. (Reproduced from Ref. [2.77] with permission courtesy of Elsevier B.V.).

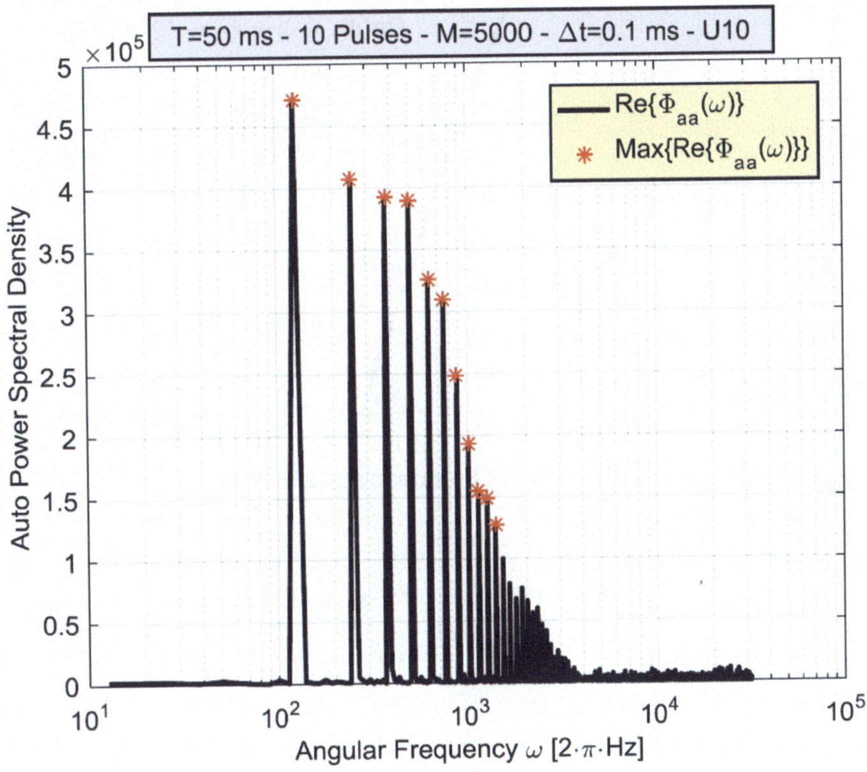

FIG. 2.67. The auto power spectral density Φbb as a function of the angular frequency from 10 pulses without the collapsing procedure. (Reproduced from Ref. [2.77] with permission courtesy of Elsevier B.V.).

The prompt neutron decay constant α can be calculated by fitting the peaks of the real part of the spectral densities with rational functions [2.20], [2.112], [2.113], as shown in Figs 2.68–2.70. The fitting functions are given in Eqs (2.29) and (2.30).

$$Re\{\Phi_{ab}(\omega)\} \cong \frac{C}{\alpha^2 + \omega^2} \qquad (2.29)$$

$$Re\{\Phi_{aa}(\omega)\} \cong \frac{C}{\alpha^2 + \omega^2} + B \qquad (2.30)$$

where C and B are constants derived from the fitting procedure. Using two detectors eliminates the noise that is present when only one detector is used (B is zero in Eq. (2.29). The values of α obtained in this experiment ranged from 958 to 1020 s^{-1}, in agreement with the value 964 s^{-1} shown in Table 2.35. As noted above, the results are based on the point kinetics assumption and require a spatial correction factor to mitigate the effect of detector position [2.103], [2.108]. Consequently, the results are more reliable from the cross power spectral density than the auto power spectral density.

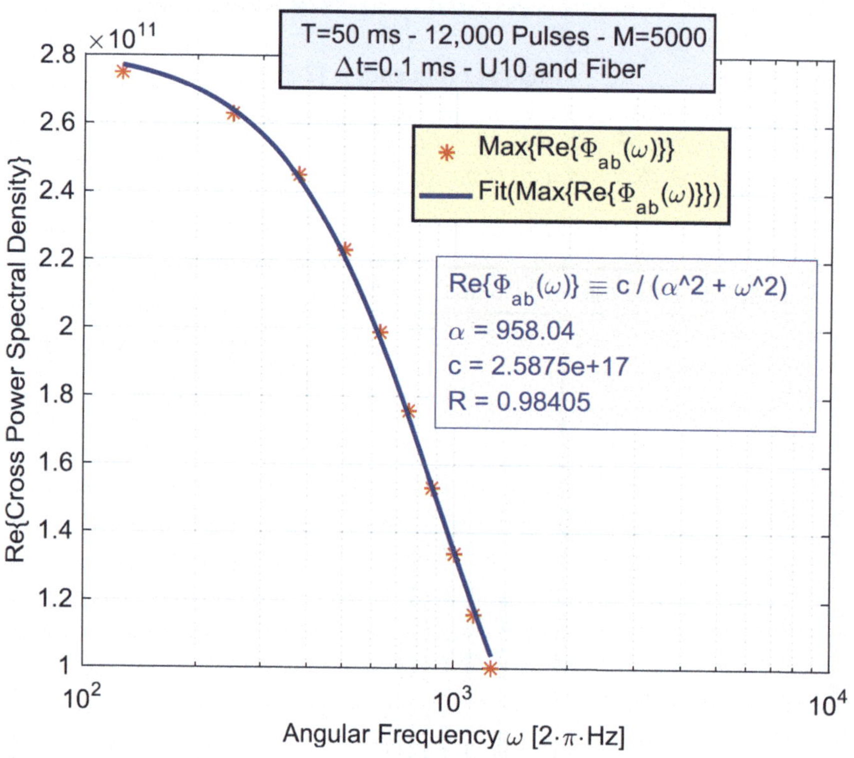

FIG. 2.68. Fitting the peaks of the real part of the cross power spectral density Φab from 12 000 pulses (Reproduced from Ref. [2.77] with permission courtesy of Elsevier B.V.).

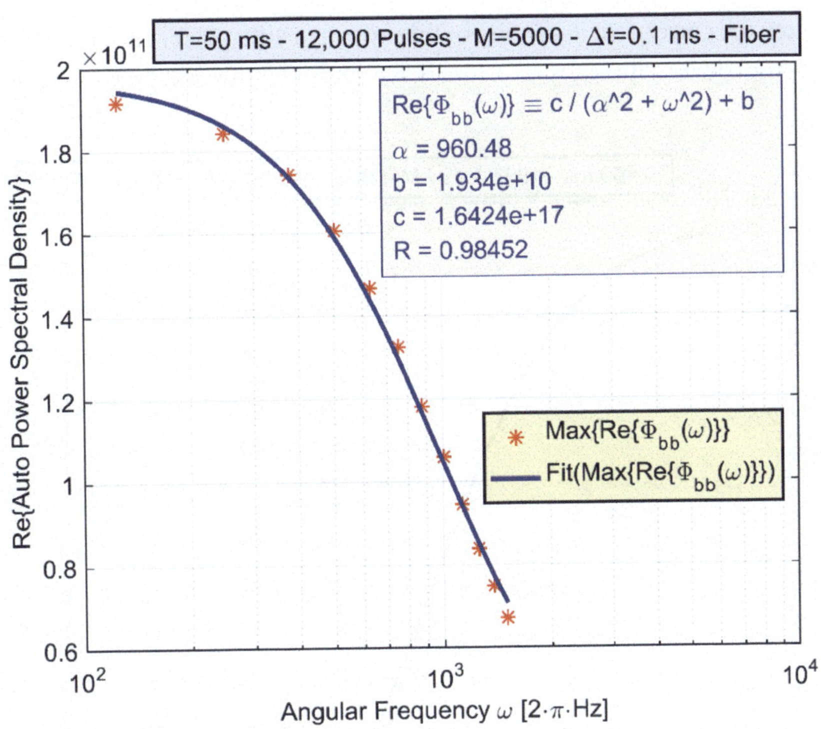

FIG. 2.69. Fitting the peaks of the auto power spectral density Φaa from 12 000 pulses (Reproduced from Ref. [2.77] with permission courtesy of Elsevier B.V.).

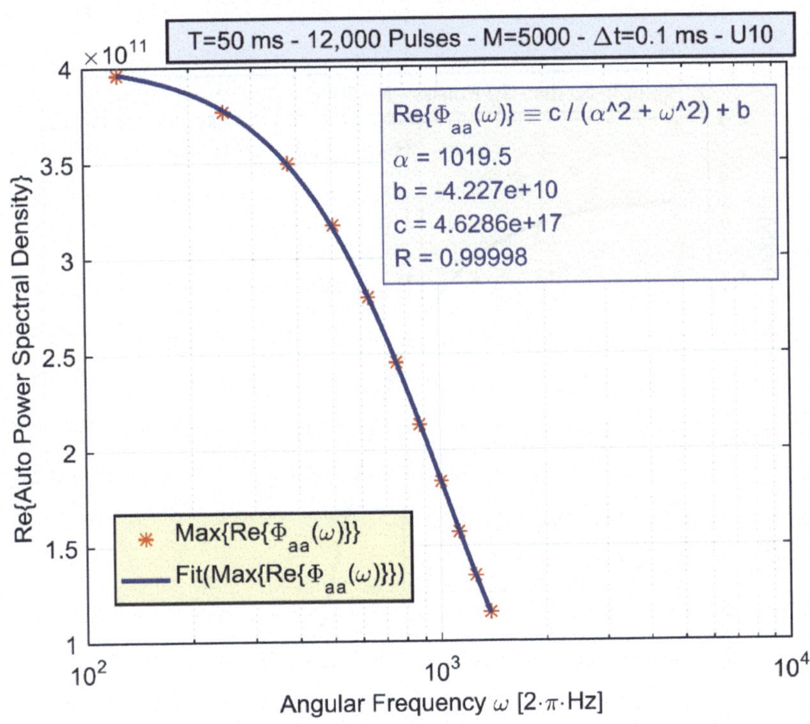

FIG. 2.70. Fitting the peaks of the auto power spectral density Φbb from 12 000 pulses (Reproduced from Ref. [2.77] with permission courtesy of Elsevier B.V.).

Figures 2.71–2.73 show the corresponding fitting of the cross and auto power spectral densities from 10 pulses without the collapsing procedure. It can be seen that this may lead to inaccurate results.

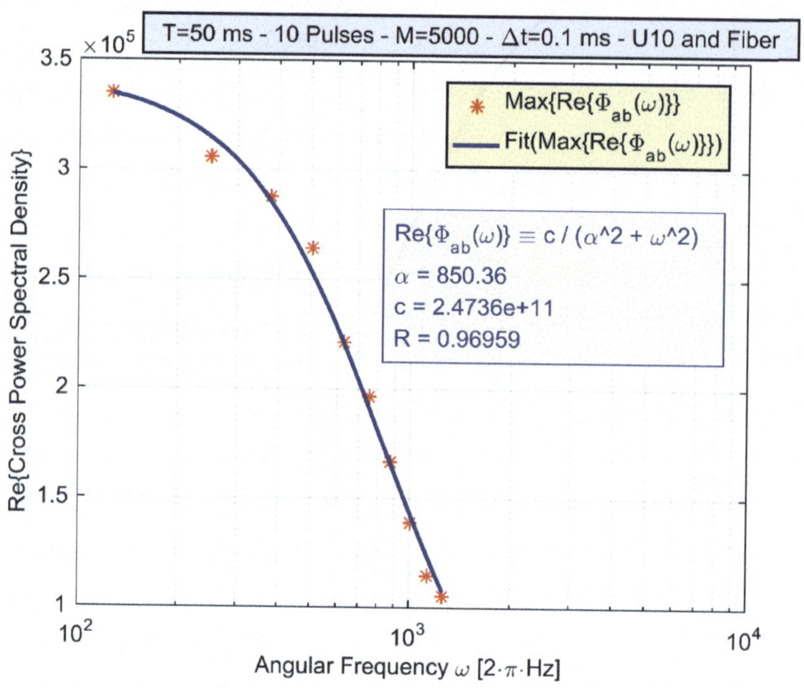

FIG. 2.71. Fitting the peaks of the real part of the cross power spectral density Φab from 10 pulses without the collapsing procedure (Reproduced from Ref. [2.77] with permission courtesy of Elsevier B.V.).

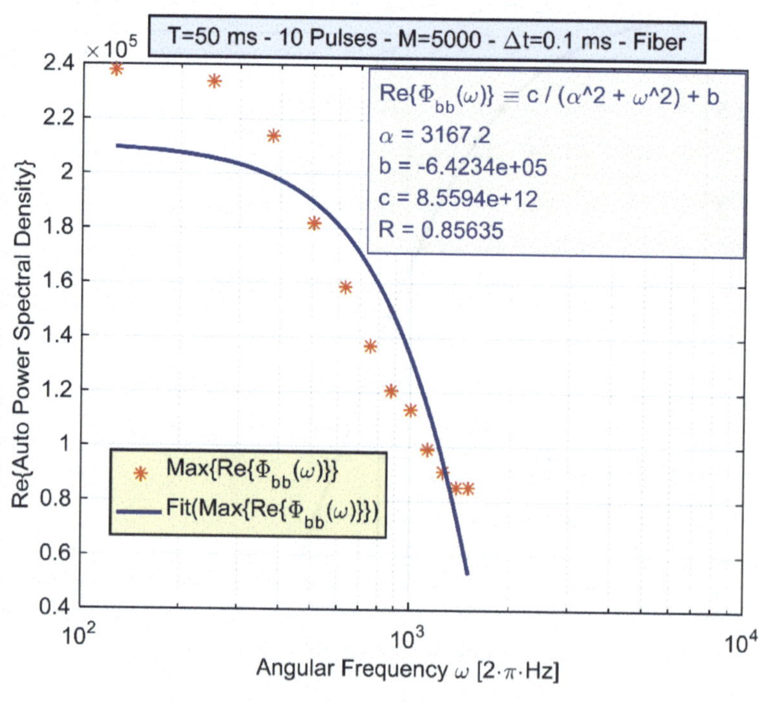

FIG. 2.72. Fitting the peaks of the auto power spectral density Φaa from 10 pulses without the collapsing procedure (Reproduced from Ref. [2.77] with permission courtesy of Elsevier B.V.).

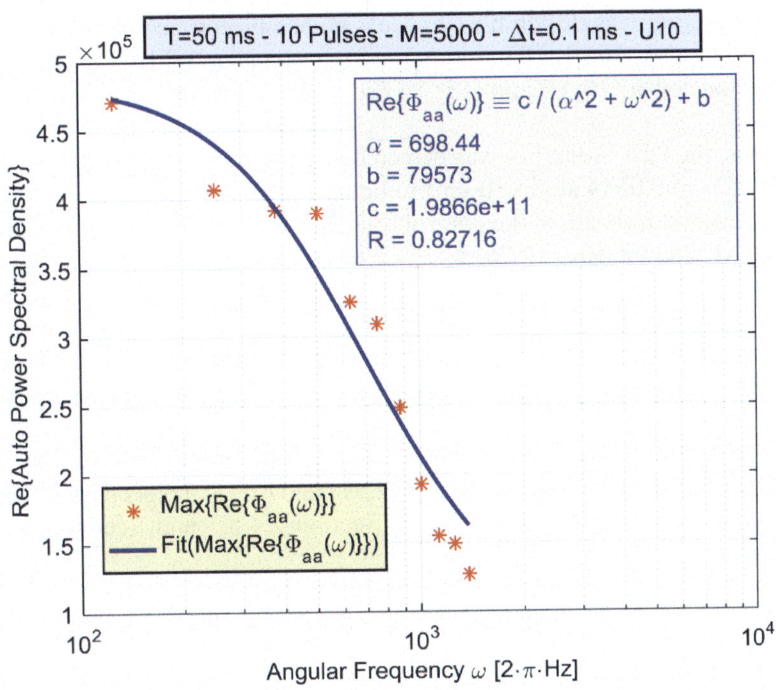

FIG. 2.73. Fitting the peaks of the auto power spectral density Φbb from 10 pulses without the collapsing procedure (Reproduced from Ref. [2.77] with permission courtesy of Elsevier B.V.).

The prompt neutron decay constant depends upon the arbitrary parameters set in the MATLAB script, the number of collapsed pulses was changed from 10 to 20 and the time bin of the neutron counts arrays c_a and c_b were varied from 0.1 to 2 ms as shown in Table 2.37.

TABLE 2.37. PROMPT NEUTRON DECAY CONSTANT OBTAINED BY DIFFERENT VALUES OF THE NUMBER OF PULSES AND TIME BIN OF THE C_A AND C_B ARRAYS [2.77]

Number of Pulses	Time bin (ms)	$\Phi_{ab}\ \alpha\ (s^{-1})$	$\Phi_{aa}\ \alpha\ (s^{-1})$	$\Phi_{bb}\ \alpha\ (s^{-1})$
10	0.1	958	1019	960
20	0.1	947	1014	958
10	0.5	958	1019	972
10	2	963	1019	987

The observed standard deviations of α according to the method of calculation were:

- cross power spectral density, 6 s^{-1};
- auto power spectral density for the BF$_3$ detector at (10, U) in the reflector zone, 2 s^{-1};
- auto power spectral density for the fibre detector in the fuel zone, 13 s^{-1}.

2.1.5.7. Calculation of the prompt neutron decay constant for the KUCA facility driven by a stationary or pulsed external neutron source

This section summarises the detailed discussion provided in Ref. [2.79]. Various methods of calculating the prompt neutron decay constant α of the KUCA [2.122], [2.27] were evaluated. The value of α was derived from timestamps of ^{10}B neutron captures in the BF$_3$ detector and processed using MATLAB scripts. The work used the KUCA I-4 core configuration shown in Figs 2.14(d) and 2.28 in which all control rods were withdrawn. The configuration is described in more detail in Refs [2.76] and [2.123].

For the experiments, the facility was driven by 100 MeV pulsed proton beam with a tungsten target, a pulse duration of 100 μs and period of 50 ms. Timestamps with a precision of 12.5 ns were collected from three detectors: BF_3 detectors located at (10, U) and (15, X) and a fibre detector located in the fuel zone.

In the MCNP simulations, the KUCA facility was driven by a stationary ^{252}Cf neutron source with the nominal fuel density reduced by 5% to 0.0544 atoms/(b·cm) to better match the experimental results [2.4], [2.76] and without modelling the aluminium sheath of the control rod cells. MCNP provided timestamps with nanosecond precision using the PTRAC and F8 cards [2.76].

The detector response in subcritical systems is significantly impacted by its position relative to the external neutron source because the neutron flux does not follow the fundamental mode. To ensure compatibility between experimental and modelled data, the ^{252}Cf source was modelled at the target position.

The two methods used to determine the prompt neutron decay constant α were based on point kinetics and on fitting the Rossi and Feynman distributions using Eqs (2.31) and (2.32), respectively. These distributions and their calculational procedures are further discussed in Section 2.1.5.5 and Section 5.6, and in Refs [2.46], [2.76], [2.97], and [2.101].

$$Rossi(\tau) = A + Be^{-\alpha\tau} \tag{2.31}$$

$$Feynman(\tau) = \frac{var\big(c(\tau)\big)}{mean\big(c(\tau)\big)} - 1 \approx \frac{B}{\alpha^2}\left(1 - \frac{e^{-\alpha\tau}}{\alpha\tau}\right) \tag{2.32}$$

where τ is the time interval between two neutron capture events, A is a constant determined by neutron captures from different fission chains, B is a constant determined by neutron captures from the same fission chain.

The parameter τ has a different physical meaning for the Rossi and Feynman distributions. For the Rossi distribution, τ is the time interval between two consecutive neutron captures. For the Feynman distribution, τ is the time gate, an arbitrary time interval used to construct the neutron counts array, c, from the detector timestamps.

MCNP scored 131 049 ^{10}B(n, α)^{7}Li reactions in the detector at (10, U) and 352 534 reactions in the detector at (15, X) using the PTRAC and F8 cards. PTRAC requires analogue capture mode, but this may have little impact on the Rossi and Feynman distributions [2.124]. PTRAC has been used for Rossi and Feynman distributions with the SILENE reactor [2.125]. The MVP-III Monte Carlo code is also able to derive α from the Feynman distribution [2.126] but was not used in this study.

The simplest method to determine α is by fitting the detector count rate as a function of time during the decay of prompt neutrons [2.76]. A single exponential function is fitted in the linear section of the curve as plotted with a logarithmic y-axis. Alternatively, α can be found by using a pulsed external neutron source and fitting the Rossi distribution using Eq. (2.33) or the Feynman distribution using Eq. (2.34) [2.35].

$$
\begin{aligned}
Rossi\ (\tau) \cong\ & \frac{C_1}{(\alpha_p + \alpha_d)(\alpha_p - \alpha_d)}\left(1 - \frac{C_2\lambda^2}{\alpha_p^2}\right)\frac{\alpha_p}{2}e^{-\alpha_p\tau} \\
& + \frac{C_1}{(\alpha_p + \alpha_d)(\alpha_d - \alpha_p)}\left(1 - \frac{C_2\lambda^2}{\alpha_d^2}\right)\frac{\alpha_d}{2}e^{-\alpha_d\tau} \\
& + C_3 \sum_{n=1}^{\infty} \cos \omega_n\tau \frac{4(\lambda^2 + \omega_n^2)}{n^2\pi^2(\alpha_p^2 + \omega_n^2)(\alpha_d^2 + \omega_n^2)}\left(\sin \frac{\omega_n W}{2}\right)^2
\end{aligned}
\tag{2.33}
$$

$$\text{Feynman } (\tau) \cong \frac{C_1}{(\alpha_p + \alpha_d)(\alpha_p - \alpha_d)}(1 - \frac{C_2\lambda^2}{\alpha_p^2})(1 - \frac{1 - e^{-\alpha_p\tau}}{\alpha_p\tau})$$

$$+ \frac{C_1}{(\alpha_p + \alpha_d)(\alpha_d - \alpha_p)}(1 - \frac{C_2\lambda^2}{\alpha_d^2})(1 - \frac{1 - e^{-\alpha_d\tau}}{\alpha_d\tau})$$

$$+ \quad \frac{C_3}{\tau}\sum_{n=1}^{\infty}\frac{1}{\omega_n^2}(\sin\frac{\omega_n\tau}{2})^2\frac{4(\lambda^2 + \omega_n^2)}{n^2\pi^2(\alpha_p^2 + \omega_n^2)(\alpha_d^2 + \omega_n^2)}(\sin\frac{\omega_n W}{2})^2 \tag{2.34}$$

where T is the accelerator pulse period, W the pulse width, C_i (with i = 1, 2, or 3) is a constant, λ is the delayed neutron decay constant, and α_p and α_d are the two roots of the in-hour equation with 1 group of delayed neutrons, and ω_n is defined by Eq. (2.35). α_p is the prompt neutron decay constant. In both equations, the magnitude of the external neutron source affects only C_3. Therefore, for a given facility configuration and k_{eff} value, the sinusoid from the summation is amplified by a stronger external neutron source.

$$\omega_n = \frac{2\pi n}{T} \tag{2.35}$$

In this work, the summation upper limit was set at 1000 because the higher terms are insignificant. The Rossi distribution was fitted using the GNUPLOT software.

The prompt neutron decay constant α can also be determined by fitting the cross power spectral density or the auto power spectral density as discussed in Section 2.1.5.5 and Ref. [2.77]. The cross power spectral density method is less sensitive to detector position because it uses two detectors.

The MAρTA method proposed in Refs [2.45] and [2.127] in another method to determine α. The detector count rate P is proportional to the core power and can be written as a combination of exponential functions as shown in Eq. (2.36).

$$P(t) \cong b_0 + b_1 e^{-\alpha_1 t} + b_2 e^{-\alpha_2 t} \tag{2.36}$$

where b_i (with i = 0, 1, 2) and α_j (with j = 1, 2) are fitting parameters. The count rate P is derived from the detector timestamps. The values for b_i and α_j can then be used with Eq. (2.37) to calculate the time derivative of the core power.

$$P'(t) = -\alpha_1 b_1 e^{-\alpha_1 t} - \alpha_2 b_2 e^{-\alpha_2 t} \tag{2.37}$$

The point kinetics analytical framework assumes that Eq. (2.38) holds at the end of the pulse period T.

$$C_i = \frac{P(T)\beta_{eff}}{\Lambda_{eff}\lambda_i} \tag{2.38}$$

where C_i is the concentration of delayed neutrons precursors for the i group (with i = 1, 2, ... 6) of delayed neutrons, β_{eff} is the delayed neutron fraction, Λ_{eff} is the neutron generation time, and λ_i is the precursors decay constant for the i group of delayed neutrons. β_{eff} and Λ_{eff} were obtained using MCNP without modelling the external neutron source, λ_i were obtained from nuclear data libraries.

MAρTA links the core power from Eq. (2.36), its time derivative from Eq. (2.37), and the delayed neutron precursors at the end of the pulse period from Eq. (2.38) as given in Eq. (2.39).

$$P'(t) - \omega_1 P(t) - \sum_{i=1}^{6}\frac{\lambda_i\omega_1(t)C_i}{\omega_1(t) + \lambda_i} = 0 \tag{2.39}$$

where ω_1 is the reactor period. Equation (2.39) assumes that the delayed neutron precursors are time-invariant. Equation (2.39) was solved using MATLAB to obtain ω_1 and the core reactivity was found from the in-hour equation shown in Eq. (2.40), which has 7 roots, assuming 6 groups of delayed neutrons.

$$\rho = \omega_1(T)\Lambda_{eff} + \sum_{i=1}^{6} \frac{\omega_1(T)\beta_i}{\omega_1(T) + \lambda_i} \tag{2.40}$$

The prompt neutron decay constant was then calculated using Eq. (2.41).

$$\alpha = \omega_7(T) = \frac{\rho - \beta_{eff}}{\Lambda_{eff}} \tag{2.41}$$

The Rossi, Feynman, the cross power spectral density, and the auto power spectral density distribution methods are noise methods [2.101]. An underlying assumption is that after a detector scores one neutron, it will score $\nu-1$ neutrons from the same fission chain and ν neutrons from an unrelated fission chain, where ν is the number of fission neutrons per fission event.

The Rossi and the Feynman distribution methods can be used with both fixed external sources such as ^{252}Cf or pulsed neutron sources. The cross power spectral density method and the auto power spectral density method are frequency domain analyses that require a pulsed neutron source. In MCNP simulations using a fixed source, the average neutron counts as a function of time can be obtained from the F4 tally results or from the timestamps from the PTRAC and F8 cards [2.97].

The count rate fitting and MAρTA methods are not noise methods. They are applied to the count rate as a function of time and so do not require the detector timestamps. They can be only used for subcritical assemblies with a pulsed external neutron source.

The work described here used MCNP with timestamps from the PTRAC and F8 cards, but not the F4 card. To increase the neutron count, the simulation assumed an enlarged detector volume compared to the experimental setup; 2.5 cm radius and 38 cm height compared to 0.635 cm radius (10, U) and 1.27 cm radius (15, X), both with 30 cm height. The MCNP simulation scored 40 437 and 107 509 ^{10}B(n, α)^{7}Li reactions in the (10, U) and (15, X) detector positions, respectively.

See Table 2.38 gives the kinetic parameters k_{eff}, β_{eff} and ρ obtained using the ENDF/B-VII, ENDF/B-VIII.β3, and JENDL-4.0 nuclear data libraries and without modelling the external neutron source. The effective neutron multiplication factors calculated using ENDF/B were 150–300 pcm higher than the values calculated using JENDL-4.0. The delayed neutron fraction and prompt neutron decay constant were insensitive to the choice of nuclear data library.

TABLE 2.38. KINETIC PARAMETERS OBTAINED FROM MCNP6 CRITICALITY SIMULATIONS USING DIFFERENT NUCLEAR DATA LIBRARIES [2.79]

Library	$S(\alpha,\beta)$	k_{eff}	β_{eff} (pcm)	ρ (μs)
ENDF/B-VIII.β3	ENDF/B-VIII.β3	$0.97449 \pm 3\text{E-}5$	809 ± 3	32.57 ± 0.03
ENDF/B-VIII.β3	ENDF/B-VII.1	$0.97581 \pm 4\text{E-}5$	812 ± 5	32.65 ± 0.03
ENDF/B-VII.1	ENDF/B-VII.1	$0.97549 \pm 4\text{E-}5$	819 ± 5	33.02 ± 0.03
JENDL-4.0	ENDF/B-VII.1	$0.97305 \pm 4\text{E-}5$	820 ± 5	33.39 ± 0.04

The Rossi and Feynman distributions plots for the KUCA facility driven by a ^{252}Cf external neutron source, and the derivation of the α value are discussed in Section 2.1.5.5. [2.76]. An additional chart showing the MCNP-derived Feynman distribution for the (10, U) BF$_3$ detector is shown in Fig. 2.74.

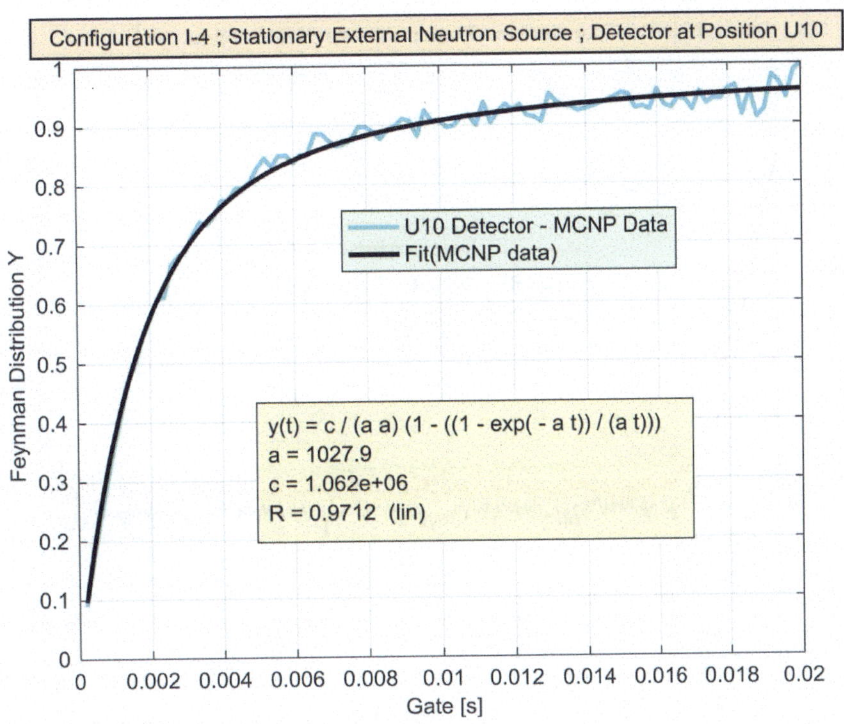

FIG. 2.74. Feynman distribution for the BF3 detector at position (10, U) (Reproduced from Ref. [2.78] with permission courtesy of Taylor & Francis.).

Figures 2.75–2.77 show the neutron count, Rossi and Feynman distributions obtained by processing the experimental data with MATLAB scripts for KUCA driven by the pulsed neutron source. The detector (15, X) was located in the reflector zone where the thermal neutron flux was higher and had the highest count rate. The Rossi distribution peaks during the neutron pulses occur due to the capture of neutrons from the same fission chain. The Feynman distribution oscillates due to the summation term of Eq. (2.34), and because the enhancement of the variance to mean ratio of the c array in Eq. (2.32) during neutron pulses causes it to dip.

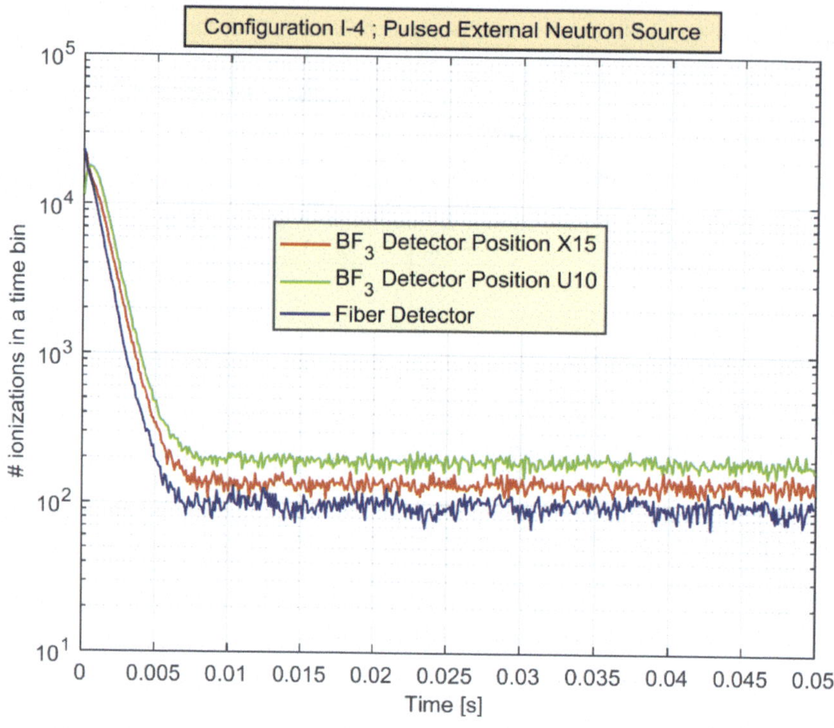

FIG. 2.75. Neutron count as a function of time for three different detectors (Reproduced from Ref. [2.79] with permission courtesy of Taylor & Francis.).

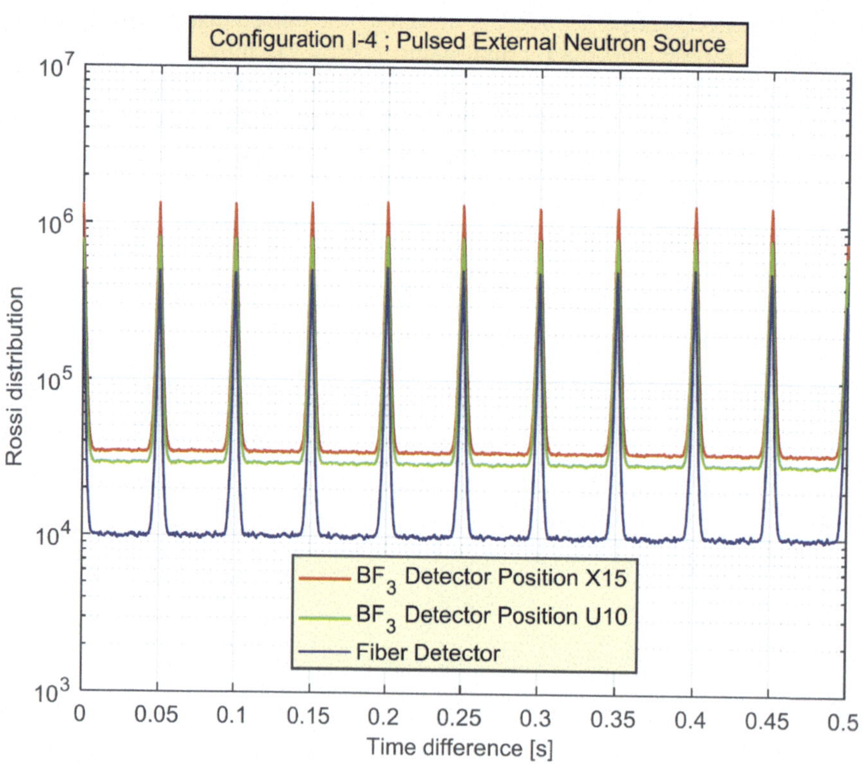

FIG. 2.76. Rossi distribution for three different detectors (Reproduced from Ref. [2.79] with permission courtesy of Taylor & Francis.).

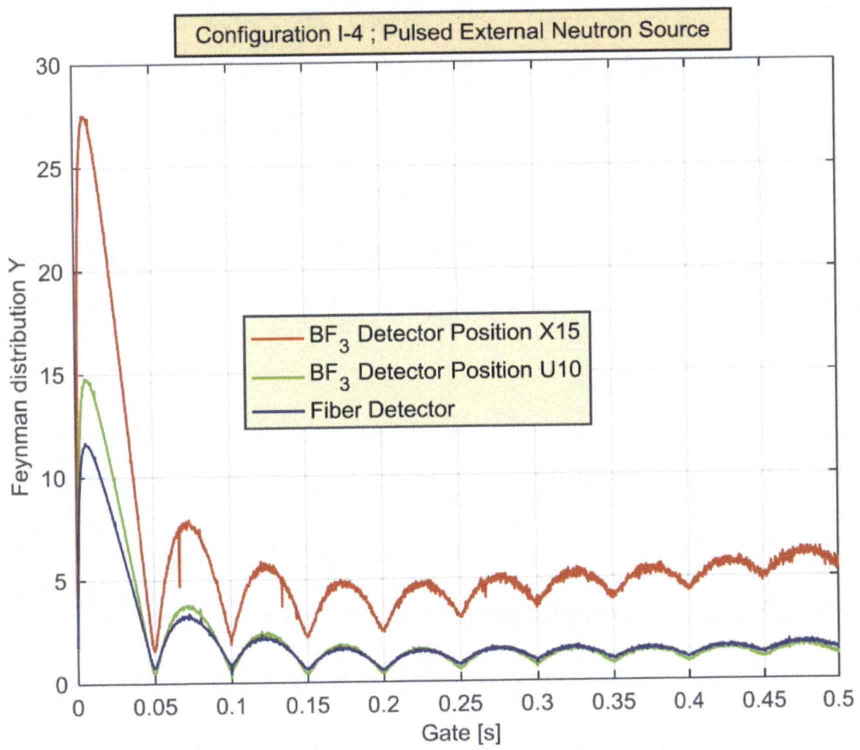

FIG. 2.77. Feynman distribution for three different detectors (Reproduced from Ref. [2.79] with permission courtesy of Taylor & Francis.).

Figure 2.78 shows the Feynman distributions for the (10, U) BF₃ detector calculated by Argonne National Laboratory and Kyoto University using the same experimental timestamps but different scripts.

Figures 2.79 and 2.80 show the fitting of this data using Eq. (2.34) and GNUPLOT software. Figures 2.81 and 2.82 show the fitting of the Rossi distributions for the (10, U) and (15, X) BF₃ detectors. Figures 2.83 and 2.84 show the fitting of the neutron count distributions using Eq. (2.36) and the MAρTA method. The fitting of the cross power spectral density and auto power spectral density distributions, from the same experimental timestamps of this study, is shown in Section 2.1.5.6.

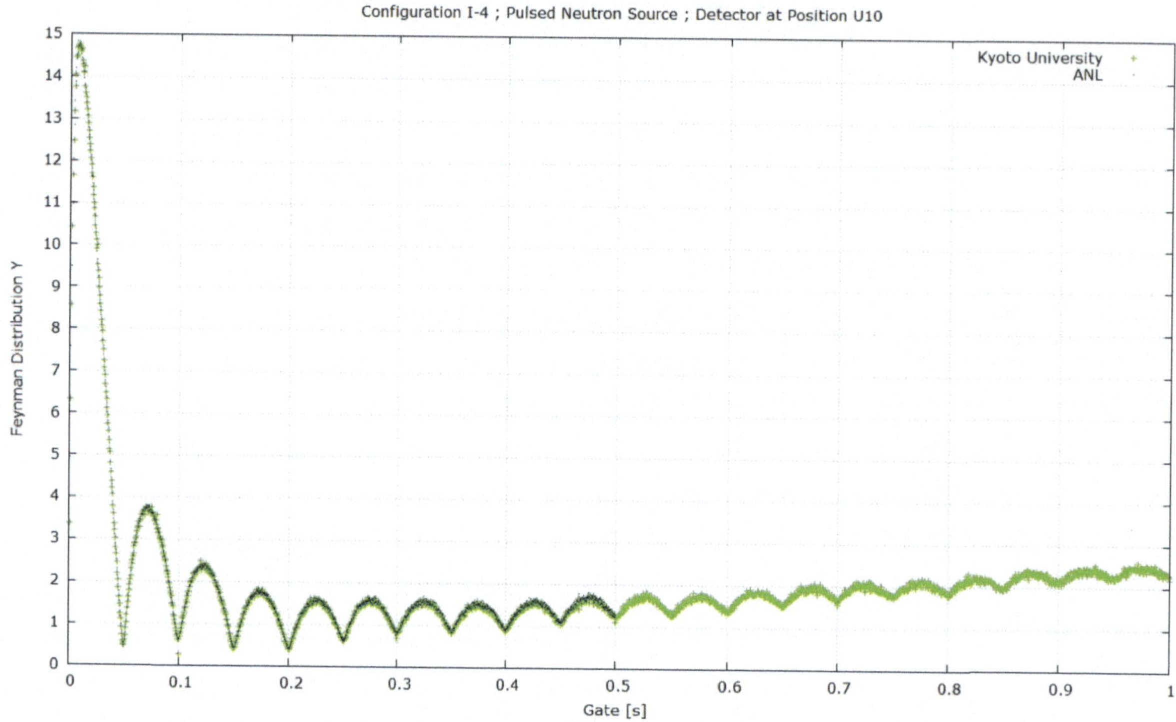

FIG. 2.78. Feynman distribution for the BF3 detector at (10, U) produced by Kyoto University and Argonne National Laboratory (Reproduced from Ref. [2.79] with permission courtesy of Taylor & Francis.).

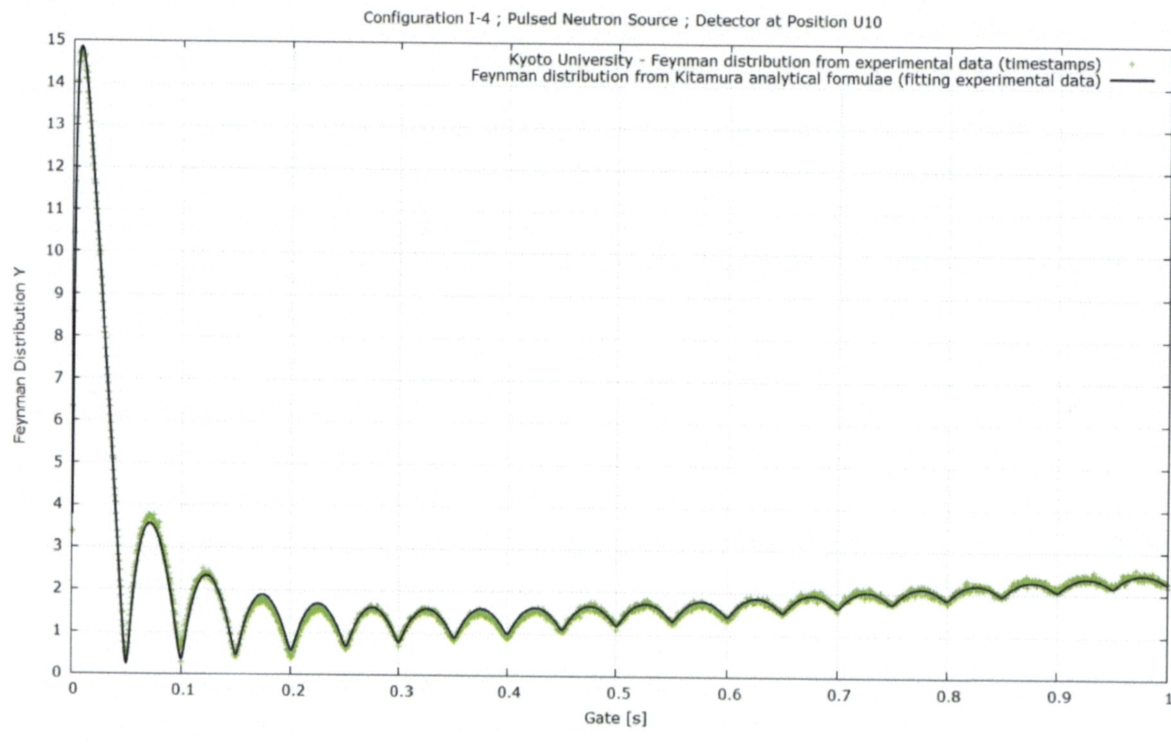

FIG. 2.79. Fitting of the Feynman distribution calculated by Kyoto University (Fig. 2.77) with Eq. (2.34) for the BF₃ detector at (10, U) (Reproduced from Ref. [2.79] with permission courtesy of Taylor & Francis.).

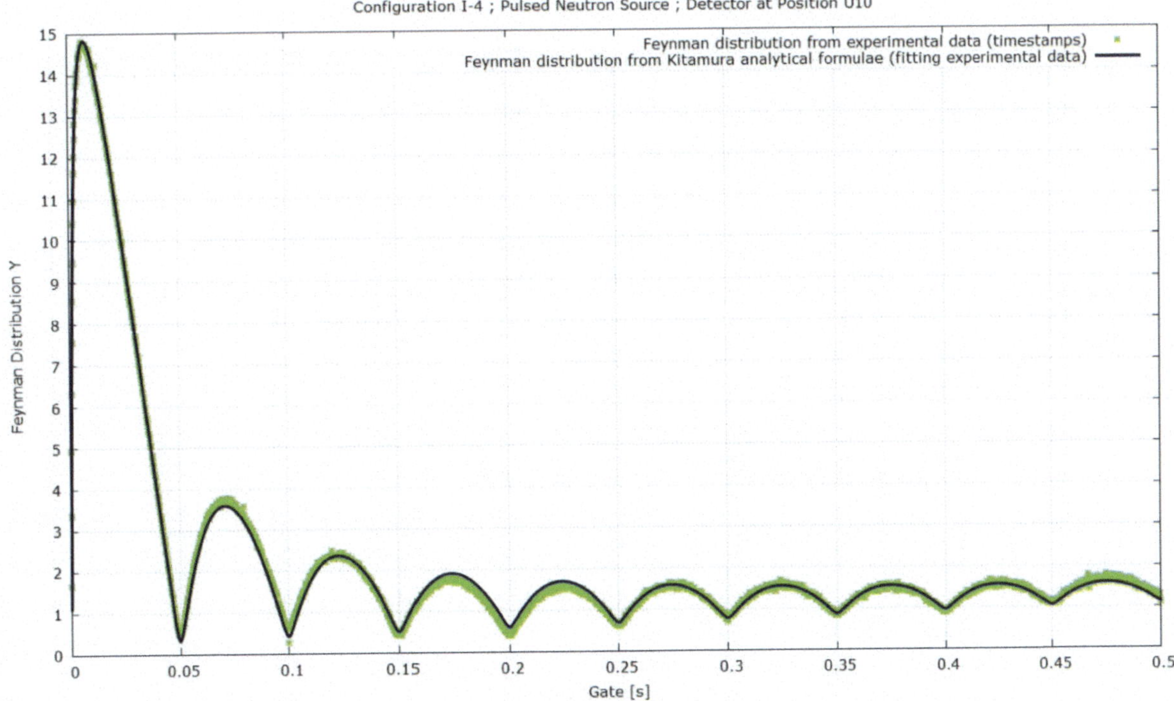

FIG. 2.80. Fitting of the Feynman distribution calculated by Argonne National Laboratory (Fig. 2.77) with Eq. (2.34) for the BF_3 detector at (10, U) (Reproduced from Ref. [2.79] with permission courtesy of Taylor & Francis.).

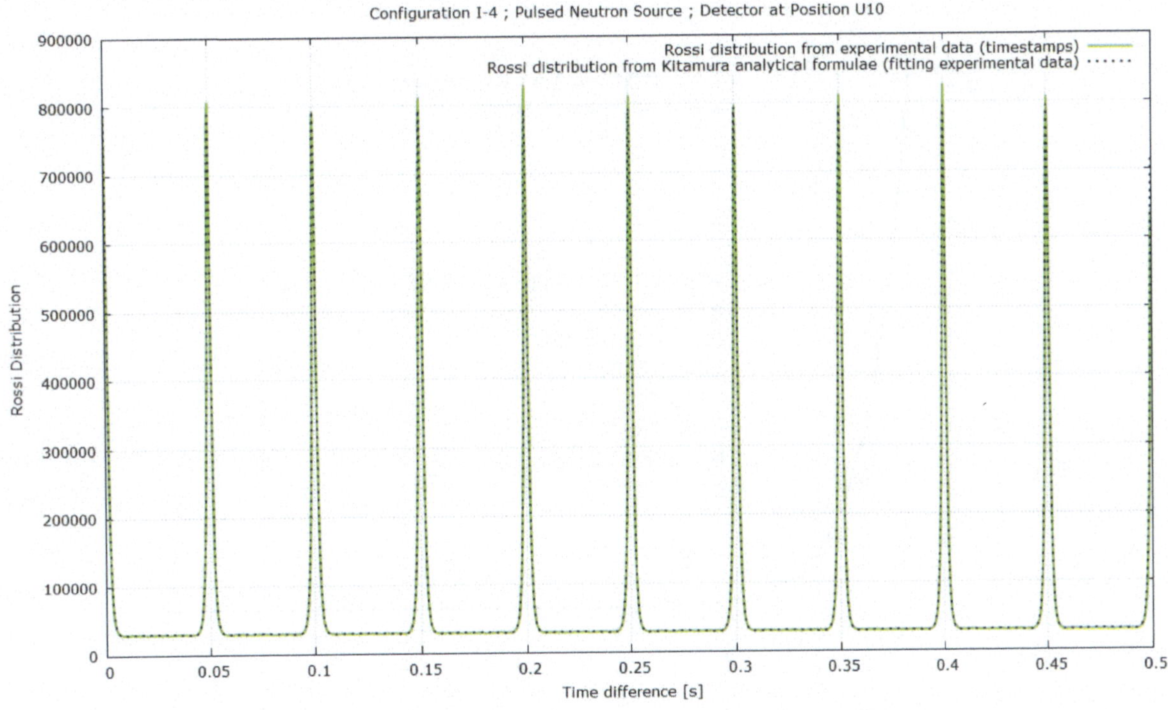

FIG. 2.81. Fitting of the Rossi distribution calculated by Argonne National Laboratory with Eq. (2.33) for the BF_3 detector at (10, U) (Reproduced from Ref. [2.79] with permission courtesy of Taylor & Francis.).

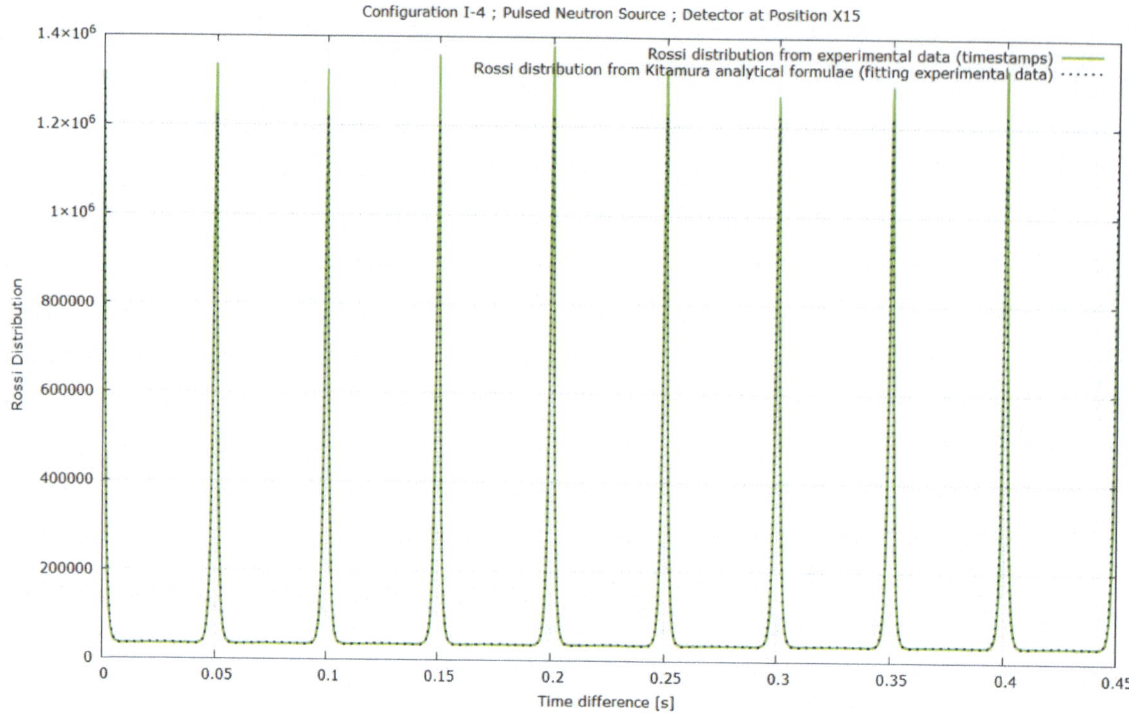

FIG. 2.82. Fitting of the Rossi distribution calculated by Argonne National Laboratory with Eq. (2.33) for the BF$_3$ detector at (15, X) (Reproduced from Ref. [2.79] with permission courtesy of Taylor & Francis.).

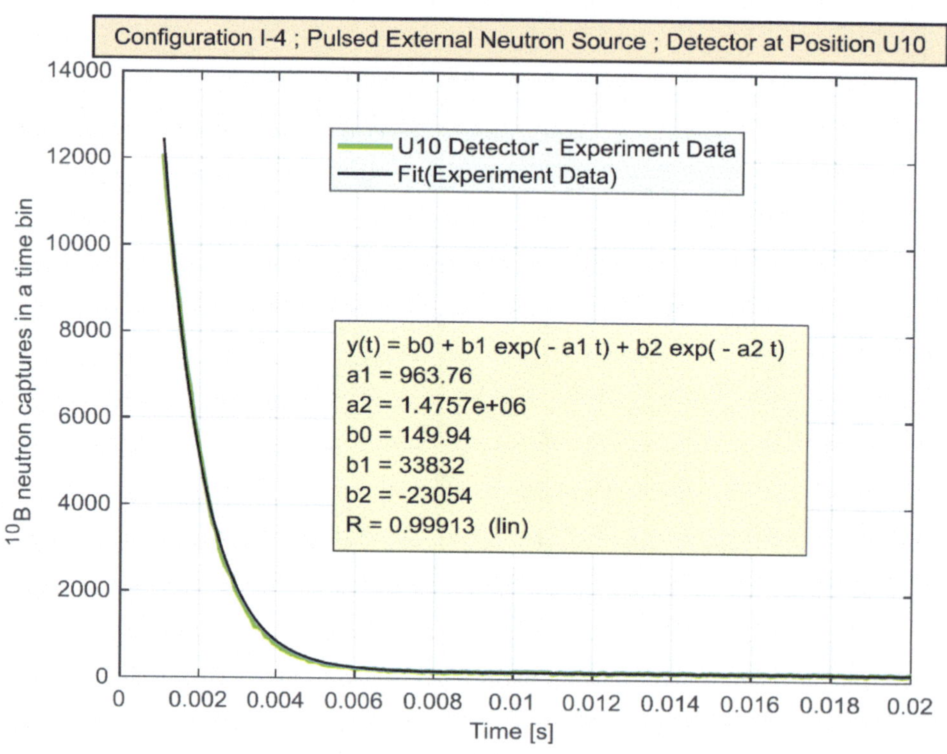

FIG. 2.83. Fitting of the neutron count for the MApTA method and the BF3 detector at (10, U) (Reproduced from Ref. [2.79] with permission courtesy of Taylor & Francis.).

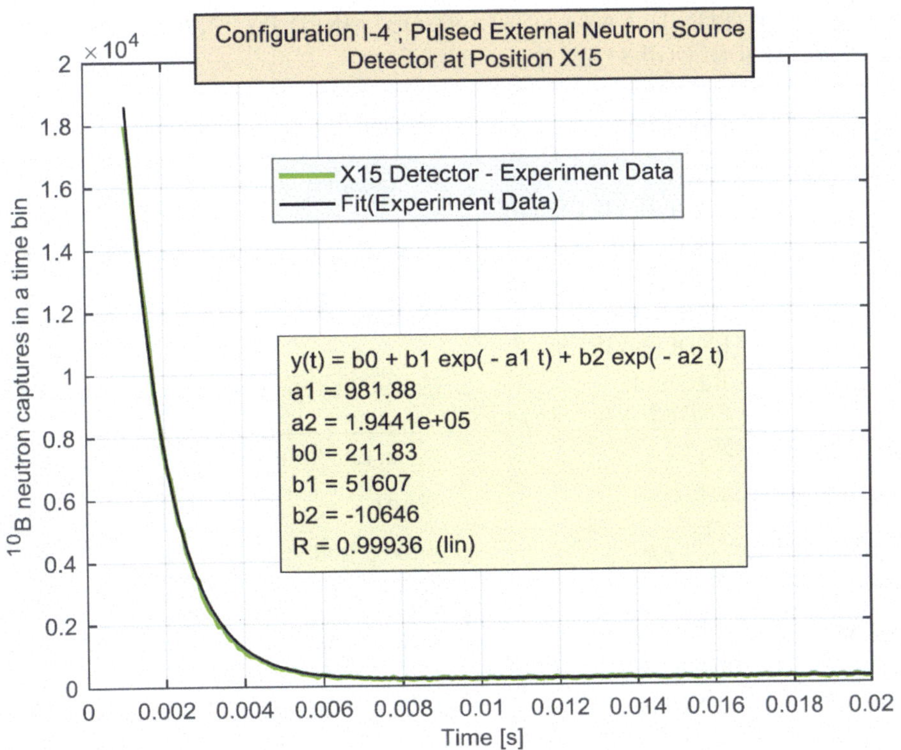

FIG. 2.84. Fitting of the neutron count for the MApTA method and the BF3 detector located at (15, X) position (Reproduced from Ref. [2.79] with permission courtesy of Taylor & Francis.).

Table 2.39 presents the values of the prompt neutron decay constant α for the KUCA I-4 configuration, calculated using different methods. The average α values for the $(15, X)$ and $(10, U)$ detectors were (977 ± 20) s^{-1} and (979 ± 28) s^{-1}, respectively, and the average k_{eff} values were (0.97615 ± 0.00065) and (0.97597 ± 0.00086). All methods agreed on the α value within a 3% uncertainty.

It was noted that the MCNP computation times could exceed several days because the detector volume was small relative to the core volume. Once the timestamps had been calculated, all methods listed in Table 2.39 to obtain the α value required only a few minutes computing time.

TABLE 2.39. PROMPT NEUTRON DECAY CONSTANT OBTAINED BY EXPERIMENTS OR SIMULATIONS USING DIFFERENT METHODOLOGIES [a] [2.79]

Neutron source	Timestamps source	Data library	Method [b]	Detector	α (s^{-1})	k_{eff}
Fission [c]	MCNP	JENDL-4.0 ENDF/B-VII.1 S(α,β)	$\alpha = (\rho\text{-}\beta_{eff})/\rho$	–	1075 ± 2	0.97305 ± 0.00004
^{252}Cf	MCNP	JENDL-4.0 ENDF/B-VII.1 S(α,β)	Rossi fitting	(15, X)	970 ± 10 [d]	0.97638
^{252}Cf	MCNP	JENDL-4.0 ENDF/B-VII.1 S(α,β)	Feynman fitting	(15, X)	981 ± 6 [d]	0.97603
Proton [e]	MCNP	JENDL-4.0 ENDF/B-VII.1 S(α,β)	Count rate fitting	(15, X)	996 ± 4 [d]	0.97556
Proton [e]	Experiment	–	Count rate fitting	(15, X)	982 ± 4 [d]	0.976
Proton [e]	Experiment	–	MAρTA	(15, X)	975 ± 2	0.97622
Proton [e]	Experiment	–	Rossi fitting	(15, X)	938 ± 50	0.9774
Proton [e]	Experiment	–	Feynman fitting	(15, X)	1000 ± 18	0.97543
^{252}Cf	MCNP	JENDL-4.0 ENDF/B-VII.1 S(α,β)	Rossi fitting	(10, U)	947 ± 30 [d]	0.97712
^{252}Cf	MCNP	JENDL-4.0 ENDF/B-VII.1 S(α,β)	Feynman fitting	(10, U)	1028 ± 15	0.97454
Proton [e]	MCNP	JENDL-4.0 ENDF/B-VII.1 S(α,β)	Count rate fitting	(10, U)	987 ± 6 [d]	0.97584
Proton [e]	Experiment	–	Count rate fitting	(10, U)	964 ± 4 [d]	0.97657
Proton [e]	Experiment	–	MAρTA	(10, U)	975 ± 2	0.97622
Proton [e]	Experiment	–	Rossi fitting	(10, U)	975 ± 11	0.97622
Proton [e]	Experiment	–	Feynman fitting	(10, U)	957 ± 19	0.9768
Proton [e]	Experiment	–	CPSD [f] fitting	(10, U) & Fibre	958 ± 2 [g]	0.97677
Proton [e]	Experiment	–	APSD [h] fitting	(10, U)	1019 ± 1 [g]	0.97483

[a] Additional values for α are discussed in Refs [2.76], [2.77], which also show additional plots on the alpha value calculations; [b] All criticality simulations used Eq. (2.41); [c] criticality simulation without external neutron source; [d] Reference [2.76]; [e] Proton beam interacting with target; [f] CPSD: cross power spectral density; [g] Reference [2.77]; [h] APSD: auto power spectral density.

2.1.5.8. Dead-time and spatial corrections for the KUCA subcritical assembly experiments

A neutron detector is affected by dead-time, δ, being the minimum time interval between two successive neutron captures that can be counted by the data acquisition system. Non-paralysable and paralysable dead-time models can be used to correct the experimental results for the dead-time effect [2.97]. In the paralysable model, the dead-time starts with each neutron count. In the non-paralysable model, the dead-time starts at a neutron count and restarts with the next neutron count occurring after a time interval larger than the dead-time. Equations (2.42) and (2.43) show the relationship between the corrected (real) neutron rate c with the measured count rate m for the non-paralysable and paralysable dead-time models, respectively.

$$c \approx \frac{m}{1 - m\delta} \tag{2.42}$$

$$m \approx c \cdot \exp(-c \cdot \delta) \tag{2.43}$$

The KUCA subcritical assembly [2.122], [2.27], [2.46] was used for the investigation of dead-time and spatial corrections. MCNP simulations assumed that the facility was driven by a external californium neutron source, and the experiments used spallation neutrons from a 100 MeV proton accelerator operating in pulsed mode.

The BF$_3$ detector timestamps of neutron captures in MCNP [2.80] simulations were collected using the PTRAC and F8 tally cards according to the procedure in Refs [2.3] and [2.6]. They were then compared with analytical formulae using the non-paralysable and paralysable models of dead-time correction.

The detector timestamps from MCNP neutron transport simulations follow a Poisson distribution. Figure 2.85 compares the number of MCNP neutron capture pairs as a function of the time interval in between them with the curve obtained using Eq. (2.44) [2.129].

$$\Delta t_i = \frac{\log\left(\frac{1}{\xi_i}\right)}{N} \tag{2.44}$$

where Δt_i is the time interval in between the i-th pair of consecutive neutron captures, ξ_i is a pseudorandom number for the i-th pair of consecutive neutron captures, and N is the total number of consecutive neutron captures. The good agreement confirms the accuracy of the MCNP simulations.

Dead-time corrections for the BF$_3$ detector timestamps obtained experimental work were made by MATLAB scripts based on the backward extrapolation method [2.128], [2.116]. The experimental reactivity was calculated from the dead-time corrected timestamps by the area method [2.116], with spatial effects corrected using the Bell and Glasstone spatial correction factor [2.108].

Figures 2.86 and 2.87 compare the count rate obtained from MCNP simulations and MATLAB scripts with the measured count rate m obtained by Eqs (2.42) and (2.43) for the detectors at position (15, X) and (10, U), respectively. The black curve represents the real (corrected) count rate from MCNP simulation, which is insensitive to the data acquisition system dead-time. The red and blue curves are the results of MATLAB scripts processing the MCNP timestamps using the non-paralysable or the paralysable dead-time models, respectively, using the rules from Ref. [2.116]. The green and cyan curves were plotted from Eqs (2.43) and (2.44), respectively, using the c value from the MCNP simulations. The agreement of the MCNP results with the analytical formulae validated the MCNP simulations and the MATLAB scripts. The agreement between computational and analytical results for the detector at position (10, U) was affected by statistical fluctuations because of the lower number of simulated neutron captures than for the detector at position (15, X). Position (10, U) is farther from the fuel zone and thus has a lower neutron flux. The timestamps of both detectors were obtained by a single MCNP simulation with a fixed number of source neutrons.

The experimental data were corrected using the backward extrapolation method, in which the accelerator pulse period was divided into an arbitrary number of time bins, with some of the neutron counts in each time bin suppressed by applying a fictitious dead-time value. This procedure makes it possible to calculate the corrected (real) counts with no dead-time and also the dead-time of the data acquisition system, as discussed Ref. [2.116].

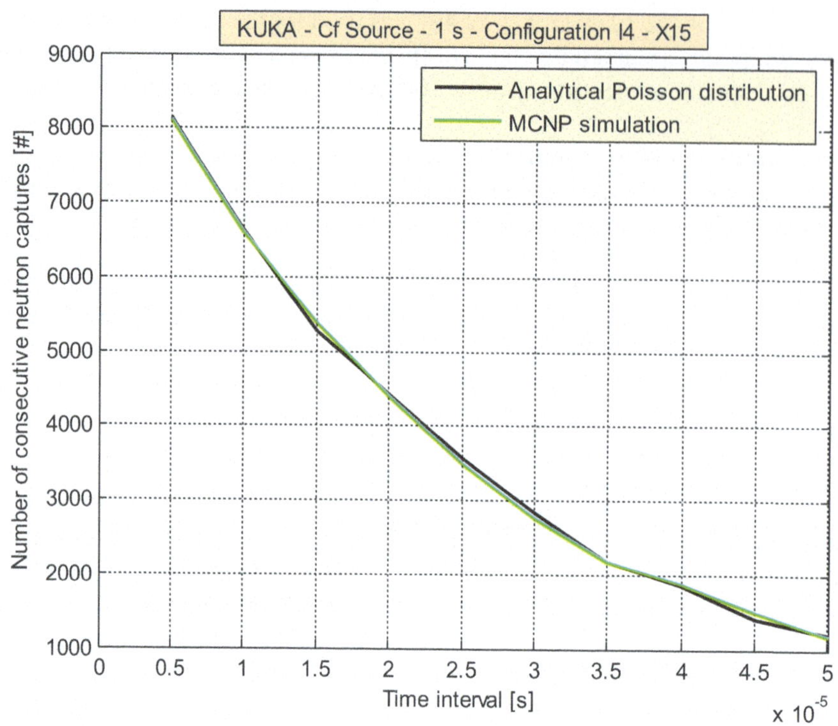

FIG. 2.85. Number of consecutive neutron captures as a function of their time interval (Reproduced from Ref. [2.79] with permission courtesy Sociedad Nuclear Mexicana, A.C.).

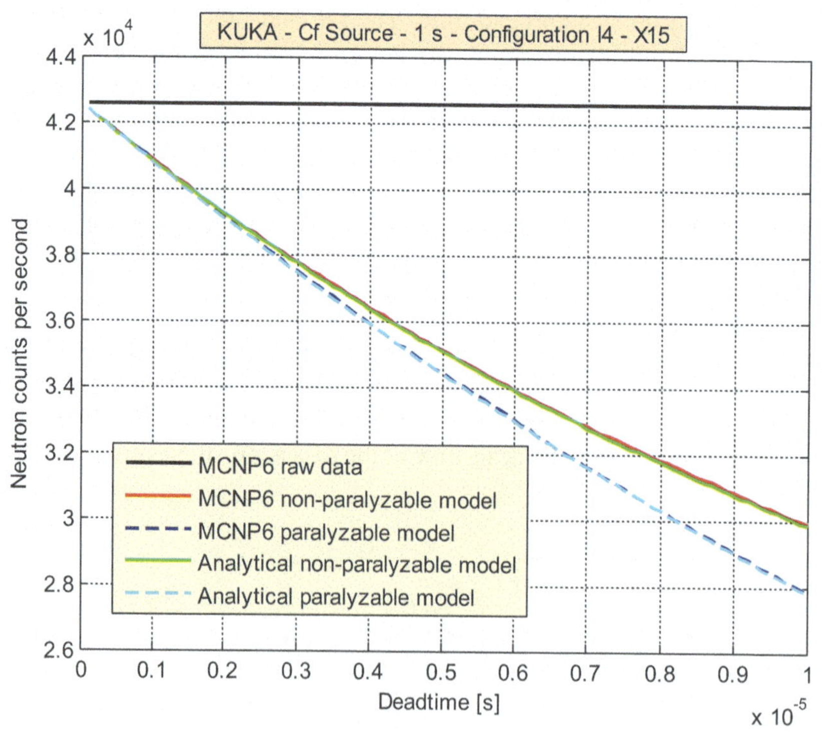

FIG. 2.86. Comparison of neutron count rates from MCNP and from analytical formulae as a function of the data acquisition system dead-time for the detector at position (15, X) (Reproduced from Ref. [2.79] with permission courtesy Sociedad Nuclear Mexicana, A.C.).

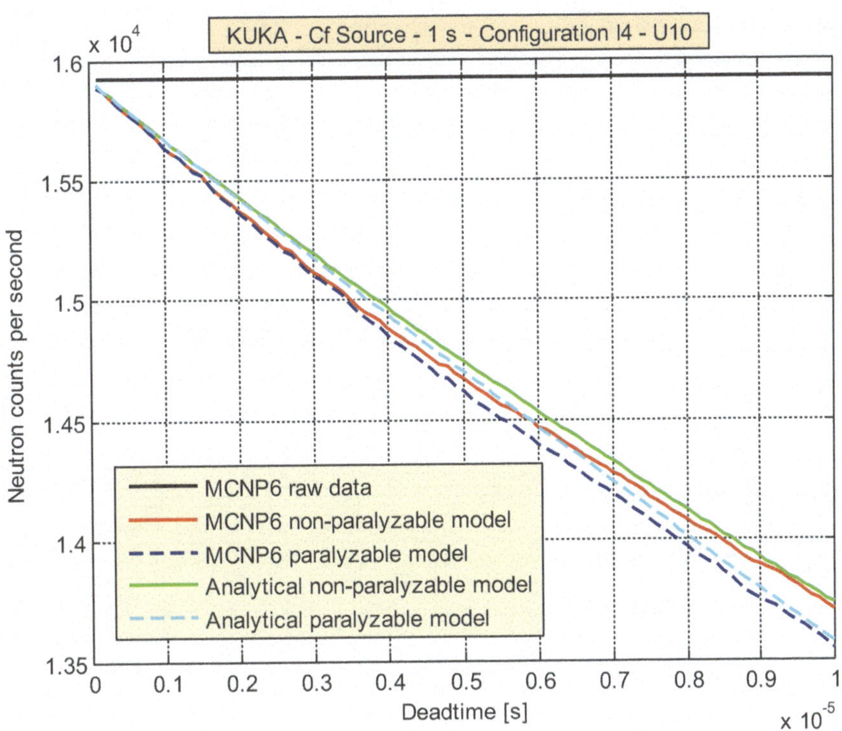

FIG. 2.87. Comparison of neutron count rates from MCNP and from analytical formulae as a function of the data acquisition system dead-time for the detector at position (10, U) (Reproduced from Ref. [2.79] with permission courtesy Sociedad Nuclear Mexicana, A.C.).

These experiments used a pulsed proton beam with energy 100 MeV, pulse duration 100 μs and repetition period of 50 ms. Figures 2.88 and 2.89 compare the measured (raw) neutron counts (m) (green curves) and the deat-time corrected counts (c) (red curves) during the pulse period for the detectors located at position (10, U) and (15, X). The dead-time correction used the backward extrapolation method and with the non-paralysable dead-time model. The neutron counts were normalized to the average number of neutron counts per pulse.

Figures 2.90 and 2.91 show the relative difference between c and m for the detectors at (10, U) and (15, X), respectively. When the count rate is low and is primarily attributable to delayed neutrons, the values of c and m are approximately equal. At high count rates, when source and prompt neutrons dominate the reaction rate, c is approximately 10% greater than m. The data in Figs 2.88–2.91 were obtained using a tungsten target and with all control rods withdrawn.

Figure 2.92 shows the effective multiplication factor k_{eff} obtained by the area method for various experimental configurations using tungsten or Pb-Bi targets and with different control rod positions. MCNP criticality simulations were used to calculate the effective delayed neutron fraction for the area method. For each configuration, the coloured columns show k_{eff} values derived from the BF$_3$ detectors at positions (15, X) and (10, U), and the (LiF+ZnS) fibre detector shown in Fig. 2.28, grouped into "original" using the measured data m, and "corrected" using the dead-time corrected data c. The difference between the measured and corrected the k_{eff} values is shown in Fig. 2.93. It can be seen that the dead-time correction lowers the k_{eff} value by 50–500 pcm. The BF$_3$ detector at position (15, X) showed the highest values because it was closer to the fuel zone and was loaded with boron, which enhanced the neutron count rate. Figure 2.94 shows the k_{eff} value was insensitive to the arbitrary number of time bins used in the backward extrapolation method.

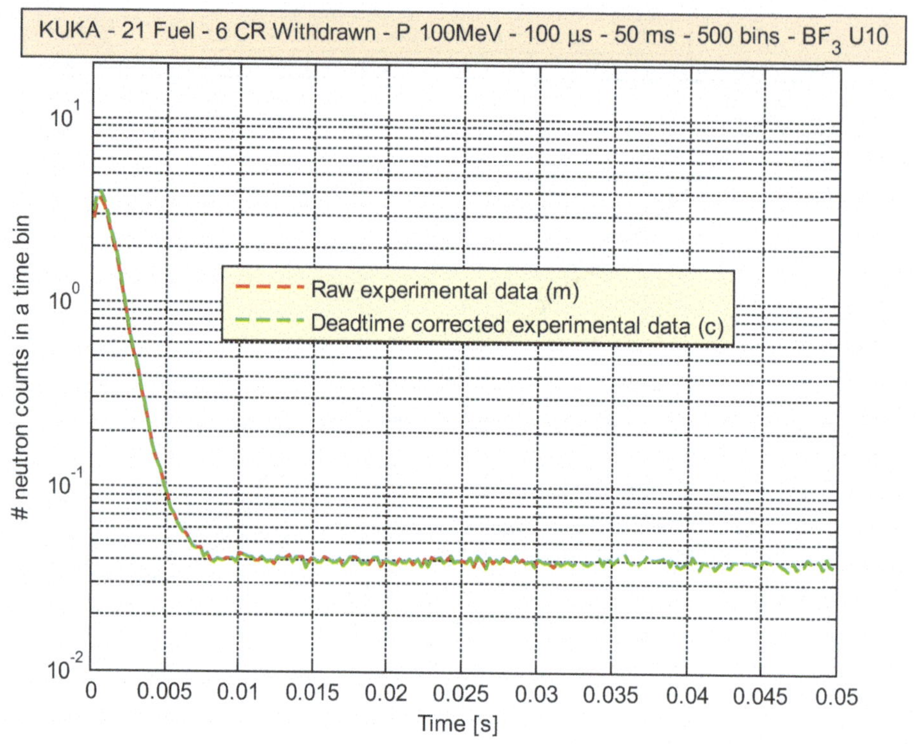

FIG. 2.88. Normalized neutron counts during the pulse period for the detector at position (10, U). (Reproduced from Ref. [2.79] with permission courtesy Sociedad Nuclear Mexicana, A.C.).

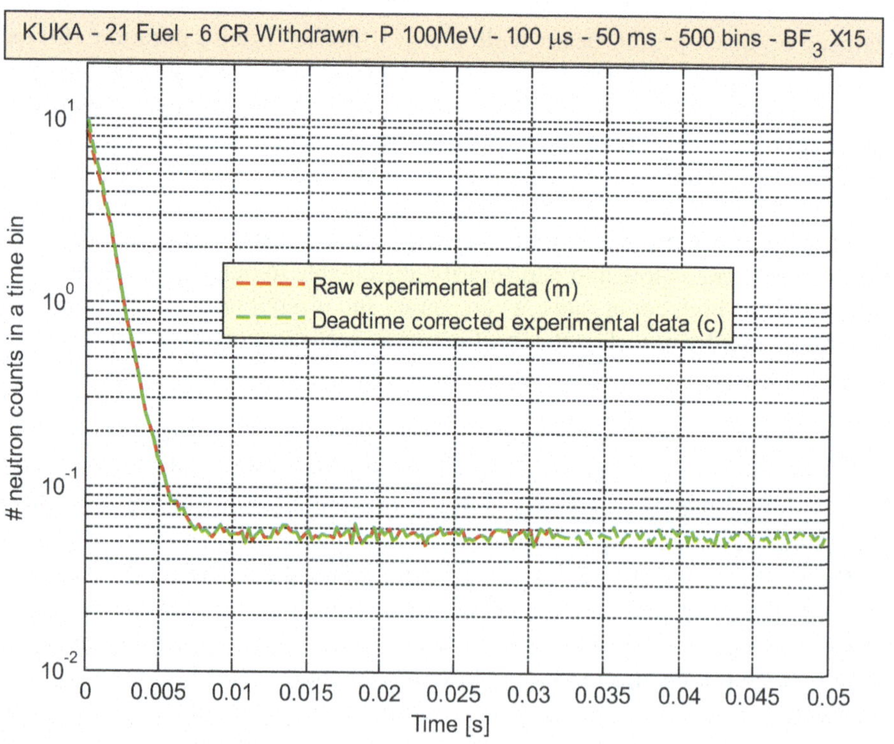

FIG. 2.89. Normalized neutron counts during the pulse period for the detector at position (15, X) (Reproduced from Ref. [2.79] with permission courtesy Sociedad Nuclear Mexicana, A.C.).

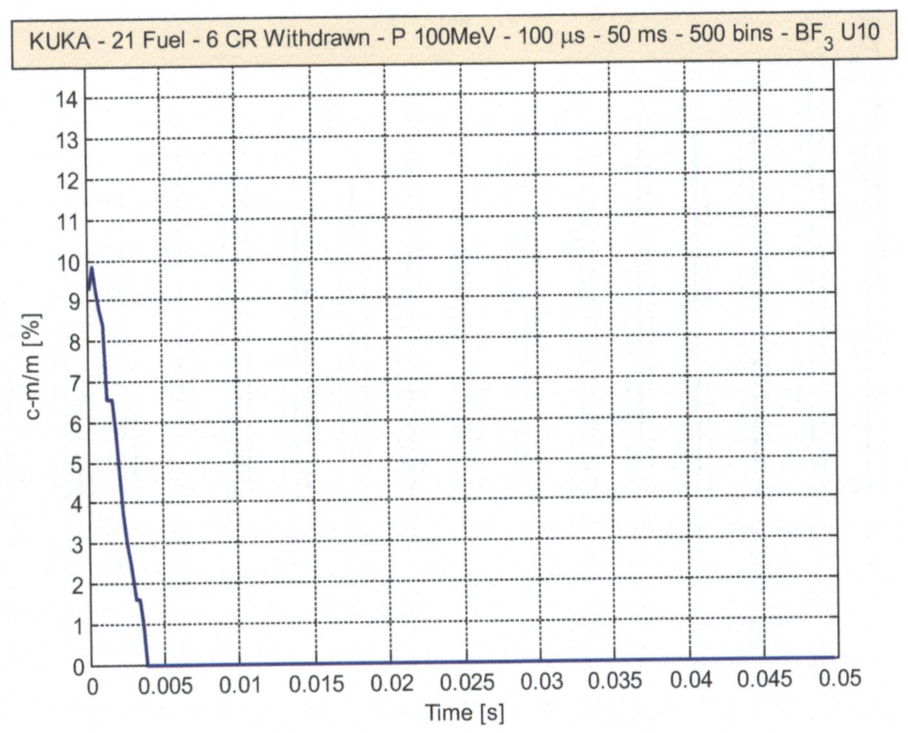

FIG. 2.90. Relative difference between corrected (c) and measured (m) neutron counts for the detector at position (10, U) (Reproduced from Ref. [2.79] with permission courtesy Sociedad Nuclear Mexicana, A.C.).

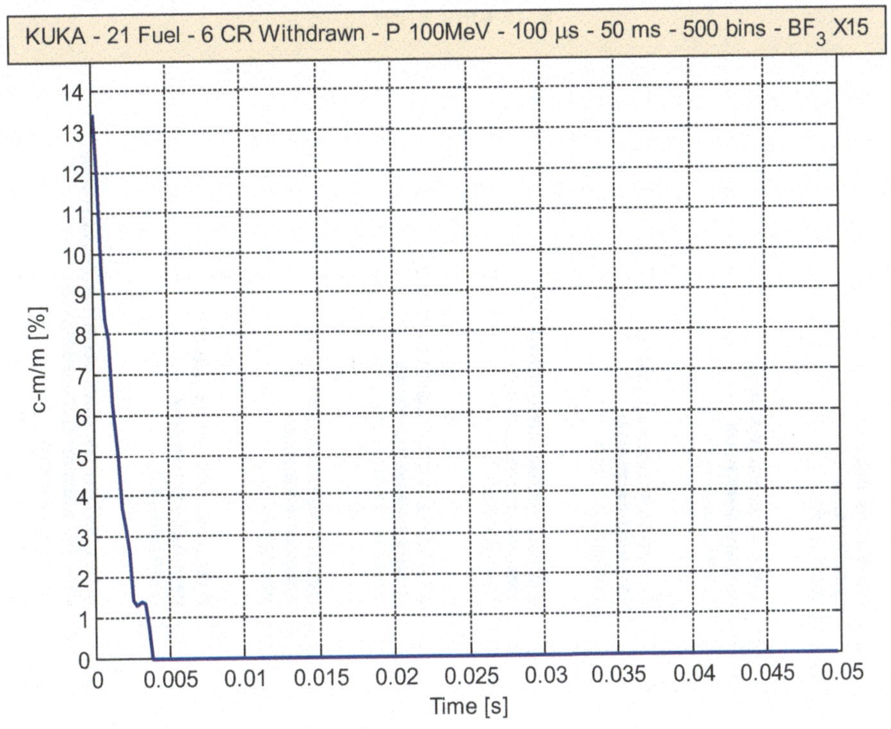

FIG. 2.91. Relative difference between corrected (c) and measured (m) neutron counts for the detector at position (15, X) (Reproduced from Ref. [2.79] with permission courtesy Sociedad Nuclear Mexicana, A.C.).

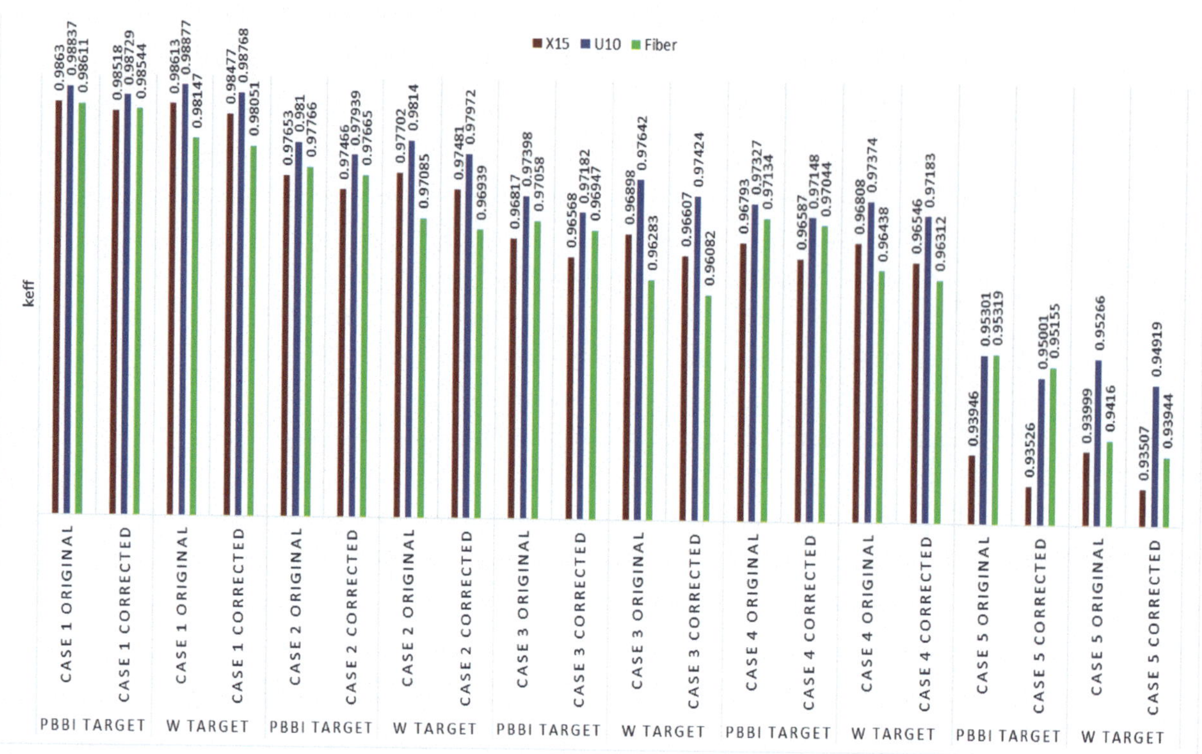

FIG. 2.92. Effective multiplication factor k_{eff} obtained by the area method from the measured experimental data (original columns) and from the backward extrapolation method and the non-paralyzable dead-time model applied to the experimental data (corrected columns) (Reproduced from Ref. [2.79] with permission courtesy Sociedad Nuclear Mexicana, A.C.).

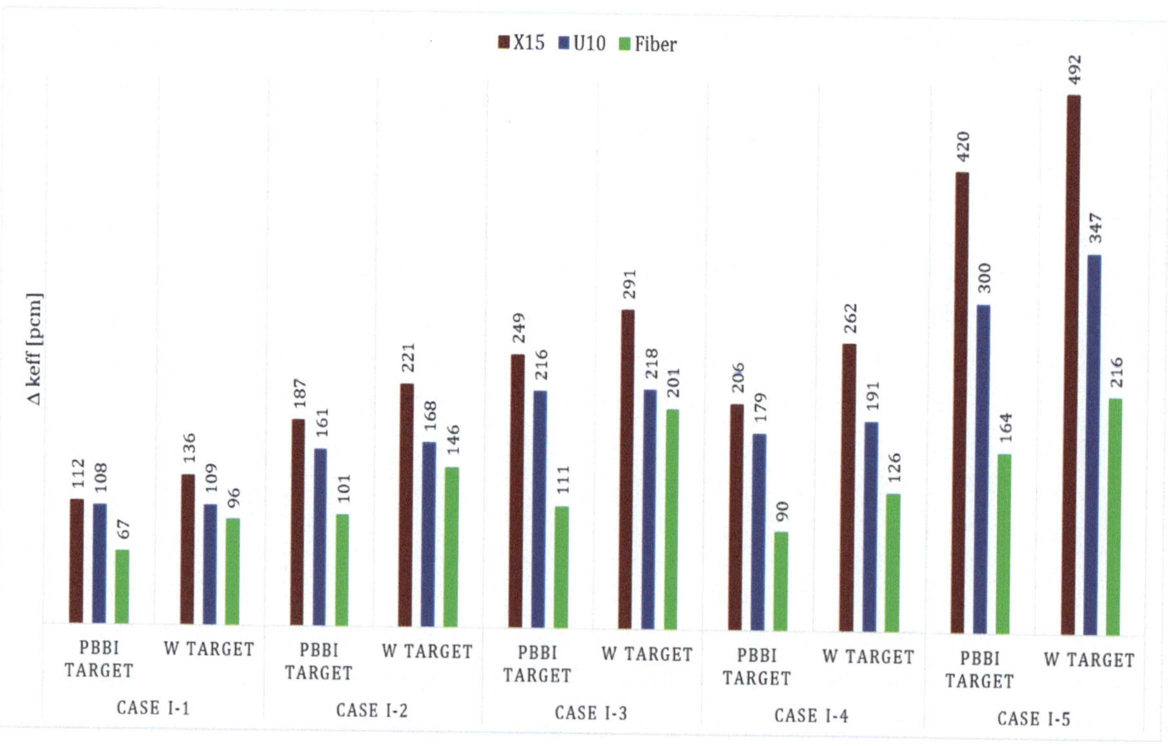

FIG. 2.93. Difference between the effective multiplication factor k_{eff} obtained by the area method from the measured experimental data (original columns) and from the backward extrapolation method and the non-paralyzable dead-time model applied to the experimental data (corrected columns) (Reproduced from Ref. [2.79] with permission courtesy Sociedad Nuclear Mexicana, A.C.).

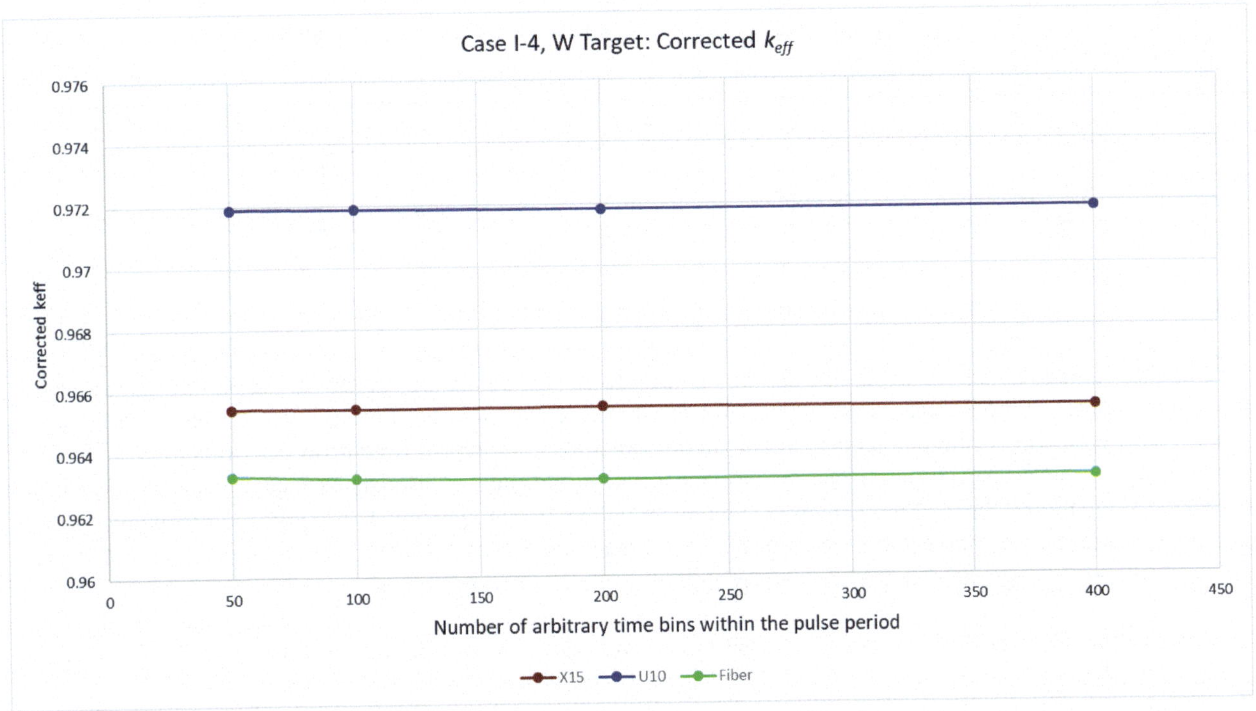

FIG. 2.94. Corrected k_{eff} value as a function of the number of arbitrary time bins used in the backward extrapolation method for the BF_3 detectors at positions (15, X) and (10, U) and the (LiF+ZnS) fiber detector (Reproduced from Ref. [2.79] with permission courtesy Sociedad Nuclear Mexicana, A.C.).

Experimental results need to also be corrected for the effect of neutron spectrum according to the detector position [2.4] and type [2.9]. This correction is achieved using MCNP as discussed in Refs [2.108] and [2.130]. Table 2.40 shows k_{eff} calculated using data from the detector at position (10, U) and an MCNP criticality simulation for Case I-4 which had a tungsten target and all control rods withdrawn. The dead-time and neutron spectrum effect corrections were less than 150 pcm.

TABLE 2.40. EXPERIMENTAL AND COMPUTATIONAL EFFECTIVE MULTIPLICATION FACTORS [2.79]

Configuration with W target	Detector position	Experimental k_{eff} area method	Experimental k_{eff} with dead-time correction	Experimental k_{eff} with dead-time and spatial corrections	MCNP k_{eff} criticality simulation
Case I-4	(10, U)	0.97374	0.97183	0.97157	0.97305±5

2.1.5.9. Application of the MAρTA method to reactivity estimation

The MAρTA algorithm was applied to the Phase I experiments at KUCA which compared different target options. The results were published in Ref. [2.45] and are summarized below. Additional calculations focussed on subcriticality level and using the Phase II experimental configurations were performed in the frame of the CRP.

For the PNS analysis a large series of pulses was collected, with a period of 50 ms and a signal collection time gate of 10 μs. It was assumed that in a single period the system response would be periodic, which is important initial condition for MAρTA. MAρTA requires delayed data and the generation time, Λ. Delayed data from the Phase 1 experiments were used as summarized in Table 2.41, and the value for the prompt neutron generation time of 30.5 μs was adopted from Ref [2.6].

TABLE 2.41. 6-FAMILIES DELAYED DATA ADOPTED FOR THE INTERPRETATION OF KUCA PHASE I EXPERIMENTS WITH THE MAρTA ALGORITHM

i	1	2	3	4	5	6
β_i (pcm)	24	134	130	371	110	38
λ_i (s^{-1})	0.01249	0.03182	0.10938	0.31700	1.35393	8.63759

β_{tot} = 807 pcm.

The use of a single set of delayed data and prompt generation time is an approximation. The various configurations in Phase I might have modified the energy spectrum and impacted the values of these parameters. However, only one set of data was available for the Phase I experiments. Likewise, for Phase II for which a value of the delayed neutron fraction was provided without additional information. Therefore, the family-wise data were obtained keeping the same proportion of the delayed neutron fractions among families and modifying their sum. The same prompt neutron generation time was used as for Phase I. Table 2.42 shows the value of the total β for the Phase II cases. The mean values from the Monte Carlo simulations were adopted, and the uncertainty estimation on the reactivity reconstruction accounts for the statistical error on the signals, although the error associated with β was not included in the error propagation.

TABLE 2.42. VALUE OF THE DELAYED NEUTRON FRACTION FOR THE KUCA PHASE II CONFIGURATIONS TAKEN FROM REF. [2.6] EVALUATED WITH MCNP6.1 ADOPTING DATA FROM JENDL-4.0.

i	Phase II-1	Phase II-2	Phase II-3	Phase II-4	Phase II-5	Phase II-6
β_i (pcm)	785 ± 4	785 ± 4	783 ± 5	788 ± 5	816 ± 5	806 ± 5

The Phase I KUCA experiments compared various target configurations (Fig. 2.14), as well as different positions of the controlling elements, with a limited modification on the number of fuel rods in the core. In each case, three detectors at different positions were used: one in the fuel region and two in the polyethylene reflector.

The response of the system clearly showed a point like behaviour, as can be seen from the three detector signals for case I-I shown in Fig. 2.95.

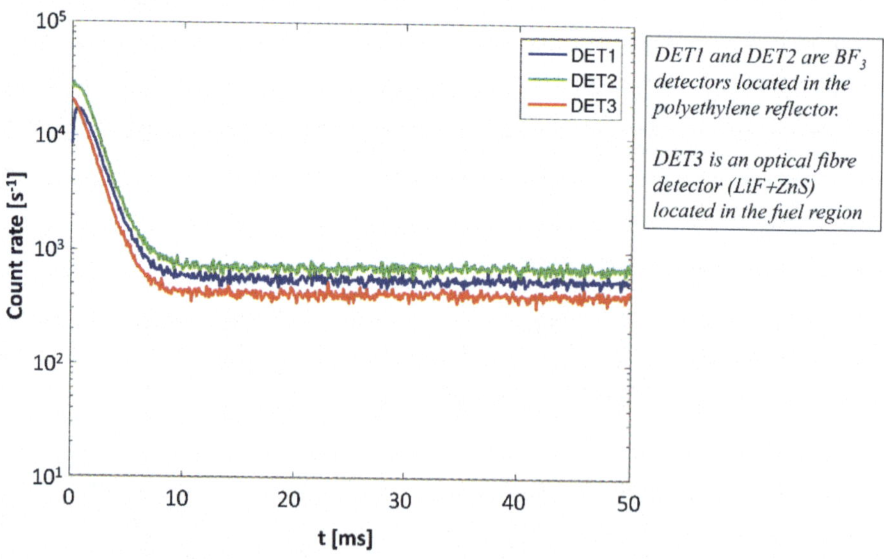

FIG. 2.95. Detector signals over one period of 50 ms for Case I-I with a Pb-Bi target. (Reproduced from Ref [2.45] with permission courtesy Elsevier B.V.).

A moving mean filter was used to process the signals [2.131] and a specific algorithm was used for the evaluation of the time derivative [2.132]. The MAρTA algorithm was used to reconstruct the reactivity value for each time instant within the period, as shown in Fig. 2.96.

Figure 2.96 suggests the correct procedure to extract an "optimized" reactivity value: as the system reactivity is constant during the period, an average is performed on the instantaneous values obtained with MAρTA, avoiding the first part of the period where non-point like effects influence the quality of the prediction. The optimal number of points to be used in the average was found to be N_{av} = 300 points. Tables 2.43–2.45 give the results of applying this procedure to the Phase I results. The results from the area method are also shown to highlight the general consistency of the results and the reduced uncertainty of the MAρTA values.

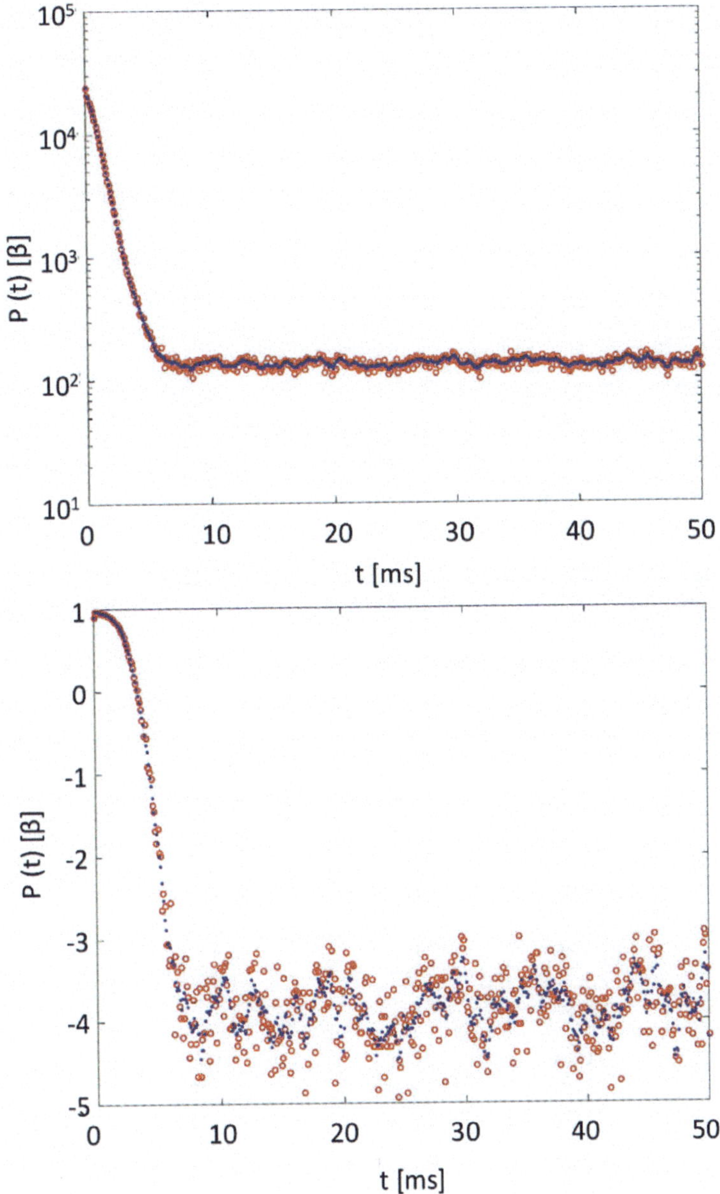

FIG. 2.96. Signal (top) and reactivity prediction (bottom) for DET3 of case II with W target. Blue dots denote the filtered signal and the corresponding reactivity prediction, while open red circles are associated to the unfiltered signal (Reproduced from Ref. [2.45] with permission courtesy of Elsevier B.V.).

The difference between the MAρTA and area ratio method shows a negative bias that can be attributed to the value adopted for the effective delayed neutron fraction, because the reactivity in the area ratio method needs to be converted from dollars to pcm, while the reactivity in MAρTA is already in pcm, but is dependent on the accuracy of β, see Ref. [2.133].

In Case V which had a larger subcriticality level, there was a larger uncertainty due to the greater noise in the detector signals. This could also be seen in the analysis of Phase II configurations that were deeply subcritical.

TABLE 2.43. AVERAGE VALUES OF THE REACTIVITY $\bar{\rho}$ AND CORRESPONDING UNCERTAINTY; FOR REACTOR CONFIGURATIONS WITH A Pb-Bi TARGET [2.45]

	Detector	$\bar{\rho}$ (β)	$\sigma_{\bar{\rho}}$ [a] ($10^{-2}\beta$)	ρ_{AM} [b] (β)	$\sigma_{\rho_{AM}}$ ($10^{-2}\beta$)	$\bar{\rho} - \rho_{AM}$ ($10^{-2}\beta$)
Case I	DET1	-1.55	0.5	-1.44	2.2	-10.6
	DET2	-1.83	0.6	-1.72	2.5	-11.2
	DET3	-1.82	0.6	-1.72	2.5	-10.0
Case II	DET1	-2.53	1.0	-2.39	3.6	-13.9
	DET2	-3.07	1.1	-2.92	4.3	-14.8
	DET3	-2.91	0.9	-2.79	4.1	-11.5
Case III	DET1	-3.41	1.2	-3.25	4.8	-16.2
	DET2	-4.18	1.5	-4.01	5.9	-16.6
	DET3	-3.80	1.6	-3.69	5.6	-11.1
Case IV	DET1	-3.57	1.3	-3.39	5.1	-17.5
	DET2	-4.27	1.6	-4.09	6.1	-17.5
	DET3	-3.75	1.5	-3.62	5.5	-12.5
Case V	DET1	-6.29	2.6	-6.08	9.5	-21.3
	DET2	-8.08	2.6	-7.88	11.4	-20.0
	DET3	-6.16	2.6	-6.09	9.2	-7.5

[a] $\sigma = \sqrt{var}$ evaluated with a sample size $N_{av} = 300$; [b] Area Ratio Method

TABLE 2.44. AVERAGE VALUES OF THE REACTIVITY $\bar{\rho}$ AND CORRESPONDING UNCERTAINTY FOR REACTOR CONFIGURATIONS WITH A W-Be TARGET [2.45]

	Detector	$\bar{\rho}$ (β)	$\sigma_{\bar{\rho}}$ [a] ($10^{-2}\beta$)	ρ_{AM} [b] (β)	$\sigma_{\rho_{AM}}$ ($10^{-2}\beta$)	$\bar{\rho} - \rho_{AM}$ ($10^{-2}\beta$)
Case I	DET1	-1.52	0.5	-1.42	2.2	-10.3
	DET2	-1.91	0.6	-1.81	2.6	-10.7
	DET3	-1.92	0.5	-1.83	2.6	-9.8
Case II	DET1	-2.46	0.8	-2.33	3.5	-13.0
	DET2	-3.19	1.2	-3.05	4.5	-13.4
	DET3	-2.86	1.1	-2.76	4.1	-10.0
Case III	DET1	-3.25	1.3	-3.10	4.8	-14.5
	DET2	-4.37	1.7	-4.22	6.3	-15.2
	DET3	-3.73	1.7	-3.64	5.6	-8.6
Case IV	DET1	-3.50	1.4	-3.33	5.1	-16.7
	DET2	-4.48	1.8	-4.32	6.4	-16.8
	DET3	-3.71	1.4	-3.61	5.3	-10.8
Case V	DET1	-6.40	3.2	-6.20	10.3	-19.9
	DET2	-8.94	5.0	-8.74	14.1	-20.5
	DET3	-6.13	3.1	-6.11	9.5	-1.9

[a] $\sigma = \sqrt{var}$ evaluated with a sample size $N_{av} = 300$; [b] Area Ratio Method

TABLE 2.45. AVERAGE VALUES OF THE REACTIVITY $\bar{\rho}$ AND CORRESPONDING UNCERTAINTY FOR REACTOR CONFIGURATIONS WITH W TARGET. [2.45].

	Detector	$\bar{\rho}$ (β)	$\sigma_{\bar{\rho}}$ [a] ($10^{-2}\beta$)	ρ_{AM} [b] (β)	$\sigma_{\rho_{AM}}$ ($10^{-2}\beta$)	$\bar{\rho} - \rho_{AM}$ ($10^{-2}\beta$)
CaseI	DET1	-1.50	0.2	-1.40	2.1	-9.9
	DET2	-1.83	0.4	-1.73	2.5	-10.4
	DET3	-2.47	0.5	-2.34	3.3	-12.8
CaseII	DET1	-2.45	0.6	-2.33	3.3	-12.0
	DET2	-3.05	0.8	-2.93	4.3	-12.5
	DET3	-3.85	1.2	-3.72	5.5	-12.8
CaseIII	DET1	-3.12	0.9	-2.97	4.3	-14.9
	DET2	-4.07	1.3	-3.91	5.7	-16.0
	DET3	-4.88	2.0	-4.74	7.2	-14.4
CaseIV	DET1	-3.5,0	1.1	-3.32	4.8	-17.3
	DET2	-4.25	1.5	-4.08	5.9	-17.0
	DET3	-4.70	2.0	-4.55	6.8	-15.3
CaseV	DET1	-6.33	2.6	-6.13	9.7	-20.4
	DET2	-8.13	4.2	-7.91	12.5	-21.7
	DET3	-7.78	4.2	-7.72	12.3	-6.4

[a] $\sigma = \sqrt{\text{var}}$ evaluated with a sample size $N_{av} = 300$; [b] Area Ratio Method

2.1.5.10. Concluding remarks

This section is a summary of the detailed conclusions in Refs [2.27], [2.76], [2.77], and [2.79]. Investigations into the accuracy of experiments and calculations in subcritical states were carried out at KUCA using 100 MeV protons from the FFAG accelerator with various target materials and supported by computer modelling. The following observations and conclusions were made:

— There was a large discrepancy between the calculated and experimental values for neutron yield;
— There was good agreement between calculation and experiment for the neutron multiplication and subcritical multiplication factors;
— The dependence of the kinetic parameters on the target materials, the prompt neutron decay constant and subcriticality was shown;
— Further studies of solid targets with high-energy protons would be beneficial, including with respect to the accuracy of reactor physics parameters.

Advances were made with the modelling and use of the Sjöstrand, Rossi, and Feynman distributions. In the Sjöstrand method, the delayed neutron equilibrium was calculated by splitting the integral of the reaction rate from a single pulse. For the noise methods, the MCNP6 source code was modified to account for the source time distribution and to increase the timestamp precision to less than a femtosecond, and the PARTISN source code was modified to transport delayed neutrons.

A novel method to calculate the prompt neutron decay constant from the cross and auto power spectral densities of detectors operating in pulse mode at low neutron count rates was defined. A MATLAB script was used to redistribute the detector timestamps arising from 12 000 particle accelerator pulses into a time interval covering 10 particle accelerator pulses. The results were similar to those obtained by detectors operating in current mode.

The non-zero imaginary part of the cross power spectral density revealed that higher order modes contribute to the neutron flux in the detectors. Therefore, the value of the prompt neutron decay constant is dependent upon detector position requiring a spatial correction factor. Measuring the prompt neutron decay constant using two detectors located at different positions mitigates this effect.

Alternative methods for calculating the prompt neutron decay constant were compared for the KUCA facility driven either by a stationary ^{252}Cf source or by a pulsed spallation neutron source. Fitting of the Rossi and Feynman distributions was used for both the stationary neutron source and the pulsed neutron source. In addition, fitting the peaks of the cross power spectral density and auto power spectral density distributions, fitting the count rate as a function of time, and the MAρTA method were applied to the pulsed neutron source configuration. Detector timestamps were required for the Rossi and Feynman distribution methods. These were generated experimentally using detectors in pulse mode and computationally using MCNP with the PTRAC and F8 cards. The cross power spectral density, auto power spectral density, count rate fitting, and MAρTA methods do not require the detector timestamps because they use the average neutron counts in time bins. These methods can be used with detectors operating in current mode or with MCNP simulations using the F4 tally. The values obtained by the different methods agreed within 3%.

The MCNP simulations with a stationary neutron source were validated with analytical formulae for dead-time corrections. The experimental data with pulsed neutron source were processed by MATLAB script that corrected for dead-time using the non-paralysable dead-time model. The dead-time correction reduced the effective multiplication factor determined by the area method by 50 to 500 pcm. Corrections were also made for the neutron spectrum effect arising from the type and position of the detector. The corrected experimental data were in good agreement with the MCNP results.

2.1.6. Analyses: Phase II – Study on subcriticality measurements

2.1.6.1. Modelling

(a) KUCA (Japan)

Numerical calculations were performed by the Monte Carlo transport code, MCNP6.1, together with JENDL-4.0 for transport and with JENDL/HE-2007 for high-energy protons and spallation neutrons. Here, in MCNP6.1, since the effects of reactivity by neutron detectors (optical fibre, fission chamber and uncompensated ion chamber detectors) and control (safety) rods are not negligible, neutron detectors and control (safety) rods were included in the simulated geometry and transport calculations.

(b) SNU (Republic of Korea)

McCARD eigenvalue calculation is done with 10^5 histories per cycle, 100 inactive cycles, 500 active cycles and JENDL-4.0 cross-section libraries.

The prompt neutron decay constant α can be directly measured by injecting a short burst of neutrons in the system, called the PNS experiment [2.134], [2.49]. Since Simmons and King [2.134] applied an exponential regression to neutron detector signals from the PNS experiment, this α measurement method has been popularly employed because it can provide α results independent of the positioning and energy characteristics of the detector and neutron source [2.134], [2.135], [2.136] by reducing higher mode contaminations on the exponential fitting [2.137], [2.138].

In this section, α for case II-6 was calculated by modelling the PNS experiment using time-dependent Monte Carlo neutron transport analyses [2.139]. MCNPX2.6.0 was used to determine the energy and direction distribution of the spallation neutrons at the target [2.141], with the assumption that azimuthal angular distributions of the source neutrons were isotropic. Their polar distribution was found by counting neutron flux at equally spaced angular bins of $15°$ on the target surface. The MCNPX2.6.0 calculation used 10^7 histories using la150h proton libraries. The neutrons angular and energy distributions were inputs to McCARD.

Time-dependent reaction rates at the three optical fibre detector positions were calculated using McCARD time-dependent Monte Carlo simulations, and α was estimated by fitting the results with Eq. (2.45).

$$R_D(r,t) = C_1 \cdot \exp[-\alpha_{est}(r|t_s) \cdot (t - t_s)] + C_2, \tag{2.45}$$

where R_D is the estimated reaction rate in the detector at position r, and $\alpha_{est}(r|t_s)$ is an estimate of α from that detector, C_1 and C_2 are fitting constants, t the time after the neutron burst, and t_s the beginning time of the fitting interval. R_D was estimated using 100 time-dependent Monte Carlo simulations, using 10^7 histories calculated at 0.1 ms intervals up to 50 ms.

(c) Politecnico di Torino (Italy)

The MAρTA method was developed as a flexible tool to estimate reactivities from flux measurements in time-dependent conditions carried out in both source-free and source-driven multiplying systems. This method was applied to the interpretation of pulsed experiments in the KUCA. A statistical analysis was performed to determine the uncertainty of the estimations. The reactivity estimations were compared with the predictions from the classical area ratio method. The analysis showed that the MAρTA method can be successfully applied to the study of pulsed experiments in subcritical systems.

(d) FDS (China)

Research on the influence caused by the flux distortions on the amplified source method (ASM) in a subcritical assembly and the elimination of the influence by detector improvements was carried out. The overall programme of work was performed in collaboration with the Kyoto University using the SuperMC (Super Multi-functional Calculation Program for Nuclear Design and Safety Evaluation) software [2.142].

Starting from the KUCA reference core shown in Fig. 2.16 that reached criticality, the number of normal fuel rods (F) was changed to analyse the deviation and the correction of the ASM results for core, as shown in Fig. 2.97.

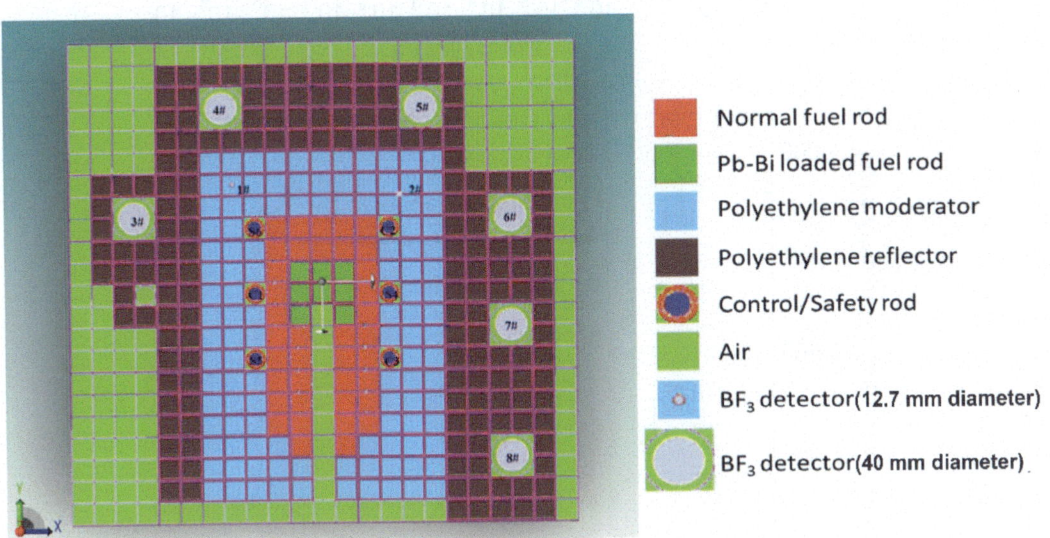

FIG. 2.97. Transverse section view of the subcritical reactor [2.140].

The reference core shown in Fig. 2.16 consists of 36 normal fuel rods (F), 8 Pb-Bi loaded fuel rods (f) and 2 partial fuel rods (16). Different operation states (perturbed states) were obtained by adjusting the insertion depth of the control rods. Control rod position 0 cm means that all control rods were completely inserted; control rod position 86 cm means that all control rods were completely withdrawn from the reactor core; control rod position n cm means that all the control rods were withdrawn n cm from the core.

The subcriticality, neutron flux distribution and detector count rate were calculated by SuperMC with the Hybrid Evaluated Nuclear Data Library (HENDL). The automatic modelling program, SuperMC/MCAM [2.143], [2.144] is an advanced interface between CAD systems and Monte Carlo codes and was used to build the core model. The virtual reality-based simulation system, SuperMC/RVIS, was used to analyse the function $f(x, y, z)$ distribution.

2.1.6.2. Experimental and numerical results and discussion

This section is a summary of the detailed information in Ref. [2.31]. Neutron spectra were computed using MCNP assuming the injection of 100 MeV protons onto a Pb-Bi target, as shown in Figs 2.98 (a) and (b). Figure 2.23 shows examples of the time evolution of the counting rate using the PNS method. Figure 2.24 provides examples of the neutron noise data from the Feynman-α method.

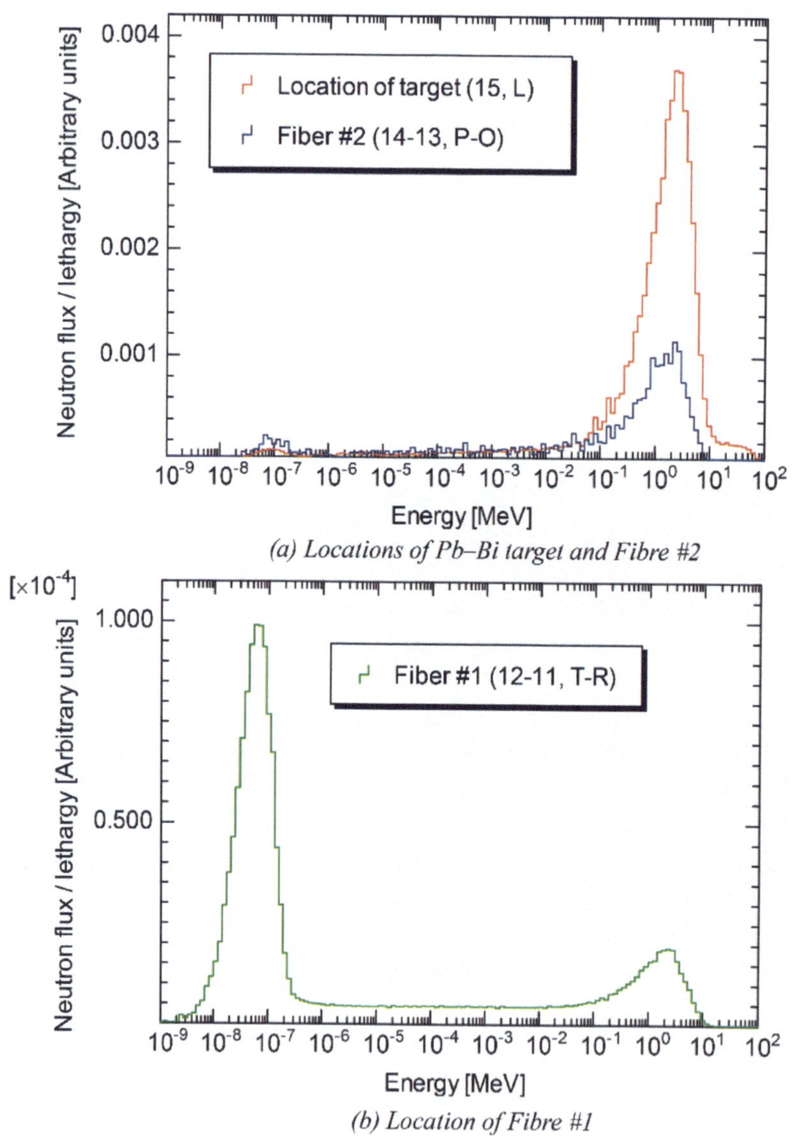

(a) Locations of Pb–Bi target and Fibre #2

(b) Location of Fibre #1

FIG. 2.98. Comparison between neutron spectra at Pb-Bi target, Fibres #1 and #2 (Reproduced from Ref. [2.31] with permission courtesy of Taylor & Francis).

Eigenvalue calculations were used to calculate the uncertainty for the reactivities of control rods C1, C2 and C3. Experiment and calculation agreed to within 3%, as shown in Table 2.46, with a total number of 10^8 histories and a statistical error of less than 5 pcm. As can be seen from the table, the excess reactivity calculated by McCARD agrees with the experimental values within the 95% confidence interval, and the

McCARD estimates of the control rod reactivities agree within the 99% confidence interval. The difference in effective multiplication factor between the McCARD calculation and experiment at the critical state was about 170 pcm.

TABLE 2.46. COMPARISON BETWEEN MEASURED AND CALCULATED REACTIVITIES [2.31]

Reactivity	Experiment (pcm)	MCNP6.1 (KUCA)		McCARD (SNU)	
		Calculation (pcm)	C/E [a]	Calculation (pcm)	C/E [a]
Excess	149 ± 3	149 ± 11	1.00 ± 0.08	172 ± 18	1.15 ± 0.11
C1	549 ± 3	567 ± 11	1.03 ± 0.02	577 ± 17	1.05 ± 0.03
C2	194 ± 1	189 ± 11	0.98 ± 0.06	193 ± 18	0.99 ± 0.09
C3	483 ± 2	498 ± 11	1.03 ± 0.02	536 ± 18	1.11 ± 0.03

a Calculation / Experiment

Table 2.47 compares the effective multiplication factors from McCARD and MCNP6.1 for the six subcritical configurations. The differences between the McCARD and MCNP6.1 results were attributed to the cross-section libraries JENDL-4.0 and ENDF/B-VII.0.

TABLE 2.47. EFFECTIVE MULTIPLICATION FACTORS OF SUBCRITICAL SYSTEMS OF PHASE II [2.89]

Case	McCARD[a] (SD) (SNU)	MCNP6.1[b] (SD) (from KUCA)
II-1	0.99021 (0.00012)	0.99219 (0.00040)
II-2	0.98512 (0.00012)	0.98723 (0.00040)
II-3	0.97734 (0.00012)	0.97950 (0.00040)
II-4	0.95556 (0.00012)	0.95784 (0.00040)
II-5	0.91088 (0.00012)	0.91355 (0.00040)
II-6	0.89725 (0.00011)	0.90006 (0.00040)

a JENDL-4.0; b ENDF/B-VII.0

2.1.6.3. Prompt neutron decay constant

The PNS and the Feynman-α methods (Eqs (2.1) and (2.4)) were used to calculate the prompt neutron decay constant using the counts from the three optical fibres in the core shown in Fig. 2.16. Table 2.48 shows a small difference in the neutron spectrum results of Fibres #1 and #2, see Figs 2.98(a) and (b), and for the subcriticality results, for subcriticality ranging from 1160 to 2483 pcm (cases II-1 to II-3), see Figs 2.23 and 2.24. The minor difference between Fibres #1 and #2 was attributed to the one-point reactor approximation being valid at shallow levels of subcriticality. For the deeper subcriticality of cases II-4 to II-6 (4812 to 11 556 pcm), a larger difference was observed, although the dependency of the prompt neutron decay constant on detector position and neutron spectrum was small. Comparing Fibre #3 with Fibres #1 and #2 a strong influence of spallation neutrons at the Pb-Bi target on the prompt neutron decay constant was apparent for both the PNS and Feynman-α methods.

TABLE 2.48. PROMPT NEUTRON DECAY CONSTANTS FROM THE PNS METHOD [2.6]

Case	Neutron decay constants α (s⁻¹)			McCARD α-iteration (from SNU)
	Fibre #1	Fibre #2	Fibre #3	
Case II-1	398.0 ± 2.5	326.8 ± 2.7	494.5 ± 10.2	380.0 ± 3.0
Case II-2	506.7 ± 3.0	452.6 ± 2.8	685.4 ± 10.8	497.0 ± 3.5
Case II-3	672.6 ± 3.5	631.9 ± 1.8	1019.7 ± 8.3	668.8 ± 3.8
Case II-4	982.5 ± 4.4	971.3 ± 3.1	1378.0 ± 18.4	1034.9 ± 5.0
Case II-5	1665.4 ± 8.7	1680.7 ± 5.1	1827.7 ± 22.8	1789.6 ± 6.8
Case II-6	1910.9 ± 7.4	1930.7 ± 3.3	2061.0 ± 37.8	2010.9 ± 7.3

The time-dependent reaction rates for the three optical fibre detectors are shown in Fig. 2.99, and the variation of α_{est} values with t_s for a fixed fitting interval of 800 μs is shown in Fig. 2.100. The α values converge to a value of ~2010 with convergence rates that depend on the detector position. Consequently, different values for α may be measured depending on detector positioning and source type if the slope fitting is applied to a time interval where the higher mode effects are insufficiently attenuated.

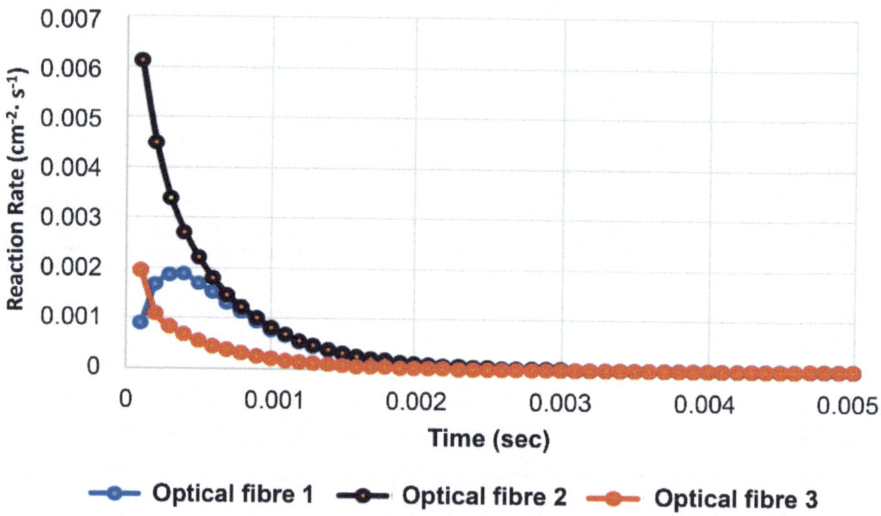

FIG. 2.99. Reaction rate from time-dependent Monte Carlo data [2.89].

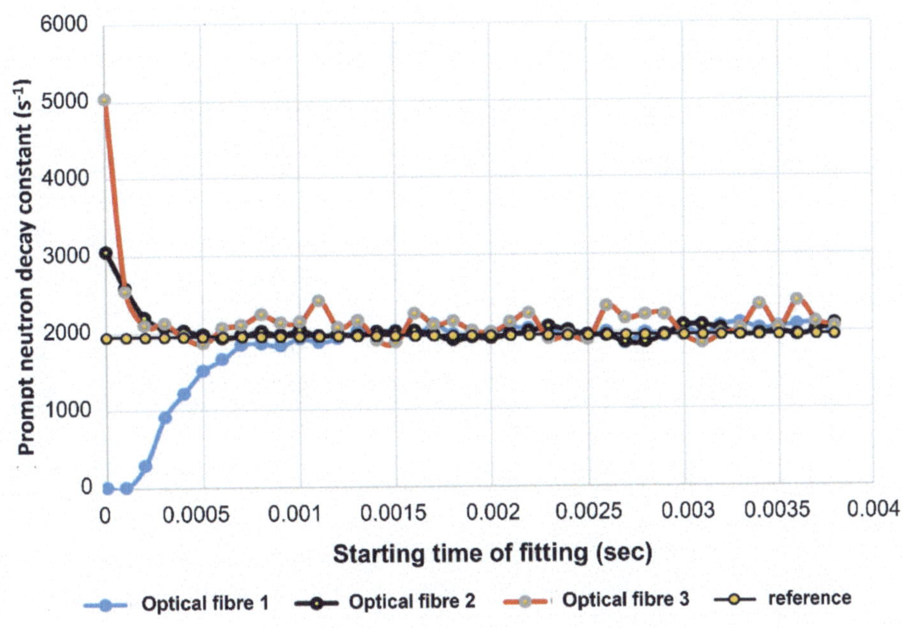

FIG. 2.100. Alpha estimates from time-dependent Monte Carlo data [2.89].

2.1.6.4. Subcriticality evaluation

The following is a summary of the detailed discussion in Ref. [2.31]. For the core configurations shown in Fig. 2.21, the experimental subcriticality in dollar units was derived using the extrapolated area ratio method using the prompt and delayed neutron components [2.145] and by the Feynman-α method. The conversion coefficient β_{eff} to convert from dollar to pcm values was obtained using MCNP6.1 with JENDL-4.0.

For cores with subcriticality between 1160 and 2483 pcm, Table 2.49 shows that the subcriticality value for the Fibre #2 location obtained from the PNS and the Feynman-α methods showed good agreement with the calculated value. The C/E error was around 10%. The difference between Fibres #1 and #2, was attributed to the detector locations. For deeper subcriticalities, between 4812 and 11556 pcm, both Fibre #1 and Fibre #2 had worse agreement with the calculated value for both the PNS and the Feynman-α methods. This was attributed to the increased effect of spallation neutrons on the neutron flux with deep subcriticality with the observation that the measurement method is not appropriate to a subcriticality level greater than 10^4 pcm. Fibre #3 showed a large influence from spallation neutrons with the two measurement methods, although the Feynman-α method had good agreement for cases II-5 and II-6. Tables 2.43 and 2.44 show the kinetic parameters of ρ and Λ, respectively. In case of the deep subcriticality, the values were obtained using MCNP. Time dependent MCNP calculations are needed to evaluate and better understand the kinetic parameters [2.146].

TABLE 2.49. MEASURED SUBCRITICALITY BY AREA RATIO AND FEYNMAN-ALPHA METHODS. [2.31]

Case	Reference (pcm)	Subcriticality ρ by area ratio method			Subcriticality ρ by Feynman-α method		
		Fibre #1 (pcm)	Fibre #2 (pcm)	Fibre #3 (pcm)	Fibre #1 (pcm)	Fibre #2 (pcm)	Fibre #3 (pcm)
II-1	1160 ± 5 [a]	1224 ± 10	1153 ± 8	2108 ± 59	1491 ± 57	1283 ± 36	2445 ± 105
II-2	1684 ± 6 [a]	1880 ± 15	1604 ± 11	4490 ± 109	2024 ± 28	1814 ± 24	3250 ± 41
II-3	2483 ± 6 [a]	2824 ± 22	2250 ± 13	10 347 ± 116	2881 ± 15	2677 ± 11	4203 ± 29
II-4	4812 ± 6 [b]	5656 ± 62	4177 ± 31	22 939 ± 1279	4336 ± 38	4384 ± 35	5740 ± 77
II-5	9895 ± 6 [b]	12 819 ± 137	8187 ± 54	14 985 ± 1136	8087 ± 61	8316 ± 50	10 110 ± 113
II-6	11 556 ± 6 [b]	16 312 ± 160	9738 ± 64	19 109 ± 2563	9633 ± 88	10 175 ± 68	12 096 ± 150

[a] ρ_{exp}; [b] ρ_{MCNP}

2.1.6.5. Improvement of the amplified source method and multiple detectors with ringed arrangement

ASM has been widely used in critical reactors for determination of small reactivity variations, and is a potential technique for subcriticality monitoring, because it uses relatively simple relationships between detector counts of two different core states [2.147], [2.148] as shown in Eq. (2.46).

$$\rho_l = \frac{R_{ref}}{R_l} \rho_{ref}$$

(2.46)

where ρ_{ref} and R_{ref} are the subcriticality and detector count rate of the reference state. ρ_l and R_l are the subcriticality and detector count rate of the l-th state. According to this equation, if one knows the ρ_r in a reference state with its corresponding detector count rate R_r, one can deduce ρ_l with the detector count rate R_l in l-th subcritical state.

However, the ASM is based on a point kinetics model; in (deep) subcritical reactors driven by an external neutron source, it is necessary to take correction factors, just like the modified source multiplication (MSM) mentioned [2.147], [2.148], [2.149]. The MSM method is difficult to be applied in online monitoring the reactivity of subcritical assembly. It is needed to calculate the correction factors.

Thus, the ASM can be regarded as a good choice for monitoring reactivity if some improvements are applied to reduce the external source effect, the higher harmonic modes effect, and the neutron flux spatial effect. Previous research work indicates that the external source effect is weak in the ASM method; the main effect comes from the other two items. Therefore, the objective of this study is to determine how to eliminate higher harmonic modes effect and the neutron flux spatial effect. First, the ASM was simulated on the KUCA assembly with SuperMC. Second, the calibration curve was set up to reduce the higher harmonic modes effect. Third, the detector position effect to the ASM was studied to reduce the neutron flux spatial effect. Finally, a multidetector ringed arrangement method was proposed to improve the ASM.

Previous work recommended that the reference state should be as close as possible to criticality [2.150]. Accordingly, a configuration with the control rods at 80 cm and $k_{eff} = 0.99721$ was selected as the reference state. The 'measured' subcriticality of l-th state can then be calculated with Eq. (2.46), while the 'raw' subcriticality of l-th state was calculated by SuperMC directly. Subcriticalities of different states and the detector count rate of detector 1# are listed in Table 2.50.

TABLE 2.50. SUBCRITICALITIES AND DETECTOR#1 COUNT RATES FOR DIFFERENT STATES [2.140]

State	Position of control rods (cm)	Raw subcriticality ρ_r (pcm)	Detector count rate (s⁻¹)	'Measured' subcriticality ρ_m (pcm)
State 1	50	2211.86	2.547420E-02	898.78
State 2	60	1902.52	2.995040E-02	764.45
State 3	65	1515.63	3.538900E-02	646.97
State 4	70	1120.41	4.438020E-02	515.90
State 5	80	279.78	8.183410E-02	279.78

It can be seen from Table 2.50 that the raw subcriticalities are not equal to the 'measured' subcriticalities, but there is a one-to-one correspondence relationship between them, as shown in Fig. 2.101. When applying the MSM method there are three correct factors that need to be considered [2.148], namely C_l^{ext}, C_l^{im}, C_l^{sp}; where C_l^{ext} is related to the extraction of the fundamental mode, C_l^{im} corrects the effect induced by the disturbance of neutron importance field, and C_l^{sp} corrects a spatial effect caused by the disturbance of the fundamental mode. The deviation between raw subcriticalities and 'measured' subcriticalities is mainly induced by extraction of the fundamental mode.

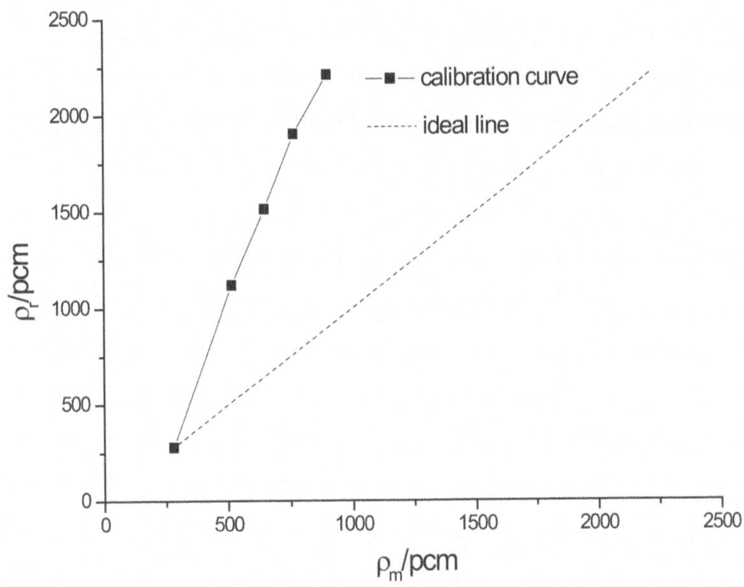

FIG. 2.101. Calibration curve of ρ_m and ρ_r at different states of detector 1# [2.140].

Therefore, by using the calibration curve of ρ_m and ρ_r, one can correct the effect induced by the extraction of the fundamental mode. By setting up the calibration curve between ρ_m and ρ_r before operating reactor, the subcriticality of any state at any time can be obtained. The calibration curve of ρ_m and ρ_r for other detectors at different states of is shown in Fig. 2.102.

As mentioned above, all the calibration curves were obtained with all control rods inserted or withdrawn at the same time, so the applicability of the calibration curve needs to be verified when the subcriticality is changed for other reasons, such as control rod ejection, change of burn-up, and temperature change of coolant or fuel. During the operation of the reactor, the control rods may not always move at the same time; there may be rod-ejection or a rod-drop accident. In order to study the influence induced by flux distortion, three typical abnormal states were simulated as listed in Table 2.51. Abnormal state 1 is the state at which control rod C1 inserts completely into the core while all other control rods remain at 80 cm. Abnormal state 2 is the state at which control rods S4 and S6 eject out of core while all other control rods remain at 0 cm. Abnormal state 3 is the state at which control rod S6 eject out of core while all other control rods remain at 0 cm.

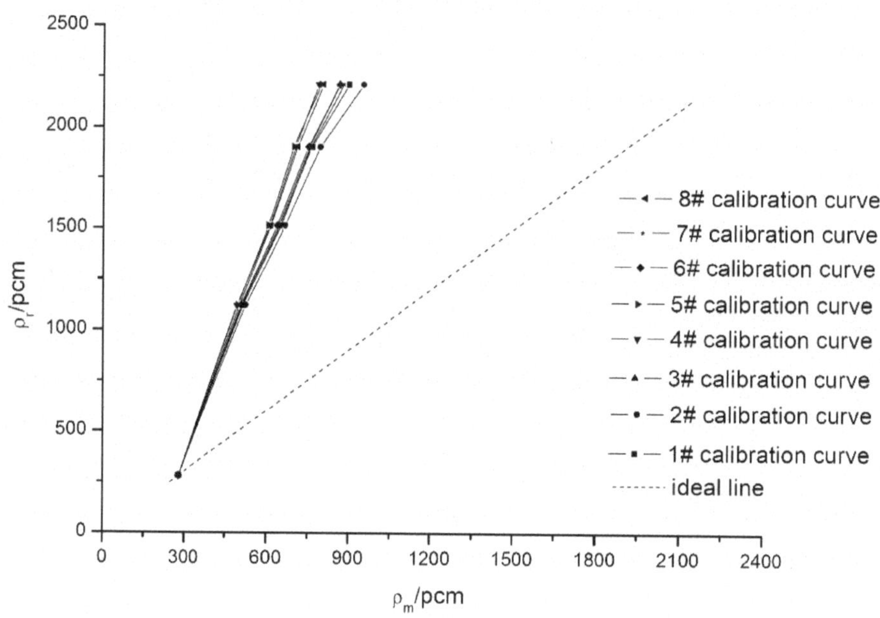

FIG. 2.102. Calibration curve of ρ_m and ρ_r at different states of different detectors [2.140].

TABLE 2.51. THREE TYPICAL ABNORMAL STATES. [2.140]

State	Control rod position	
	Control Rods at 0 cm (in)	Control Rods at 80 cm (out)
Abnormal state 1	C1	C2, C3, S4, S5, S6
Abnormal state 2	C1, C2, C3, S5	S4, S6
Abnormal state 3	C1, C2, C3, S4, S5	S6

The count rate of detector 1#, the raw subcriticality ρ_r, and the 'measured' subcriticality ρ_m of the three typical abnormal states are listed in Table 2.52. The comparison of abnormal state results and calibration curves are shown in Fig. 2.103.

TABLE 2.52. COUNT RATE OF DETECTOR #1, RAW SUBCRITICALITY ρ_r, AND 'MEASURED' SUBCRITICALITY ρ_m OF THREE TYPICAL ABNORMAL STATES [2.140]

State	ρ_r (pcm)	Count rate (s^{-1})	ρ_m (pcm)
Abnormal state 1	272.74	8.38295E-02	273.12
Abnormal state 2	1479.57	4.33980E-02	524.57
Abnormal state 3	2008.55	3.53554E-02	647.58

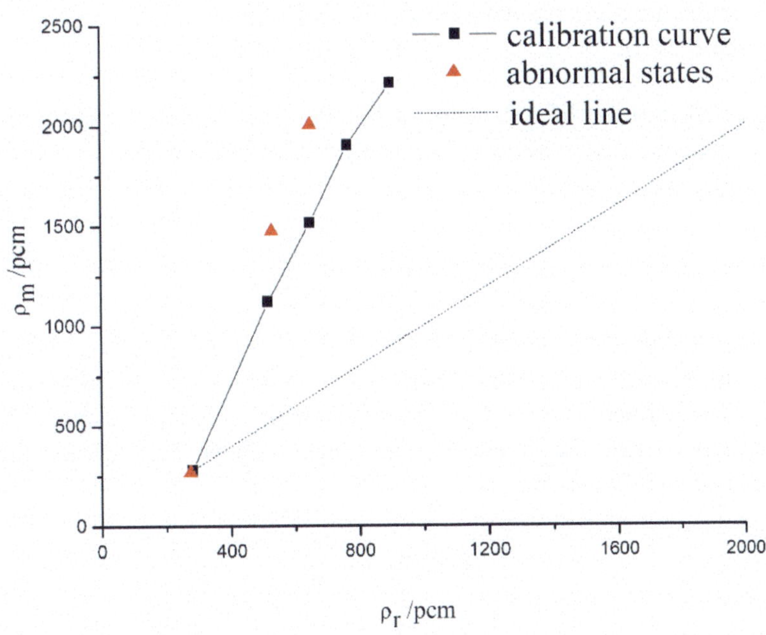

FIG. 2.103. Comparison of abnormal states results and calibration curves [2.140].

Figure 2.103 shows that ρ_m of abnormal states were not near or on the calibration curve, it means the subcriticality cannot be measured exactly using only one detector with a calibration curve.

There are similar results as detector 1# simulating with other detectors to measure subcriticality. The 'measured' subcriticality and the corresponding deviation at two states are listed in Table 2.53. These two states have similar raw subcriticality but with different control rod positions, namely the subcriticality is changed by different reasons with different flux distortion.

TABLE 2.53. THE MEASUREMENT SUBCRITICALITY OF NORMAL STATE 3 AND ABNORMAL STATE 2 GIVEN BY DIFFERENT DETECTORS WITH THE CALIBRATION CURVE [2.140]

State	ρ_r(pcm)	ρ_m (pcm)	Deviation	ρ_m (pcm)	Deviation	ρ_m (pcm)	Deviation
Detector			1#		2#		3#
Normal state 3	1515.63	1487.37	1.86%	1473.26	2.80%	1511.98	0.24%
Abnormal state 2	1479.57	1110.91	24.92%	1654.41	11.82%	1260.84	14.78%
Detector			4#		5#		6#
Normal state 3	1515.63	1542.58	1.78%	1520.14	0.30%	1500.20	1.02%
Abnormal state 2	1479.57	1295.14	12.47%	1484.31	0.32%	1427.87	3.49%
Detector			7#		8#		
Normal state 3	1515.63	1514.56	0.07%	1520.60	0.33%		
Abnormal state 2	1479.57	1351.95	8.63%	1579.83	6.78%		

From Table 2.53, one can see all the deviations of normal state 3 are less than 5% while the maximum deviation of abnormal state 2 is 24.92%. These results clearly shown even the two states have the similar raw subcriticality, there are big deviations for the same detector, because of the mechanism that led to the change of the reactivity is different.

In (deep) subcritical reactors driven by an external source, the neutron flux is composed of not only fundamental mode, but also higher harmonic modes. So, the accuracy of the ASM is affected by the position of detectors. Just as mentioned above, the determined subcriticality results obtained with different detectors are very different (see Table 2.53).

In order to find the optimal position of a detector, a function $f(x, y, z, \Sigma_{\text{det}})$ is proposed to estimate the deviations between the real subcriticality and the subcriticality determined by the ASM in a perturbed state. The function $f(x, y, z, \Sigma_{\text{det}})$ is defined as:

$$f(x, y, z, \Sigma_{det}) = \frac{\rho_l R_l}{\rho_r R_r} \tag{2.47}$$

Where ρ_l is the raw subcriticality of the perturbed state. The function $f(x, y, z)$ can indicate the measurement accuracy when the detector placed in a certain position (x, y, z) with a certain neutron energy response Σ_{det}. The detector count rate R is defined as $R = \langle \Sigma_{det}, \Phi \rangle$. If the neutron energy response Σ_{det} is assumed to be independent of the neutron energy, the detector count rate R will be proportional to the neutron flux Φ. Therefore, Eq. (2.47) can be simplified as:

$$f(x, y, z) = \frac{\rho_l \Phi_l}{\rho_r \Phi_r} \tag{2.48}$$

The simplified function $f(x, y, z)$ only relates to the spatial position. The value of $f(x, y, z)$ ought to be 1 if the 'measured' subcriticality ρ_l is equal to the real subcriticality ρ_r. When $f(x, y, z)$ is greater than 1, it means the 'measured' subcriticality given by the ASM is larger than the real subcriticality. Otherwise, the 'measured' subcriticality is less than the real subcriticality.

As analysed above, there ought to be a position at which the 'measured' subcriticality is nearly equal to the real subcriticality, i.e., where the function $f(x, y, z)$ is close to 1. Therefore, the spatial distribution of $f(x, y, z)$ was studied preliminarily; a detailed study of axial and radial distribution of $f(x, y, z)$ was also performed.

Firstly, as a typical normal state, state 3 (CRs position at 65 cm) was analysed. The $f(x, y, z)$ distributions on the horizontal mid-plane and the vertical mid-plane through control rod S4 are shown in Fig. 2.104. As indicated in Fig. 2.104, the values of $f(x, y, z)$ are away from 1 in most positions, which means the 'measured' subcriticalities given by the ASM are not in good agreement with the real subcriticalities in most positions. Moreover, the largest deviations are found around the perturbation position.

If subcriticality is corresponded with value of $f(x, y, z)$ in a certain position, the correct result can still be obtained by calibration even if the value is not 1. In order to identify this, a comparison of $f(x, y, z)$ between normal state 3 and abnormal state 2 was made; these two states have similar subcriticalities of 1501.20 pcm and 1439.43 pcm, respectively. The $f(x, y, z)$ distributions of these two states are shown in Figs 2.104 and 2.105, respectively.

As shown in Figs 2.104 and 2.105, the $f(x, y, z)$ distributions are very different, especially in the positions around the perturbations. The large deviation indicates that even two different states with similar subcriticalities, the results may be very different determined only by one detector. Consequently, it is hard to find out an ideal position for place detector to exactly measure the subcriticalities, even the calibration curve was applied.

The count rate of detector is not only related to the neutron flux distribution Φ but also related to the arrangement of the detector. To find the optimal position with multi detectors, the effect of axial and radial position for improving the ASM were analysed.

The axial $f(x, y, z)$ distributions at normal state 3 and abnormal state 2 are shown in Fig. 2.106. several typical radial positions were selected, one is at the centre of source zone (line 1), one is at middle of fuel zone (line 2), one is at centre of control rod C1 (line 3), and one is beyond the shielding zone (line 4).

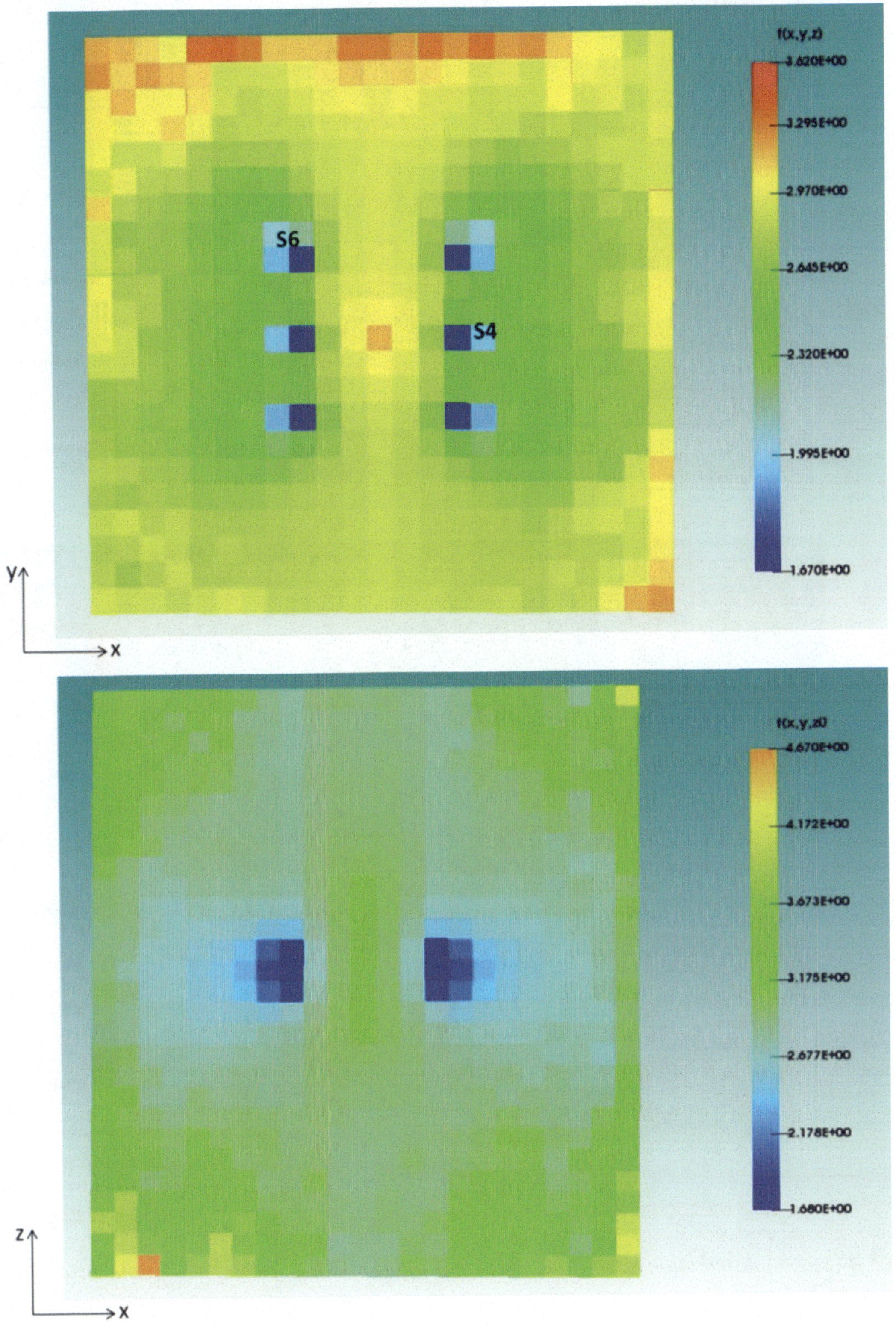

FIG. 2.104. $f(x, y, z)$ distribution on the horizontal mid-plane and the vertical mid-plane through rod S4 of state 3 [2.140].

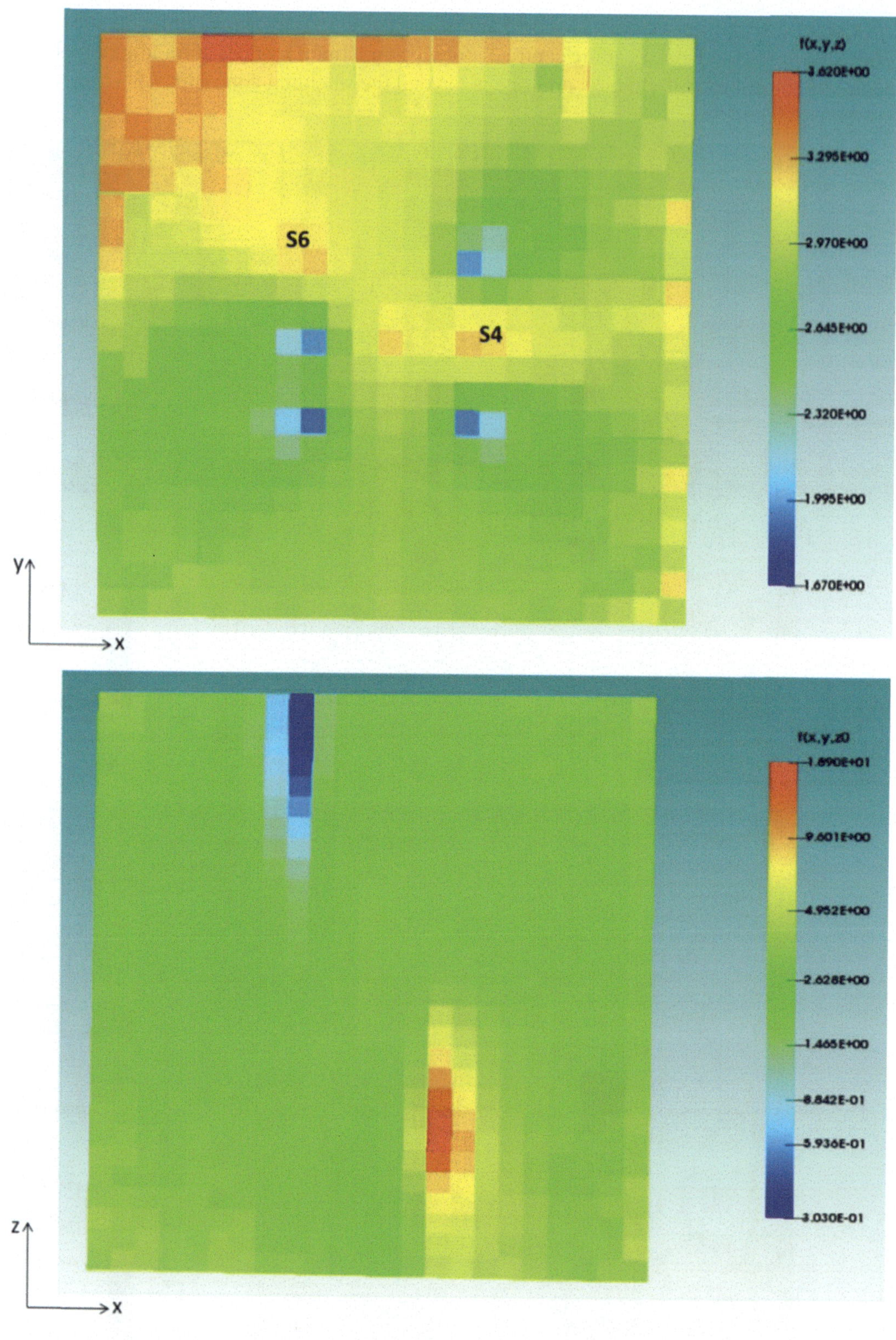

FIG. 2.105. $f(x,y,z)$ distribution on the horizontal mid-plane and the vertical mid-plane through rod S4 of abnormal state 2 [2.140].

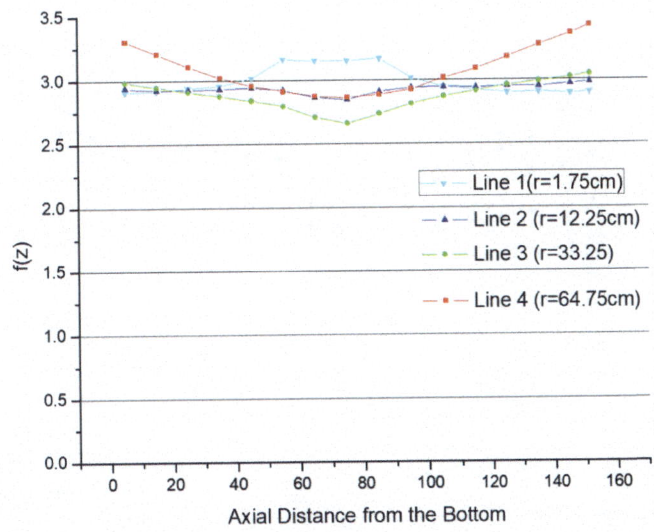

(a) Graph of f(x,y,z) along z at different radial positions at normal state 3

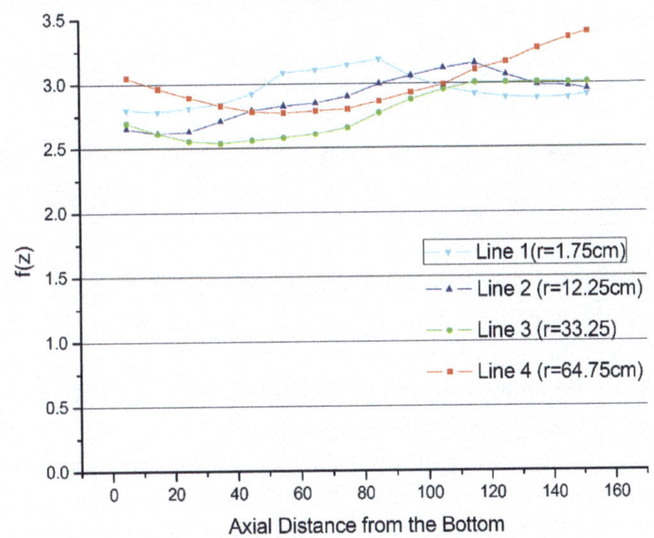

(b) Graph of $f(x,y,z)$ along z at different radial positions at abnormal state 2

FIG. 2.106. Axial shape of the function $f(x,y,z)$ at different radial positions at different states [2.140].

Figure 2.106 indicates that the $f(x,y,z)$ distributions are different at different (x,y) positions, and they are never equal to 1. Thus, it can be inferred that through axial orientation to arrange detector positions the 'measured' results cannot be improved. Meanwhile, the difference among the line 1, 2, 3, 4 shows that through radial orientation to arrange detectors positions may more important.

To investigate the effect of radial detector position, the core was divided into 20 annular regions, each of which represented a ringed detector. It was assumed that the detectors were long enough to detect all the neutrons at the axial orientation. The $f(x,y,z)$ simplified as $f(r)$ in this Section was calculated considering the average flux Φ in each annular region. The distribution of $f(r)$, where r denotes the distance from the central point (i.e., $r = \sqrt{(x^2 + y^2)}$), for normal states 1–5 and abnormal states 1–3 is shown in Fig. 2.107.

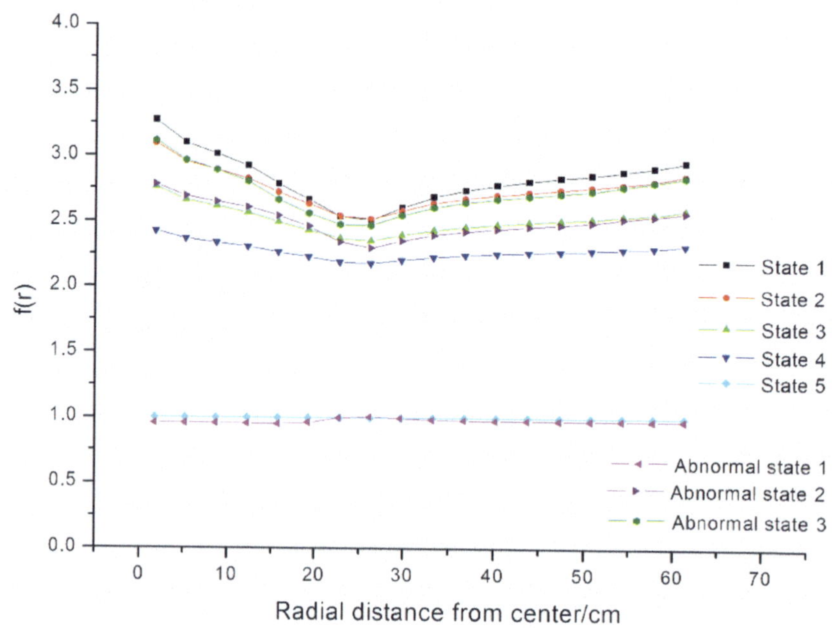

FIG. 2.107. Radial $f(r)$ under different states [2.140].

Figure 2.107 shows that the values of $f(r)$ are not equal to 1 in most cases. This means that the 'measured' subcriticality is not in good agreement with the real subcriticality given by the ASM using a ringed detector. However, some important information can still be obtained from Fig. 2.107:

- The values of $f(r)$ decrease along with increasing radial distance from the centre of core and decrease more quickly when the reactor is at a deeper subcritical state. The values of $f(r)$ reach a minimum near the point A (just beyond the moderator boundary), increase slowly till to point B (beyond the reflector boundary), and then keep unchanged. This phenomenon is caused by the high harmonic modes of the neutron flux from external source;
- The curves of normal state 3 and abnormal state 2 (with similar subcriticality) are almost overlapped although their perturbations are different. This means that the 'measured' subcriticality given by the ASM can be corrected through arranging several detectors at radial orientation.

Since the values of $f(r)$ before the point A are sensitive to the radial distance, a small position deviation may cause a large measurement error; it is therefore not practical to place the detectors near this point. Hence, a ringed detector arrangement far from the external source was proposed to correct the effect induced by the disturbance of neutron importance field and spatial effect.

In order to simulate the count rate of detectors, six virtual $^{10}BF_3$ detectors (3# to 8#) were set in the core, the position shown as in Fig. 2.97. In the simulation, the count rate R in Eq. (2.46) was the sum of the count rates of detectors 3# to 8#. The results of these simulations are shown in Fig. 2.108. Where ρ_m represents the 'measured' subcriticality given by the ASM with the sum count rate, and ρ_r represents the real subcriticality obtained by theoretical calculation.

In Fig. 2.108, the calibration curve was set up through normal states 1–5 with the real subcriticality and the 'measured' subcriticality. Abnormal states 1–3 are shown as the triangular symbols. It can be seen that the three triangular symbols are very close to the calibration curve, in contrast to the triangular symbols in Fig. 2.103 which produced by single detector.

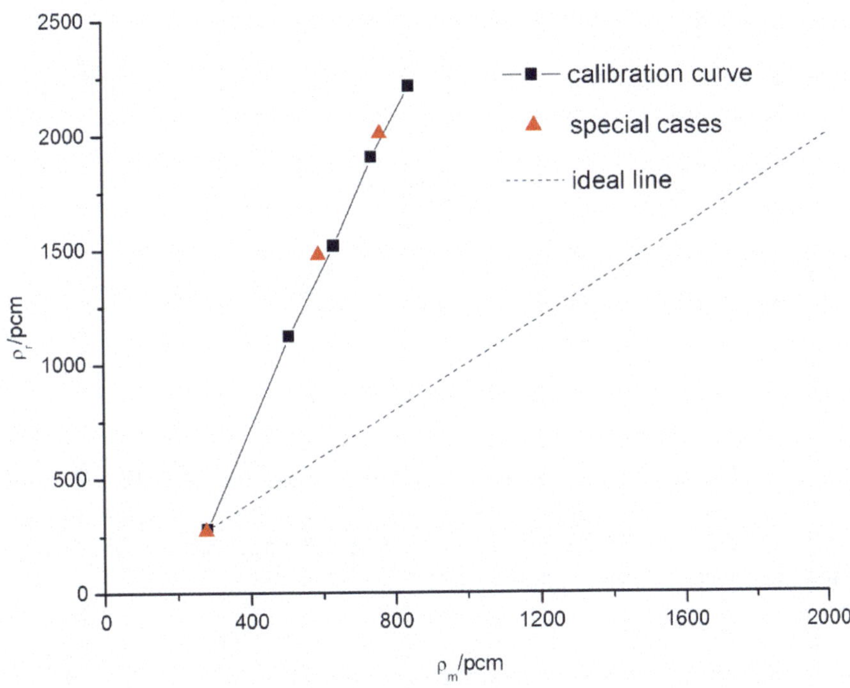

FIG. 2.108. Results given by ASM with a ringed arrangement of detectors (3# – 8#) [2.140].

The least squares method was used for the calibration curve (both single detector and ringed arrangement detectors) fitting with a cubic equation to obtain the desired parameters. The polynomial fitting curves are as follows:

(a)   For the single detector:

$$\rho_{m,1} = -1.4907 \times 10^{-6} \times \rho_m{}^3 + 1.153 \times 10^{-3} \times \rho_m{}^2 + 3.0178 \times \rho_m - 650.4278$$

(b)   For the ringed arrangement detectors:

$$\rho_{m,2} = 1.8058 \times 10^{-7} \times \rho_m{}^3 - 1.3193 \times 10^{-4} \times \rho_m{}^2 + 4.2986 \times \rho_m - 852.8250$$

ρ_m is the 'measured' subcriticality given by the ASM method; $\rho_{m,1}$ and $\rho_{m,2}$ are the subcriticalities obtained from the fitting formula.

Comparison of 'measured' subcriticality of abnormal states 1–3 given by the single detector and the ringed arrangement detectors through ASM are listed in Table 2.54. It can be seen that the deviation can reach up to 23.30% using the single detector. However, the deviation is less than 6% when using the ringed detector arrangement. This means that, although the flux is distorted, the correct results can still be obtained through the calibration curve by the ASM with ringed arrangement detectors. Consequently, the ringed arrangement detectors method is indeed an effective method to improve the results of the ASM.

TABLE 2.54. COMPARISON OF 'MEASURED' SUBCRITICALITY OF ABNORMAL STATES 1–3 GIVEN BY SINGLE DETECTOR AND RINGED ARRANGEMENT DETECTORS THROUGH ASM [2.140]

State	Raw subcriticality ρ_r (pcm)	Single detector 1# $\rho_{m,1}$ (pcm)	Deviation	Detectors 2#–8# $\rho_{m,2}$ (pcm)	Deviation
Abnormal state 1	272.74	257.56	5.57%	269.85	1.06%
Abnormal state 2	1479.57	1148.63	22.37%	1395.83	5.66%
Abnormal state 3	2008.55	1540.64	23.30%	1959.25	2.45%

2.1.6.6. Discussion on the interpretation of the subcritical measurements

This section is a summary of the discussion in Ref. [2.31]. For the α-fitting method, the prompt neutron decay constant α can be calculated from Eq. (2.49):

$$\alpha = \frac{\beta_{eff} - \rho}{\Lambda} \tag{2.49}$$

where ρ is the subcriticality in pcm units, β_{eff} is the effective delayed neutron fraction and Λ is the neutron generation time. The subcriticality $\rho_\$$ in dollar units can be expressed as shown in Eq. (2.50):

$$\rho_s = -\frac{\Lambda}{\beta_{eff}}(\alpha - \frac{\beta_{eff}}{\Lambda}) \tag{2.50}$$

Table 2.55 gives the value of Λ as calculated using Eq. (2.50). A study at KUCA [2.151] showed that the value of β_{eff} has a small effect on the value of subcriticality in pcm units converted from dollar units compared to numerical subcriticality in pcm units. This section considers the kinetic parameter, Λ, derived from α and $\rho_\$$ by varying the subcriticality.

The value of Λ for Fibre #2 was calculated from Eq. (2.50), using MCNP6.1 to determine β_{eff}, noting the good agreement between the numerical simulations and the experimental results for Fibre #2 by the PNS method. Table 2.55 compares the experimental and the numerical values of Λ showing acceptable agreement at various subcriticality levels. Fitting the experimental values of α and $\rho_\$$ for Fibres #1 and #2 using Eq. (2.50) yielded values for Λ/β_{eff} for Fibre #1 of 12.27×10^{-3}, and for Fibre #2 of 6.82×10^{-3}, as shown in Fig. 2.109. Table 2.56 compares these values with the results from MCNP6.1. The experimental fitting of counts from Fibre #2 demonstrated good agreement with the MCNP6.1 calculations by varying the subcriticality, with the relative difference between the experiments and the calculations of 15% at most. From the results in Fig. 2.109 and Tables 2.55 and 2.56, the kinetic parameters were deduced by the combination of experiments and calculations and verified by the PNS and the α-fitting methods.

TABLE 2.55. COMPARISON BETWEEN $\rho_\$$, β_{eff} AND Λ BY THE PNS METHOD AND MCNP CALCULATIONS [2.31]

| Case | $\rho_\$$ ($) | β_{eff} (pcm) | Λ ($\times10^{-5}$ s) | |
	Fibre #2 (PNS)	MCNP6.1	MCNP6.1	Fibre #2 (PNS)
II-1	1.47 ± 0.01	785 ± 4	4.88 ± 0.01	5.93 ± 0.08
II-2	2.04 ± 0.01	785 ± 4	4.77 ± 0.01	5.28 ± 0.09
II-3	2.87 ± 0.01	783 ± 5	4.64 ± 0.01	4.80 ± 0.05
II-4	5.30 ± 0.02	788 ± 5	5.16 ± 0.01	5.11 ± 0.06
II-5	10.03 ± 0.03	816 ± 5	5.43 ± 0.01	5.53 ± 0.05
II-6	12.08 ± 0.03	806 ± 5	5.50 ± 0.01	5.46 ± 0.05

TABLE 2.56. COMPARISON OF MCNP6.1 AND EXPERIMENTAL VALUES OF Λ/β_{eff} [2.31]

| Case | Λ/β_{eff} values ($\times10^{-3}$) | | |
	MCNP6.1	Fibre #1 [a]	Fibre #2 [a]
II-1	6.22 ± 0.01		
II-2	6.08 ± 0.01		
II-3	5.93 ± 0.01		
II-4	6.55 ± 0.01	12.27 ± 0.08	6.82 ± 0.05
II-5	6.65 ± 0.01		
II-6	6.82 ± 0.01		

[a] PNS method

Tables 2.53–2.56 show that the experimental benchmarks are important for the interpretation of the experimental results and derivation of the correction factors, and for determining the effect of detector position, neutron spectrum and subcriticality measurement methods on the kinetic parameters.

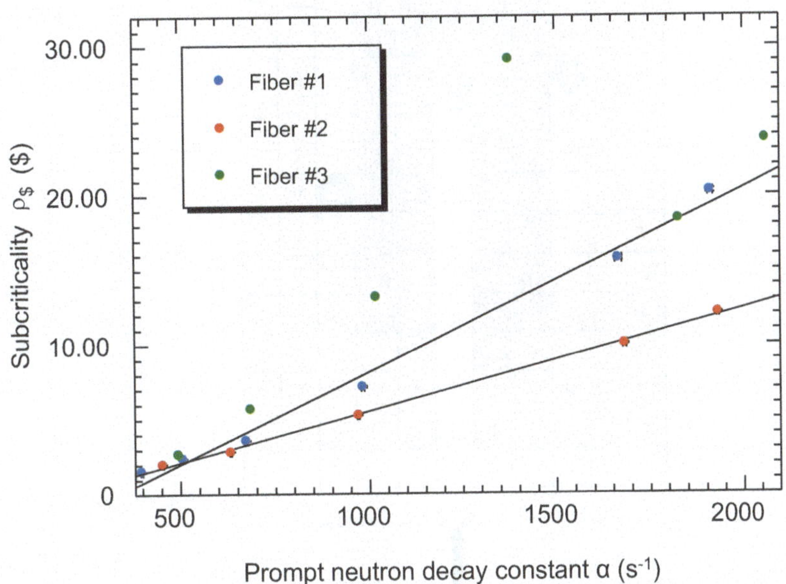

FIG. 2.109. Linear relationship between prompt neutron decay constant α and subcriticality ρ$ (Reproduced from Ref. [2.31] with permission courtesy of Elsevier B.V.).

2.1.6.7. Interpretation of measurements by the MAρTA method

The KURRI Phase II experiments focused on the subcriticality level, as stated in the documentation provided in the frame of the CRP work. Six configurations of the KUCA core were studied, with changes to the number of fuel rods and positions of the control rods (as an example, the core configuration for case II-1 is shown in Fig. 2.110). The computational analyses of these configurations using provided by KURRI, using Monte Carlo simulations, showed a wide range of values for the effective neutron multiplication factor k_{eff}, as presented in Table 2.57 [2.6]

TABLE 2.57. EFFECTIVE NEUTRON MULTIPLICATION FACTORS OF SUBCRITICAL SYSTEMS IN CASES II-1 TO II-6 [2.6]

Case	k_{eff} [a]
Critical state	1.00378 ± 0.00040
II-1	0.99219 ± 0.00040
II-2	0.98723 ± 0.00040
II-3	0.97950 ± 0.00040
II-4	0.95784 ± 0.00040
II-5	0.91355 ± 0.00040
II-6	0.90006 ± 0.00040

[a] MCNP6.1 with ENDF/B-VII.0)

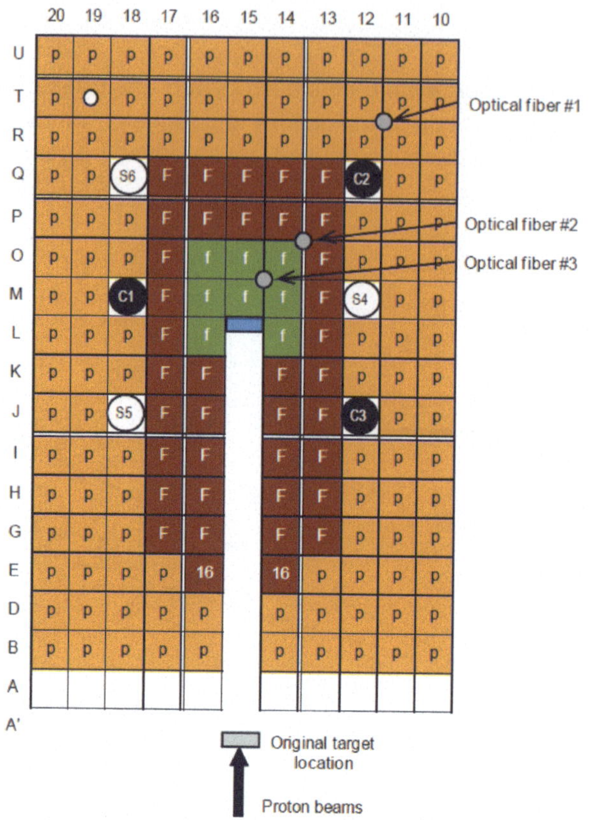

FIG. 2.110. Configuration of case II-1 in KUCA adopted for the experiments (Reproduced from Ref. [2.6] with permission courtesy of Kyoto University).

In Phase II, the measurements were performed with three optical fibre detectors, positioned on a diagonal from the polyethylene reflector towards the source location, as shown in Fig. 2.110. Optical fibre #1 is in the polyethylene reflector, while optical fibres #2 and #3 are surrounded by multiplying material; in particular, #3 is very close to the high energy neutron source. Consequently, the detectors are detecting the neutron population in very different conditions, and this directly impacts the quality of the reactivity reconstruction for the signal analysis. Ref. [2.6] shows that the results obtained from the optical fibres #1 and #2 were generally consistent, while optical fibre #3 was very much influenced by the contribution of the source neutrons with predictions for the reactivity that were far from those of the other detectors. The results are presented in Table 2.58 showing the inconsistencies associated with detector #3, as well as the reduced quality of the reconstruction for deeply subcritical configurations.

TABLE 2.58. MEASURED SUBCRITICALITY USING THE EXTRAPOLATED AREA RATIO METHOD [2.6]

| Case | $\beta_{eff}{}^a$ (pcm) | Reference (pcm) | | Subcriticality ρ (by area ratio method) | | |
				Fibre #1 (pcm)	Fibre #2 (pcm)	Fibre #3 (pcm)
II-1	785 ± 4	$\rho_{exp} =$	1160 ± 5	1224 ± 10	1153 ± 8	2108 ± 59
II-2	785 ± 4	$\rho_{exp} =$	1684 ± 6	1880 ± 15	1604 ± 11	4490 ± 109
II-3	783 ± 5	$\rho_{exp} =$	2483 ± 6	2824 ± 22	4177 ± 31	$22\,939 \pm 1279$
II-4	788 ± 5	$\rho_{MCNP} =$	4182 ± 6	5656 ± 62	8187 ± 54	$14\,985 \pm 1136$
II-5	815 ± 5	$\rho_{MCNP} =$	7895 ± 6	$12\,819 \pm 137$	9738 ± 64	$19\,109 \pm 2563$
II-6	806 ± 5	$\rho_{MCNP} =$	$11\,556 \pm 6$	$16\,312 \pm 160$	$16\,312 \pm 160$	$16\,312 \pm 160$

$^a \beta_{eff}$ from MCNP6.1 with JENDL4.0

The MAρTA algorithm has been applied to the signals provided with the same methodology applied in Phase I, including the statistical analysis and error propagation from the signals to the reactivity value averaged on the period of the source pulse. The integral parameters necessary for the calculations, provided in Ref. [2.6], have been taken as input; in particular, the different values of the effective delayed neutron fraction as reported in Table 2.57 have been adopted, keeping the same proportion among the β_i values for each family of precursors. The value of the neutron generation time Λ is taken as for the experiments in Phase I, for lack of additional information.

The results of the signal interpretation with MAρTA are reported in Table 2.59. The values obtained confirm the consistency of the results with the area method, as observed with the Phase I results. However, some notable differences can be observed:

- As experienced in the previous phase, the statistical error associated with MAρTA is in general lower than with the area method;
- In case 3, identical results were obtained for the first two detectors: it was observed that the Excel input file had identical columns for the two detectors, possibly due to a mistake in the postprocessing of the raw data;
- When moving away from criticality, the results of MAρTA departed from the area method, and both are different from the MCNPX simulation results, taken by KURRI as reference (see Table 2.58).

TABLE 2.59. SUBCRITICALITY LEVEL FOR THE PHASE II CONFIGURATIONS, EVALUATED WITH MAρTA

| Case | MAρTA | | |
	BF1 (pcm)	BF2 (pcm)	BF3 (pcm)
II-1	1240.7 ± 3.1	1296.8 ± 2.4	2457.6 ± 8.8
II-2	1822.4 ± 4.8	$1741.9 \pm 3{:}5$	4742.8 ± 18.2
II-3	2616.9 ± 3.3	2616.9 ± 3.3	8299.5 ± 16.3
II-4	4668.7 ± 14.4	4541.5 ± 13.2	16084 ± 16.2
II-5	7842.1 ± 9.6	5780.5 ± 5.2	34432 ± 94.5
II-6	9503 ± 7.2	6871.2 ± 4.2	41897 ± 65.3

The distance from the reference can be associated to the large distance from criticality of the analysed configurations, although a deeply subcritical core is known to be more point-like, therefore more consistent with the hypotheses at the basis of both interpretation methods. The reason for the large difference between the two interpretation methods, both based on point kinetics, is not known and would require analysis of more cases, with a consistency check on the experimental data to avoid the situation described above.

In the ADS experiments with spallation neutrons, the kinetic parameters were measured using both the PNS and Feynman-α methods with the optical fibre detectors, for subcriticalities ranging between 1160 and 11 556 pcm. The results confirmed the validity of the prompt neutron decay constant and the subcriticality in dollar units through the deduction of kinetic parameters. However, the effects of detector position dependency, neutron spectrum and subcriticality measurement methods still remained.

The experimental benchmarks of kinetic parameters obtained by ADS experiments with spallation neutrons are expected to play an important role in theoretical and numerical studies with respect to the detector position dependency, neutron spectrum and subcriticality measurement methods. In further studies, the experimental benchmarks could be conducive to the numerical verification of subcriticality online monitoring, to the analysis of subcriticality uncertainty and to the deterministic approach to kinetic parameters.

In addition, KUCA experimental data was analysed using SuperMC code. The main parameters were compared with SuperMC simulation. The studies could draw the following conclusions. Firstly, a calibration curve was set up of ρ_m and ρ_r to eliminate the influence of the higher harmonic modes. But the calibration curve did not give a correct result when the reactor was in an abnormal state. Secondly, a function $f(x, y, z)$ was proposed which indicates the measurement accuracy. In deeply subcritical reactors driven by an external source, the flux is composed of not only fundamental mode, but also higher harmonic modes, so the accuracy of the ASM was seriously affected by the detector positions. To find the optimal position of detector to improve the results of the ASM, a function $f(x, y, z)$ was proposed and simulated on KUCA. It was found that the measured subcriticality using a single detector was seriously affected by the perturbations. Thirdly, the effects of the axial and radial detector position were both evaluated. The results showed that the radial detector position improved the results of the ASM. In the case of radial detector position, the influence of the high harmonic modes of flux on the 'measured' subcriticality is more serious near the external source than that far from the external source. Finally, a ringed detector arrangement with calibration was proposed, and simulation work was performed on KUCA. The results showed that, compared to using a single detector, the ringed arrangement detector method with calibration has a better effect on improving the results of the ASM.

2.1.7. Analyses: Phase III – Study on reaction rates

2.1.7.1. Modelling

(a) KUCA (Japan)

The numerical analyses of reaction rates in high-energy thresholds were conducted with the Monte Carlo calculation code MCNP6.1 [2.39] coupled with JENDL/HE-2007 [2.30] , JENDL-4.0 [2.40], and JENDL/D-99 [2.50] (total number of histories: 10^8) with the combined use of MCNP6.1 and JENDL/HE-2007 for high energy protons and spallation process, JENDL-4.0 for transport and JENDL/D-99 for reaction rates.

(b) SNU (Republic of Korea)

For the reaction rate estimate at the indium wire, McCARD source-mode calculations were conducted with 10^8 histories using JENDL-4.0 libraries. The spallation neutron data at the target were obtained from the MCNPX2.6.0 proton simulations.

2.1.7.2. Experimental and numerical results

For an easy understanding of neutron irradiation, neutron spectra were obtained by MCNP calculations at several significant positions around the Pb-Bi-zoned core (Fig. 2.21(c)), as shown in Fig. 2.111: at the location of the target, the Pb-Bi-zoned fuel assembly (15,M), and two positions (14-13, P-O) and (14-13, L-K) at the boundary between Pb-Bi-zoned and normal fuel regions.

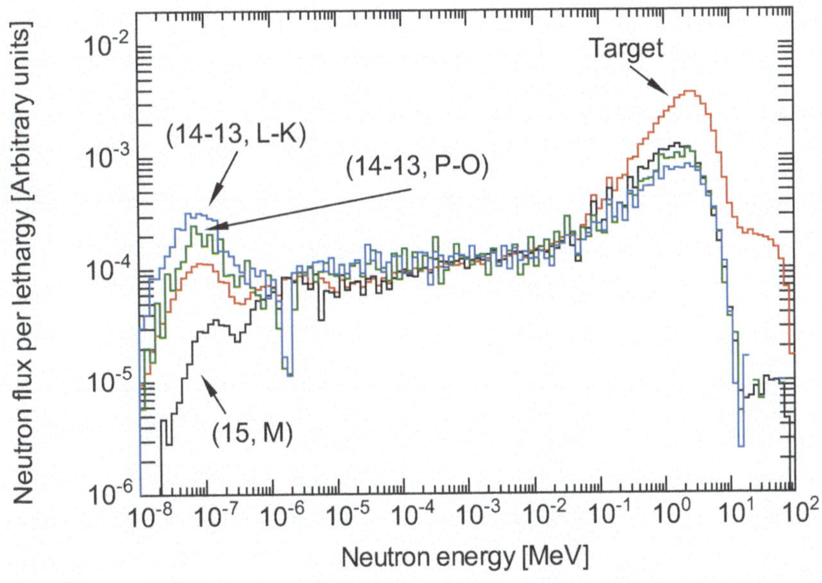

FIG. 2.111. Neutron spectra at target and several locations in Fig. 2.21(c) during injection of 100 MeV protons onto the Pb-Bi target. (Reproduced from Ref. [2.42] with permission courtesy of Taylor & Francis).

The following is a summary of the detailed discussion in Ref. [2.42]. Table 2.60 shows that measured and calculated subcriticalities were in good agreement, with a relative difference of 8%. The precision of subcriticality derived from eigenvalue calculations was key to ensuring the reliability of the MCNP6.1 calculations with a fixed-source for the ADS core driven by spallation neutrons.

TABLE 2.60. MEASURED AND CALCULATED SUBCRITICALITIES (PCM) [2.42]

Case	MCNP6.1 Calculation (pcm)	Experiment (pcm)	C/E
II-3	2483 ± 12	2302 ± 6	1.08 ± 0.01
II-4	4812 ± 12	—	—
II-5	9895 ± 12	—	—
II-6	11556 ± 12	—	—

(a) Foil activation method

Experimental values for reaction rates were calculated from the total counts of the peak energy of γ-rays, using the value of the saturation activity that is proportional to the reaction rate as given in Eqs (2.51) and (2.52):

$$D_\infty = \frac{\lambda T_c C}{\varepsilon_D \varepsilon_E (1 - e^{-\lambda T_i}) e^{-\lambda T_w}(1 - e^{-\lambda T_c})} \tag{2.51}$$

$$RR = D_\infty \frac{\rho}{M} = \frac{\lambda T_c C \rho}{\varepsilon_D \varepsilon_E (1 - e^{-\lambda T_i}) e^{-\lambda T_w} (1 - e^{-\lambda T_c}) M} \qquad (2.52)$$

where D_∞ (s^{-1}) is saturation activity, RR (s^{-1}·cm^{-3}) are the reaction rates, λ (s^{-1}) is the decay constant, T_c (s) is the measurement counting time, C (s^{-1}) is the counting rate, ε_D (%) is the detection efficiency, ε_E (%) is the emission rate, T_i (s) is the irradiation time, T_w (s) is the waiting time until the measurement starting after the irradiation, ρ (g·cm^{-3}) is the density, and M (g) is the mass of the foil. The experimental errors were estimated at 15% for the activation wire and 5% for the foils, including the statistical errors of the γ-ray counts and γ-ray peak width. Table 2.61 shows the atomic densities used in the calculations.

TABLE 2.61. ATOMIC DENSITY OF ACTIVATION FOILS FOR REACTION RATES MEASUREMENT [2.6]

Foil	Isotope	Abundance (%)	Purity (%)	Atomic density($\times 10^{24}$/cm^3)
In	^{113}In	4.3	99.99	1.64406E-03
	^{115}In	95.7	99.99	3.66790E-02
Ni	^{58}Ni	68.27	99.0	6.09388E-02
	^{60}Ni	26.10	99.0	2.41006E-02
	^{61}Ni	1.13	99.0	1.06083E-03
	^{62}Ni	3.59	99.0	3.42546E-03
	^{64}Ni	0.91	99.0	8.96354E-04
Fe	^{54}Fe	5.82	99.99	4.83829E-03
	^{56}Fe	91.18	99.99	7.79975E-02
	^{57}Fe	2.1	99.99	1.81771E-03
	^{58}Fe	0.28	99.99	2.46635E-04
^{27}Al	^{27}Al	100	99.0	5.99156E-02
^{197}Au	^{197}Au	100	99.95	5.90193E-02

i.   Indium wire distribution

Indium wire was used to obtain thermal neutron flux information based on the ^{115}In$(n, \gamma)^{116m}$In reaction rates, with the assumption that the thermal cross-sections of ^{115}In$(n, \gamma)^{116m}$In are proportional to the ^{235}U(n, f) thermal cross-sections. The accuracy of the reaction rate analyses for various subcritical states was assessed using the indium wire located at (14–13, P–A) in Figs 2.21(c)–(f) with subcriticality between 2483 and 11 556 pcm. Figures 2.112(a)–(d) show that good agreement was found between the measured ^{115}In$(n, \gamma)^{116m}$In reaction rate distributions for cases II-3 to II-6, , the MCNP fixed-source calculations. Figures 2.113–2.116 compare the McCARD calculations and experimental results in which the which the 1097.3 keV γ-ray was used to measure the ^{115}In$(n, \gamma)^{116m}$In reaction.

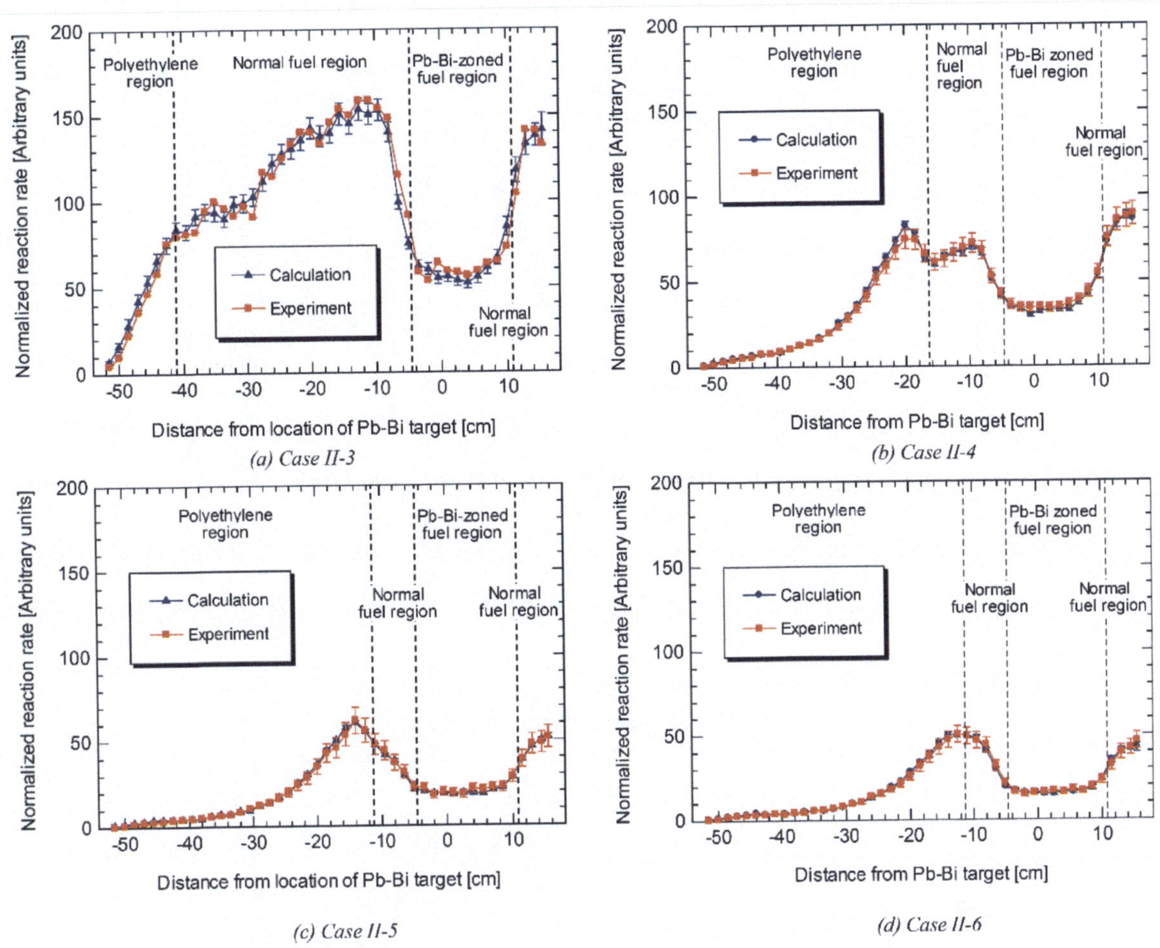

FIG. 2.112. Comparison between measured and calculated $^{115}In(n, \gamma)^{116m}In$ reaction rate distributions along (14–13, P–A) (Reproduced from Ref. [2.42] with permission courtesy of Taylor & Francis).

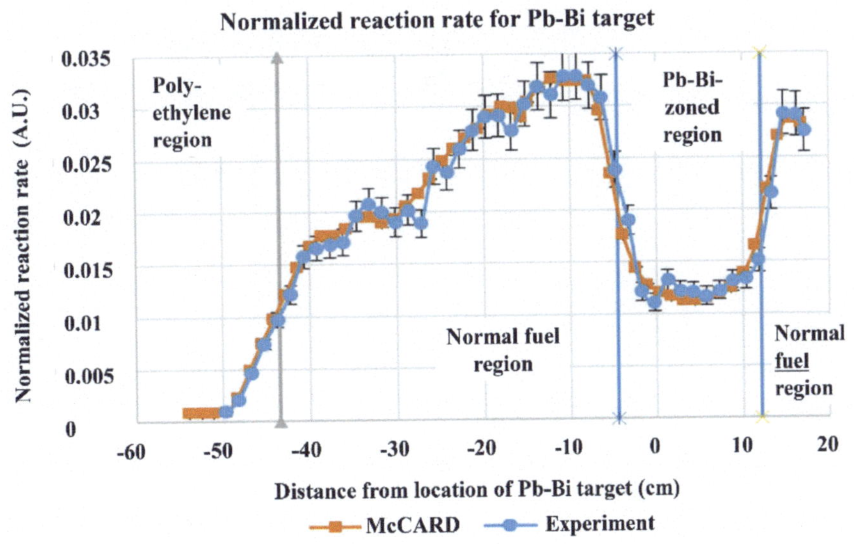

FIG. 2.113. Comparison of reaction rate distributions, Phase III case II-3 of in the Pb-Bi benchmark [2.89].

119

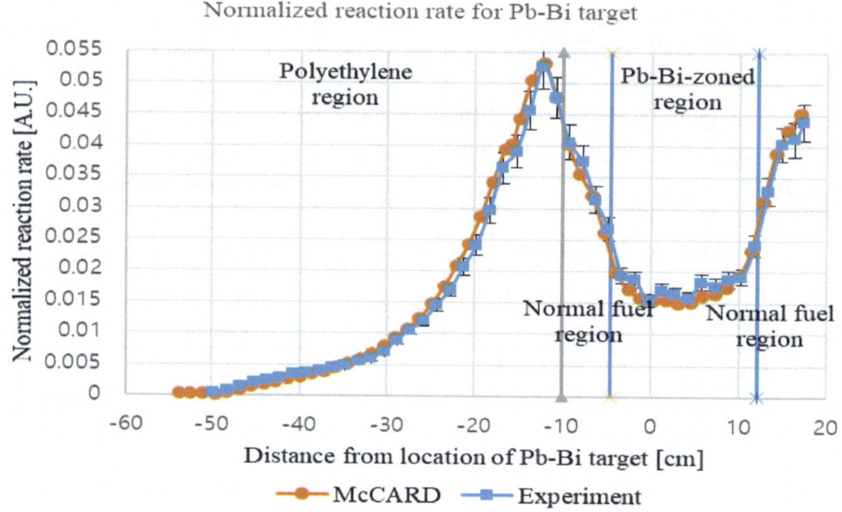

FIG. 2.114. Comparison of reaction rate distributions, Phase III case II-4 of the Pb-Bi Benchmark [2.89].

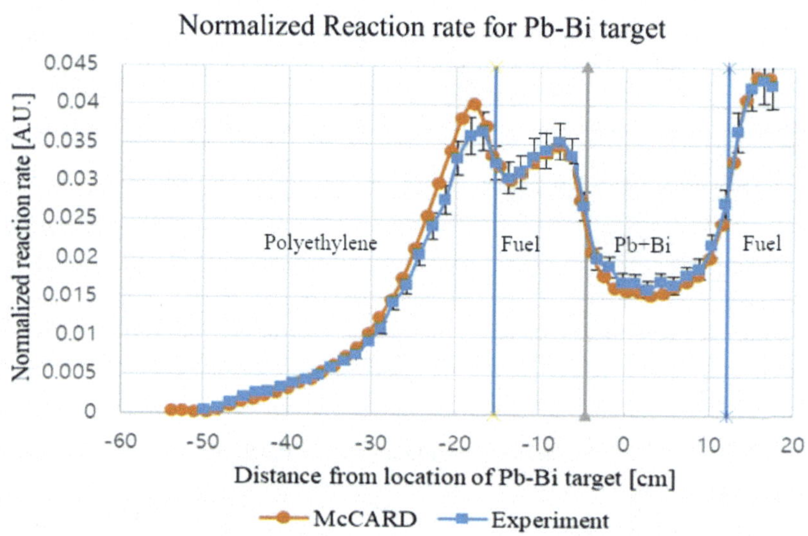

FIG. 2.115. Comparison of reaction rate distributions Phase III case II-5 of the Pb-Bi Benchmark [2.89].

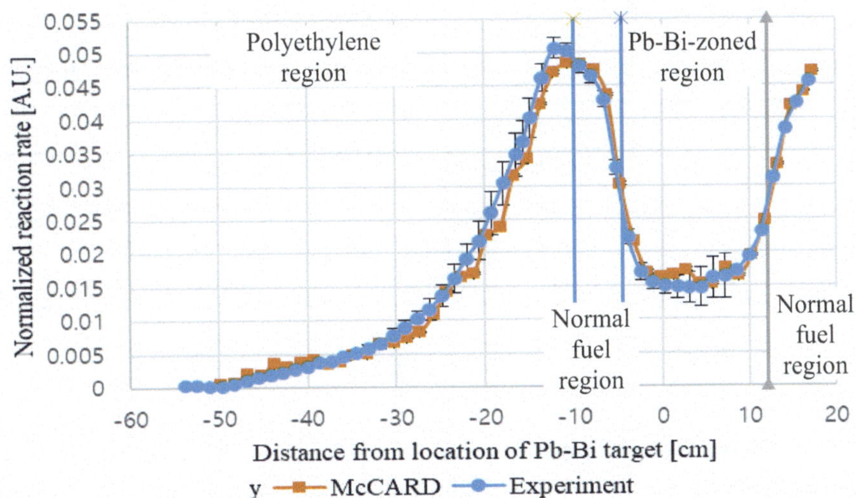

FIG. 2.116. Comparison of reaction rate distributions Phase III case II-6 of the Pb-Bi Benchmark [2.89].

ii. Subcritical multiplication factor

The subcritical multiplication factor k_s was used to investigate the effect of spallation neutrons on the ADS neutron multiplication using Eq. (2.53) [2.2]:

$$k_S = \frac{F}{F + S} \tag{2.53}$$

where F is the total number of fission neutrons and S total number of neutrons from the external neutron source. It is ^{115}In$(n, \gamma)^{116m}$In reaction rate in the indium wire in the core is related to F by the coefficient C_f and by the ^{115}In$(n, n')^{115m}$In reaction rate in the foil at the target location is related to S by the coefficient C_s as given in Eqs. (2.54) and (2.55), respectively [2.26]:

$$F = A \cdot C_f \int_0^{T_1} \int_{fuel\ region} R_{In(n,\gamma)}(x_0, y, z_0, t)\, dy\, dt \tag{2.54}$$

$$S = C_S \int_0^{T_i} R_{In(n,n')}(r_s, t)\, dt \tag{2.55}$$

where A is the constant of variable separation, C_f the coefficient of relating the cross-sections of ^{115}In$(n, \gamma)^{116m}$In and ^{235}U(n, f), C_s the coefficient relating ^{115}In$(n, n')^{115m}$In and capture rates of source neutrons, $R_{In(n,\gamma)}$ in the indium wires at position x_0 (location (14–13)), y (fuel region) and z_0 (axial centre position in fuel region), $R_{In(n,n')}$ is the inelastic scattering reaction rate of the indium foil at position r_s, and T_i is the irradiation time during the experiments. The assumption that the indium capture reaction rates are proportional to the fission neutrons, that is, that C_f can be treated as a constant, was validated. The second assumption, that 1-D reaction rates can be extended to 3-D reaction rates using the constant A shown in Eq. (2.54) was validated in previous ADS studies [2.18], [2.26].

MCNP was used to obtain values for A, C_f and C_s. Equation (2.53)–(2.55) can be rewritten as Eq. (2.56):

$$k_s = \frac{F}{F + S} = \frac{A \cdot C_f \int_0^{T_i} \int_{fuel\ region} R_{In(n,\gamma)}(x_0, y, z_0, t)\, dy\, dt}{A \cdot C_f \int_0^{T_i} \int_{fuel\ region} R_{In(n,\gamma)}(x_0, y, z_0, t)\, dy\, dt + C_s \int_0^{T_i} R_{In(n,n')}(r_s, t)\, dt} \tag{2.56}$$

121

Table 2.62 shows that good agreement was found between experiments and calculations of k_s using Eq. (2.56) within a relative difference of 1%.

TABLE 2.62. COMPARISON BETWEEN MEASURED AND MCNP6.1CALCULATED k_S VALUES [2.42]

Case	Calculation	Experiment
II-3	0.9989 ± 0.0005	0.9996 ± 0.0008
II-4	0.9997 ± 0.0005	0.9991 ± 0.0009
II-5	0.9990 ± 0.0005	0.9991 ± 0.0009
II-6	0.9985 ± 0.0005	0.9989 ± 0.0005

iii. Activation foils

Table 2.63 gives the measured reaction rates of the foils irradiated in the subcritical cores. Table 2.64 and Fig. 2.12 provide a comparison of the measured and calculated reaction rates. In case II-3 which had a subcriticality of 2483 pcm, there was good agreement for reaction rates with low or non-threshold energy, with a relative difference around 10%. However, the calculations for the high-threshold reaction rates of the ^{27}Al and ^{56}Fe foils underestimated by up to 50%. Figure 2.12 shows that the C/E value reduces with deepening subcriticality.

A comparison with earlier studies that used 14 MeV neutrons [2.17] showed that the discrepancy between measured and calculated reaction rates was larger for the ADS with spallation neutron. This was attributed to the uncertainty of high-energy threshold reaction rates and the difficulty of making a precise simulation of defocused proton beams. At KUCA, the proton beams were in a vacuum until they reached the original target location and thereafter transited the air space (15, A-L) to the Pb-Bi target at the new target location. See Figs 2.21(c)–(f).

TABLE 2.63. MEASURED REACTION RATES OF ACTIVATION FOILS IN CASES II-3 TO II-6. [2.6]

	Measured reaction rate ($s^{-1} \cdot cm^{-3}$)			
Reaction	Case II-3	Case II-4	Case II-5	Case II-6
^{197}Au$(n, \gamma)^{198}$Au (bare)	(8.88 ± 0.02)E+06	(4.88 ± 0.04)E+06	(3.51 ± 0.08)E+06	(2.53 ± 0.04)E+06
^{197}Au$(n, \gamma)^{198}$Au (Cd)	(7.84 ± 0.09)E+06	(4.46 ± 0.04)E+6	(3.11 ± 0.04)E+6	(2.30 ± 0.03)E+06
^{115}In$(n, n')^{115m}$In	(8.60 ± 0.13)E+04	(4.27 ± 0.22)E+04	(4.27 ± 0.03)E+04	(2.86 ± 0.06)E+04
^{58}Ni$(n, p)^{58}$Co	(4.90 ± 0.08)E+04	(3.18 ± 0.03)E+04	(3.23 ± 0.16)E+04	(2.00 ± 0.10)E+04
^{56}Fe$(n, p)^{56}$Mn	(1.82 ± 0.07)E+03	(1.26 ± 0.02)E+03	(1.55 ± 0.03)E+03	(1.39 ± 0.02)E+03
^{27}Al$(n, \alpha)^{24}$Na	(1.54 ± 0.05)E+03	(1.11 ± 0.02)E+03	(1.62 ± 0.03)E+03	(1.10 ± 0.01)E+03

TABLE 2.64. COMPARISON OF MEASURED AND CALCULATED REACTION RATES [2.42]

Reaction	Calculated / Experimental (C/E) Ratio			
	Case II-3	Case II-4	Case II-5	Case II-6
^{197}Au$(n, \gamma)^{198}$Au (bare)	1.14 ± 0.09	0.96 ± 0.06	0.90 ± 0.06	0.75 ± 0.07
^{197}Au$(n, \gamma)^{198}$Au (Cd)	1.10 ± 0.09	0.81 ± 0.05	0.80 ± 0.06	0.71 ± 0.07
^{115}In$(n, n')^{115m}$In	0.86 ± 0.02	0.69 ± 0.04	0.52 ± 0.01	0.47 ± 0.02
^{58}Ni$(n, p)^{58}$Co	0.99 ± 0.04	0.74 ± 0.03	0.59 ± 0.03	0.67 ± 0.04
^{56}Fe$(n, p)^{56}$Mn	0.60 ± 0.03	0.54 ± 0.03	0.47 ± 0.02	0.40 ± 0.02
^{27}Al$(n, \alpha)^{24}$Na	0.47 ± 0.01	0.26 ± 0.02	0.19 ± 0.01	0.22 ± 0.02

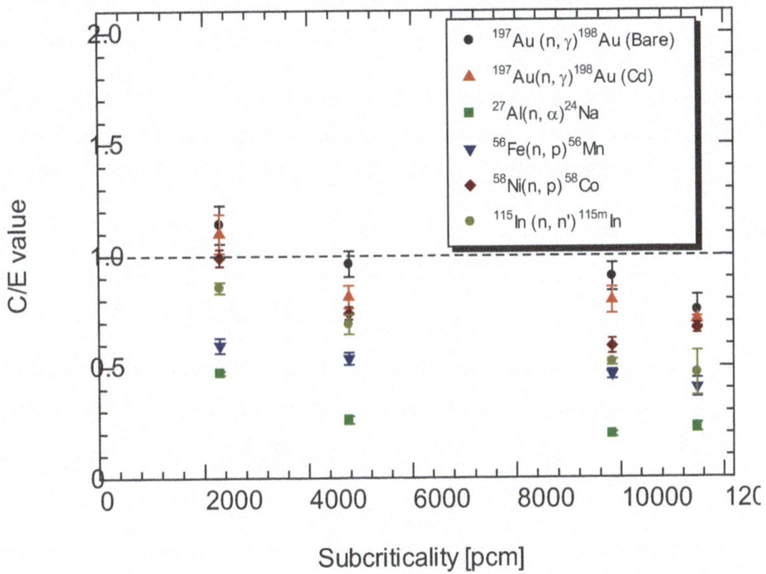

FIG. 2.117. C/E- value between measured and calculated reaction rates by varying subcriticality in cases II-3 to II-6. (Reproduced from Ref. [2.42] with permission courtesy of Taylor & Francis).

(b) Spectrum index

i.   Cd ratio

The Cd ratio is defined as the ratio of activation reaction rates of the bare and Cd-covered gold foils, $R_{Au\text{-}bare}$ and $R_{Au\text{-}Cd}$ as given in Eq. (2.57). It is interpreted as the neutron spectrum index, based on relation between absorption cross-sections of gold and cadmium.

$$Cd\ ratio = \frac{R_{Au-bare}}{R_{Au-Cd}} \tag{2.57}$$

Table 2.65 compares the experimental and calculated results for the Cd ratio using the bare and Cd-covered gold foils for subcritical core configurations. The good agreement seen in Table 2.65 confirms the accuracy of the MCNP simulated Au absorption reaction rates in both the thermal and epithermal neutron regions.

TABLE 2.65. COMPARISON BETWEEN MEASURED AND CALCULATED Cd RATIOS. [2.42]

Case	Calculation	Experiment	C/E
II-3	1.15 ± 0.09	1.11 ± 0.06	1.04 ± 0.10
II-4	1.24 ± 0.09	1.09 ± 0.02	1.13 ± 0.09
II-5	1.27 ± 0.09	1.12 ± 0.04	1.13 ± 0.09
II-6	1.16 ± 0.09	1.10 ± 0.04	1.05 ± 0.09

ii. In ratio

As discussed above, the ratio of thermal to epithermal neutrons measured by the Cd ratio was interpreted as a spectrum index of about 10%, by subtracting the unit value from the Cd ratio experimental results shown in Table 2.65. The F8/F5 ratio, the ratio of fast fission and total fission reaction rates of ^{238}U and ^{235}U, respectively, is a spectrum index that is used with fast reactors as a quantitative evaluation of the neutron spectrum. This work described here introduces a new concept of the ^{235}U fission ratio to measure thermal and fast neutrons, analogous to the Cd ratio and the F8/F5 value. The ^{235}U fission ratio in the specified Pb-Bi-zoned fuel region from (14–13, L; y_1) to (14–13, O; y_2) shown in Figs 2.21(c)–(f), is defined as shown in Eq. (2.58).

$$^{235}U\ fission\ ratio = \frac{\int_{y_1}^{y_2} R_{U-fission}^{thermal}(x_0, y, z_0)dy}{\int_{y_1}^{y_2} R_{U-fission}^{fast}(x_0, y, z_0)dy} \tag{2.58}$$

where $R_{U-fission}^{thermal}$ and $R_{U-fission}^{fast}$ indicate ^{235}U(n, f) reaction rates in the thermal (< 0.1 eV) and fast (> 0.1 MeV) neutron regions, respectively. With the assumption discussed above for the Cd ratio the numerator of Eq. (2.58) can be written:

$$\int_{y_1}^{y_2} R_{U-fission}^{thermal}(x_0, y, z_0)dy = C_{fission}^{thermal} \int_{y_1}^{y_2} R_{In(n,\gamma)}^{thermal}(x_0, y, z_0)dy \tag{2.59}$$

where $C_{fission}^{thermal}$ is the proportionality coefficient of between the cross-sections for ^{115}In$(n, \gamma)^{116m}$In and ^{235}U(n, f) in the thermal neutron region. Similarly, by a new assumption of the proportionality of the In inelastic scattering (threshold energy 0.4 MeV) and ^{235}U fast fission reactions, the denominator in Eq. (2.58) is written as follows:

$$\int_{y_1}^{y_2} R_{U-fission}^{fast}(x_0, y, z_0)dy = C_{fission}^{fast} \int_{y_1}^{y_2} R_{In(n,n')}^{fast}(x_0, y, z_0)dy \tag{2.60}$$

where $C_{fission}^{fast}$ is the proportionality coefficient of between cross-sections for ^{115}In$(n, n')^{115m}$In and ^{235}U(n, f) in the fast neutron region, noting that the assumption of proportionality in the fast neutron region is more complicated because of the characteristics of the cross-sections. Using MCNP fixed-source calculations with Eqs (2.59) and (2.60), the results of $C_{fission}^{thermal}$ and $C_{fission}^{fast}$ were found to be nearly constant around 1.11 and 6.25, respectively.

Direct experimental measurement of ^{235}U(n, f) reaction rates is complex, so it was proposed to use the indium wire distribution method as discussed above to obtain neutron flux information for both the thermal and fast energy regions simultaneously. With the assumption that the ^{235}U fission ratio in Eq. (2.58) corresponds approximately to the ratio of ^{115}In$(n, \gamma)^{116m}$In and ^{115}In$(n, n')^{115m}$In reaction rate distributions, a new spectrum index based on the In ratio can be defined by introducing the coefficients of $C_{fission}^{thermal}$ and $C_{fission}^{fast}$ as shown in Eq. (2.61):

$$In\ ratio = \frac{C_{fission}^{thermal} \int_{y_1}^{y_2} R_{In(n,\gamma)}^{thermal}(x_0, y, z_0)dy}{C_{fission}^{fast} \int_{y_1}^{y_2} R_{In(n,n')}^{fast}(x_0, y, z_0)dy} \qquad (2.61)$$

A relative difference of around 10% in the numerical results of ^{235}U fission ratio and the In ratio confirms that that the assumption in Eqs (2.59) and (2.60) is mostly valid. The measured and calculated In ratios shown in Table 2.66 were in good agreement with a relative difference of <5% in cases II-3 to II-6. Figure 2.118 shows that the ratio was constant, particularly in the Pb-Bi-zoned fuel regions, except for the region boundaries, and that fast neutrons dominate the Pb-Bi zoned fuel region. The agreement between measured and calculated In ratios supports the use of the In ratio to determine the ratio of thermal to fast neutrons as an additional neutron spectrum index for ADS.

TABLE 2.66. COMPARISON BETWEEN MEASURED AND CALCULATED INDIUM RATIOS [2.42]

Case	^{235}U fission ratio (Eq. (2.63))	In ratio (Eq. (2.66))		C/E
	Calculation	Calculation	Experiment	
II-3	$(5.16 \pm 0.08) \times 10^{-3}$	$(4.62 \pm 0.08) \times 10^{-3}$	$(4.52 \pm 0.10) \times 10^{-3}$	1.02 ± 0.03
II-4	$(2.21 \pm 0.10) \times 10^{-3}$	$(2.15 \pm 0.10) \times 10^{-3}$	$(2.17 \pm 0.10) \times 10^{-3}$	0.99 ± 0.07
II-5	$(1.79 \pm 0.07) \times 10^{-3}$	$(1.60 \pm 0.07) \times 10^{-3}$	$(1.57 \pm 0.11) \times 10^{-3}$	1.02 ± 0.08
II-6	$(1.70 \pm 0.09) \times 10^{-3}$	$(1.54 \pm 0.09) \times 10^{-3}$	$(1.48 \pm 0.10) \times 10^{-3}$	1.04 ± 0.10

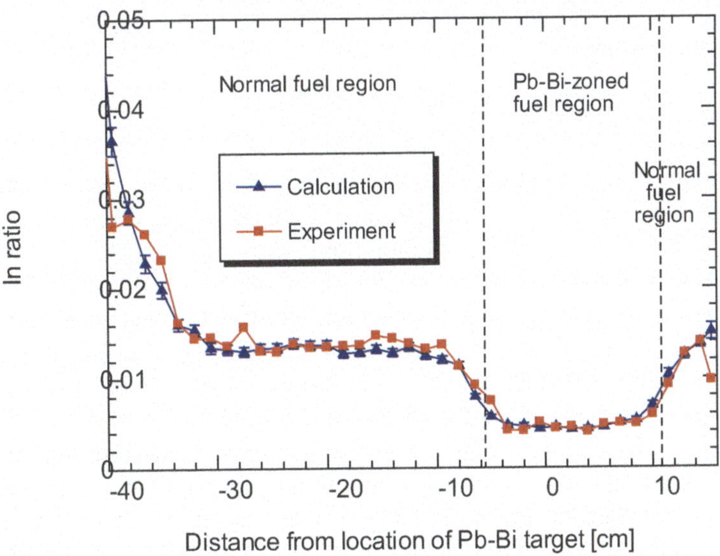

FIG. 2.118. Comparison between measured and calculated In ratios in case II-3. (Reproduced from Ref. [2.42] with permission courtesy of Taylor & Francis).

Reaction rate experiments to investigate the accuracy of the foil activation reaction rate measurement at various subcriticality levels were carried out at KUCA using spallation neutrons produced by 100 MeV protons with a Pb-Bi target. The agreement between the measured and calculated ^{115}In$(n, \gamma)^{116m}$In reaction rate distributions demonstrated the reliability of the MCNP calculations of the subcritical multiplication factor k_s. The high-threshold energy reaction rates showed significant discrepancies between experiment and calculation as well as a dependence of the reaction rates on the subcriticality level. This remains an important issue for fixed-source calculations. The In ratio was proposed as neutron spectrum index for ADS and validated by the agreement between experiment and calculation.

2.1.8. Analyses: Phase IV – Study on effective delayed neutron fraction

2.1.8.1. Modelling

(a) KUCA (Japan)

To obtain correction factors g and g^* by Eqs (5.22) and (5.23), respectively (see the discussion in Section 5.6), the calculations of reaction rates and adjoint flux were performed by MCNP6.1 [2.152] together with JENDL-4.0 [2.40] or ENDF/B-VII.0 [2.64] (uranium and boron) and JENDL/HE-2007 (all nuclides except uranium and boron); total histories for adjoint flux and reaction rates were 10^7 and 10^6, respectively.

(b) SNU (Republic of Korea)

The effective delayed neutron fraction β_{eff} was calculated by McCARD and compared with MCNP6.1 and the experimental results for the seven core configurations in Phase IV. McCARD eigenvalue calculations were conducted with 10^5 histories per cycle and JENDL-4.0 cross-section libraries. The cycle length of the adjoint weighted calculation was 10.

(c) KIT (Germany)

The 3-D (XYZ) ERANOS and PARTISN models of the case I-4 (Fig. 2.14(d)) were benchmarked against the corresponding MCNPX2.7 [2.74] model and the reference value from KUCA team (MCNP6) with respect to the criticality level and the kinetic parameters. Transport calculations were performed by using the effective macroscopic 40-group cross-sections, with 10 groups below 1 eV, generated from an ultra-fine-group European Cell Code (ECCO) [2.75] data library (1968 energy groups), using the JEFF3.1 reference data library. Concerning ERANOS, the VARIANT transport solver was used. The MCNPX2.7 calculations were performed by employing the JEFF3.1, JEFF3.2, and JENDL4.0 point-wise cross-section libraries. Results concerning the criticality level are given in Table 2.31 and show a good agreement between the deterministic and the Monte Carlo calculations.

2.1.8.2. Results for the Rossi-α method and β_{eff} estimation

This section is a summary of the detailed discussion in Ref. [2.46]. PNS experiments at KUCA were used to investigate neutron noise analyses by the Rossi-α method using the core configurations shown in Fig. 2.27(a)–(g). The key equations for this method are given in Section 5.6. Values of α to supplement the neutron noise analyses were derived from a plot of the count rate as a function of time, as shown in Fig. 2.119. Equation (2.5) was fitted to the region where the correlation probability is negligible to obtain the intensity of the second term in Eq. (5.11). The decay of the correlation probability was confirmed at 0.025 s, and intensity B was obtained by the fitting with Eq. (5.12) in the region between 0.025 and 0.125 s. Subtraction of the uncorrelated joint probability term from Eq. (5.13) and fitting with Eq. (5.14) yielded the value of the correlation term in Eq. (5.10). Figure 2.120 shows that the uncorrelated probability decreased rapidly. The correlated probability increased in the vicinity of zero rather than exponentially decreasing. This was attributed to an overestimation of the uncorrelated probability in that region due to the shape of the external neutron source changing from Gaussian to pulsed. The calculation of the uncorrelated probability is sensitive to the form of the neutron source.

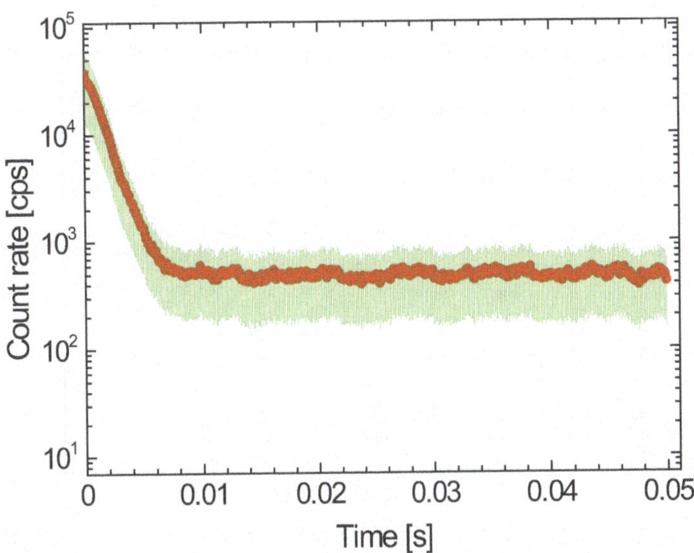

FIG. 2.119. PNS histogram of optical fibre detector in Case IV-1. (Reproduced from Ref. [2.46] with permission courtesy of Taylor & Francis).

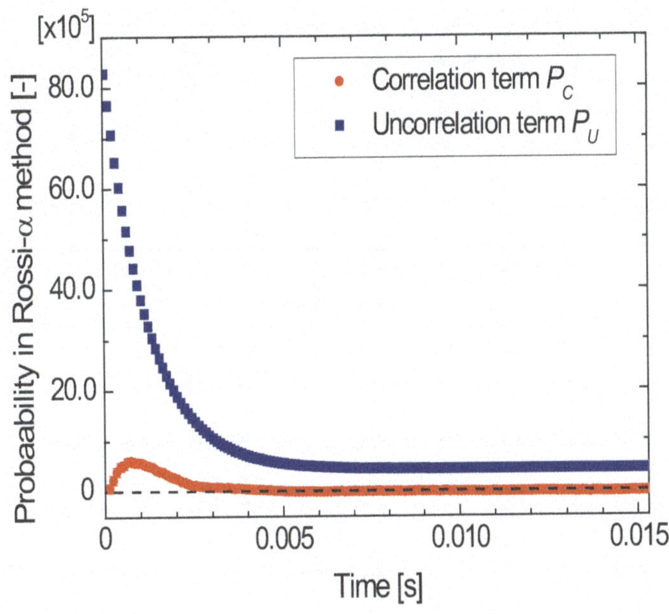

FIG. 2.120. Correlated and uncorrelated probabilities of fibre detector in case IV-1 by Rossi-α method in PNS experiments. Reproduced from Ref. [2.46] with permission courtesy of Taylor & Francis).

The α and B fitted values varied with subcriticality as shown in Figs 2.121(a) and (b), attributed to the increase of the subcriticality for α and the decrease of neutrons for B. Figure 2.121(c) shows that the optical fibre had a decreasing tendency in the intensity of correlated probability C because the fission reaction rate decreases with increasing subcriticality. The BF$_3$ detector, however, showed the increasing tendency in cases IV-4, IV-5 and IV-6. This was attributed to low accuracy of the calculated correlated probability because of the low correlated probability of the fission neutrons in BF$_3$ detectors positioned outside the core.

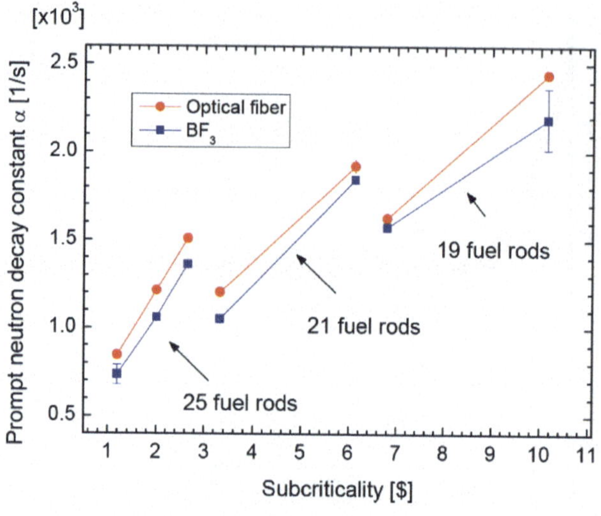

a)   Prompt neutrondecay constant α

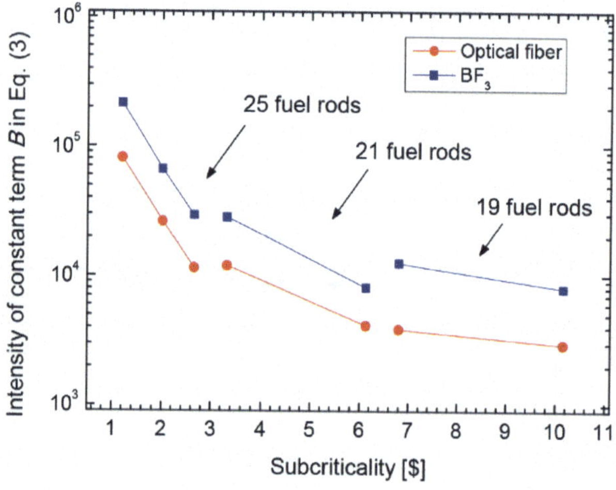

b)   Intensity of uncorrelated probability B

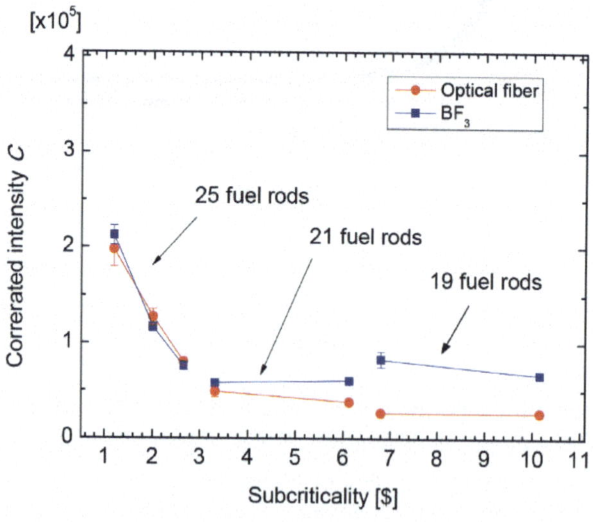

c)   Intensity of correlated probability C

FIG. 2.121. Parameters obtained by the Rossi-α method in PNS experiments by varying subcriticality (Reproduced from Ref. [2.6] with permission courtesy of Kyoto University).

Reaction rates and adjoint flux calculations using MCNP6.1 [2.39] with JENDL-4.0 [2.40] or ENDF/B-VII.0 [2.64] (uranium and boron) and JENDL/HE-2007 [2.30] (all nuclides except uranium and boron) were made to determine the g and g^* correction factors according to Eqs (5.22) and (5.23). A total of 10^7 histories were used for adjoint flux and 10^6 histories for the reaction rates, with a statistical error of < 1%. The procedure for manually obtaining the 3-D adjoint flux was: an external neutron source in the Watt spectrum of ^{235}U was set inside an HEU plate; the reaction rates for the response of $\nu\Sigma_f$, which is better suited than Σ_f to estimate the adjoint flux [2.153], were tallied over the core with NONU option to avoid the neutron multiplication; these reaction rates approximated the adjoint flux at the HEU plate; the external neutron source in the HEU plate was repositioned and the calculation repeated. Table 2.67 shows that the correction factor g was invariant with subcriticality, but g^* decreased with increasing subcriticality. There was a small difference in g and g^* depending on the cross-section library used. For the β_{eff} measurements, g and g^* were calculated using JENDL-4.0 and JENDL/HE-2007.

The β_{eff} values were determined using Eq. (5.27) with the above values for g and g^* and the fitted parameters α, B and C from the ADS experiments. Table 2.68 compares the measured and MCNP6.1 calculated values of β_{eff}. The optical fibre at the centre of the core had acceptable accuracy with approximately 13% difference in cases IV-1 to IV-7, with $k_{eff} \approx 0.93$. There was a large difference between the calculated β_{eff} and that measured with the BF$_3$ detector. This was attributed to the variation in the shape of the pulsed neutron source and the determination of correlated probability using Eq. (2.5). The optical fibre in the core was accurate because it was not affected by the higher mode in flux distribution and correlated neutrons from the same fission chain are largely detected. A detector located outside of the core the poor measurement accuracy is due to the difficulty in obtaining the correlated term according to Eq. (5.13). To estimate the measurement accuracy of β_{eff}, it was compared with the value estimated by the area ratio method and the relation between reactivity in dollar and pcm units. If k_{eff} is known, β_{eff} can be estimated by Eq. (2.62):

$$\beta_{eff}^{Area\ ratio\ method} = \frac{\rho}{\rho_s} = \frac{1}{\rho_s}\left(1 - \frac{1}{k_{eff}}\right) \tag{2.62}$$

Table 2.68 compares the measured and calculated values of β_{eff} and shows that, in cases IV-1–IV-7, the proposed methodology for measuring β_{eff} with the optical fibre detector (optical fibre column) was comparable or more accurate than using the area ratio method (area ratio column) set out in Eq. (2.62). This confirmed the validity of the measurement methodology for β_{eff} for the subcritical ADS with $k_{eff} \approx 0.93$. To obtain accurate β_{eff} values with the proposed methodology, the values for the correlated probability can be improved with using a specific shape (Gaussian distribution) of the external neutron source. The McCARD results for k_{eff} and β_{eff} are also shown in Table 2.68. The β_{eff} calculated by McCARD agrees within the 95% confidence interval with the values from the MCNP6.1 calculations and the optical fibre measurements. The values measured with the BF$_3$ detector showed an increased discrepancy as the system subcriticality deepened.

TABLE 2.67. CORRECTION FACTORS [2.46]

Case	g [a]		g^* ($\times 10^{-3}$) [b]	
	JENDL-4.0 and JENDL/HE-2007	ENDF/B-VII.0 and JENDL/HE-2007	JENDL-4.0 and JENDL/HE-2007	ENDF/B-VII.0 and JENDL/HE-2007
IV-1	1.03 ± 0.01	1.03 ± 0.01	4.29 ± 0.01	4.32 ± 0.01
IV-2	1.03 ± 0.01	1.03 ± 0.01	4.28 ± 0.01	4.31 ± 0.01
IV-3	1.04 ± 0.01	1.04 ± 0.01	4.18 ± 0.01	4.19 ± 0.01
IV-4	1.03 ± 0.01	1.03 ± 0.01	3.84 ± 0.01	3.83 ± 0.013
IV-5	1.03 ± 0.01	1.03 ± 0.01	3.74 ± 0.01	3.78 ± 0.01
IV-6	1.03 ± 0.01	1.03 ± 0.01	2.84 ± 0.023	2.84 ± 0.01
IV-7	1.03 ± 0.01	1.03 ± 0.01	2.75 ± 0.01	2.76 ± 0.01

[a] Calculated by Eq. (5.22); [b] Calculated by Eq. (5.23) (see Section 5)

TABLE 2.68. COMPARISON BETWEEN MEASURED AND CALCULATED β_{eff} VALUES [2.46]

Case	k_{eff} (SD)		Effective delayed neutron fraction β_{eff} [pcm] (SD)			
	McCARD (from SNU)	MCNP6.1	McCARD (from SNU)	MCNP6.1	Optical fibre	BF$_3$
Case 1	0.98871 (0.00013)	0.99093 (0.00009)	804 (5)	817 (11)	870 (81)	911 (29)
Case 2	0.98017 (0.00013)	0.98274 (0.00009)	812 (5)	794 (11)	876 (58)	845 (33)
Case 3	0.97373 (0.00014)	0.97627 (0.00009)	821 (5)	814 (11)	855 (58)	725 (36)
Case 4	0.97450 (0.00012)	0.97662 (0.00009)	824 (5)	801 (11)	903 (109)	804 (28)
Case 5	0.95411 (0.00013)	0.95680 (0.00009)	816 (5)	826 (12)	915 (75)	303 (7)
Case 6	0.95087 (0.00013)	0.95275 (0.00009)	820 (5)	817 (11)	921 (86)	189 (19)
Case 7	0.93123 (0.00013)	0.93358 (0.00009)	830 (5)	815 (11)	811 (72)	162 (14)

Table 2.69 shows the very good agreement between kinetic parameters calculated with ERANOS and KIN3D/PARTISN and the KUCA reference. The 3-D ERANOS results were obtained using a code extension developed at KIT.

TABLE 2.69. KINETIC PARAMETERS OF CASE I-4: ERANOS, KIN3D/PARTISN, AND REFERENCE RESULTS. [2.67]

Code	β_{eff} (SD) [pcm]	Λ [s]
ERANOS/VARIANT	829	3.027E-05
KIN3D/PARTISN	827	3.032E-05
McCARD (From Seoul National University)	824 (5)	—
MCNP6.1 (From Kyoto University)	801 (11)	—
KUCA reference	807	3.050E-05

2.1.8.3. Concluding remarks

The Rossi-α method for estimating β_{eff} was used in the PNS experiments with spallation neutrons. The values for β_{eff} that were obtained from an optical fibre detector located at the centre of the core and a BF$_3$ detector located outside the core were compared with the values calculated by MCNP6.1. The optical fibre detector results had a relative difference of about 13% in the subcritical ADS with $k_{eff} \approx 0.93$. The results from the BF$_3$ detector did not agree with the MCNP6.1 calculations because of the low accuracy of the uncorrelated probability. This was attributed to the optical fibre detector being at the centre of the core, thus minimizing the effect of higher-order flux modes. The applicability of the Rossi-α measurement methodology was demonstrated by the agreement between calculated and measured β_{eff} values in the experiments. To further improve the β_{eff} measurements, the shape of the external neutron source should be a Gaussian distribution.

2.2. KUCA ADS U–PB EXPERIMENTS

The U–Pb benchmark experiments conducted at the KUCA ADS with solid lead plates with various core configurations were published in Ref. [2.7] and are summarised in this section. The U–Pb benchmark consisted of experiments with subcritical ADS configurations using lead plates with two different external neutron sources: 14 MeV neutrons and 100 MeV protons impacting a Pb-Bi target. The U–Pb benchmark was divided into two sections, Phase I which investigated kinetic parameters and Phase II which focussed on the reaction rate distribution.

The U–Pb benchmark compared neutronics parameters calculated with Monte Carlo codes with experimental results. Monte Carlo codes were used to calculate both static and kinetic parameters including the effective multiplication factor, excess reactivity, control rod worth, reaction rate distribution, and the prompt neutron decay constant. The benchmark specifications are given in Sections 2.2.1 to 2.2.3, and their analyses are provided in Sections 2.2.4 to 2.2.6.

2.2.1. Benchmark specifications

2.2.1.1. Description

The KUCA type-A core has assemblies with a height of 1524 mm, a square cross-section with sides of 54.3 mm. The assembly had a 0.5 mm thick aluminium sheath, with a gap of 0.25 mm between the assembly and the aluminium sheath. Internally, the assembly comprised a laminated structure of thin plates, with various plate compositions and thicknesses depending on the type of assembly. The inter-assembly gap was normally 1 mm, with some regions having a 3 mm gap to enable the insertion of an indium wire for activation experiments. Six control rods were composed of anhydrous boron, with a double aluminium sheath, and an inter-sheath gap of 2.5 mm. The outer sheath was fixed at the bottom, but the inner sheath could move up and down. Figure 2.122 shows the side view of target and core configuration with the 14 MeV neutron source and Fig. 2.123 shows the side view of target and core configuration with the 100 MeV proton beam.

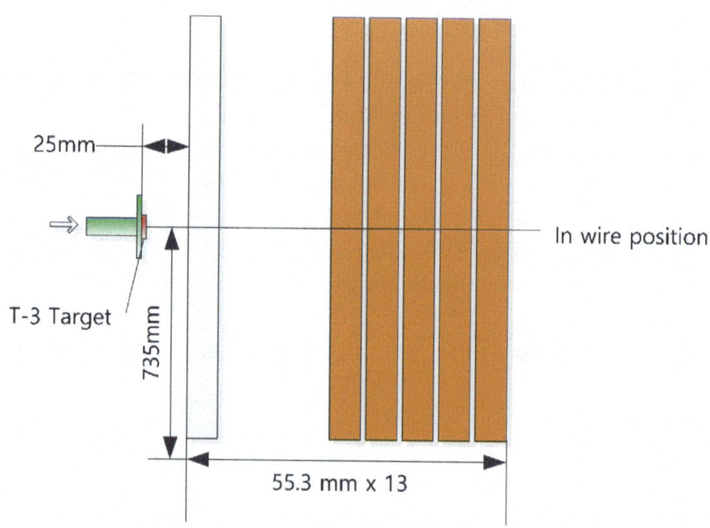

FIG. 2.122. Side view of target and core configuration with 14 MeV neutrons [2.89].

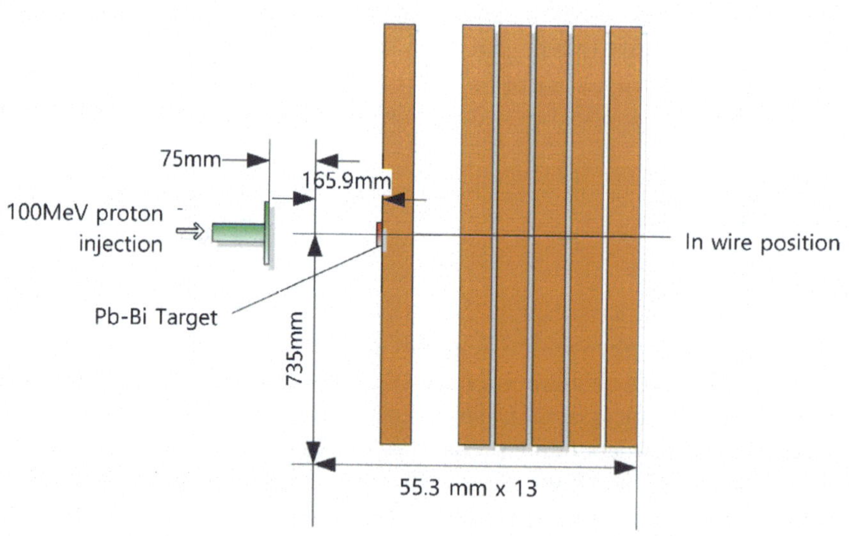

FIG. 2.123. Side view of target and core configuration with 100 MeV protons [2.89].

2.2.1.2. Core configurations

In the U–Pb benchmark, two types of fuel assemblies were used. The type-F assembly had 60 unit fuel cells with HEU and PE plates. The type-f assembly had 40 cells with HEU-Pb in the middle of the assembly, and 10 unit fuel cells with HEU–PE plates arranged above and below the HEU-Pb region. Table 2.70 shows the thickness of the plates used in the assemblies. Figures 2.124 and 2.125 show the type-F and type-f assemblies, respectively. The polyethylene moderator (p) and the polyethylene reflector (PE) are shown in Figs 2.126 and 2.127, respectively. The graphite assembly is shown in Fig. 2.128 and the polyethylene reflector with a detector in Fig. 2.129.

TABLE 2.70. THICKNESS OF PLATES [2.7]

Plate	Thickness (mm)
1/8" PE	3.086
1/4" PE	6.300
1/2" PE	12.500
19" PE	483.085
1/8" p	3.158
10" p	254.000
1/16" EUx2	3.175
1/8" Pb	3.011

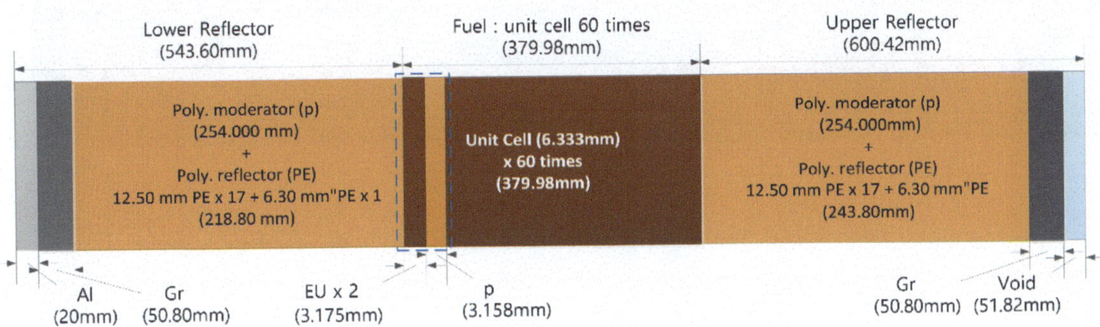

FIG. 2.124. Type-F assembly (F) in the U-Pb benchmark [2.89].

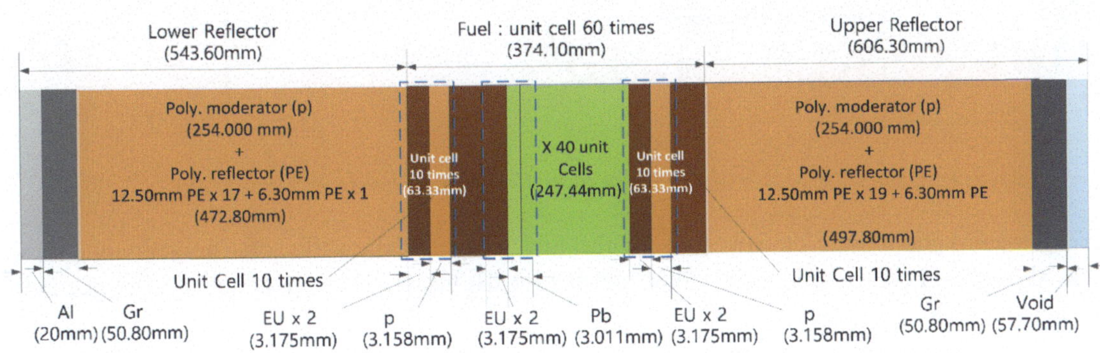

FIG. 2.125.. Type-f assembly (f) in the U–Pb benchmark [2.89].

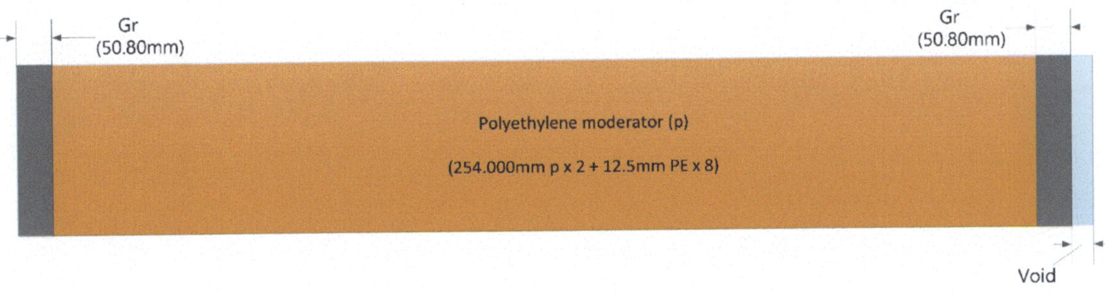

FIG. 2.126. Polyethylene moderator (p) in the U–Pb benchmark [2.89].

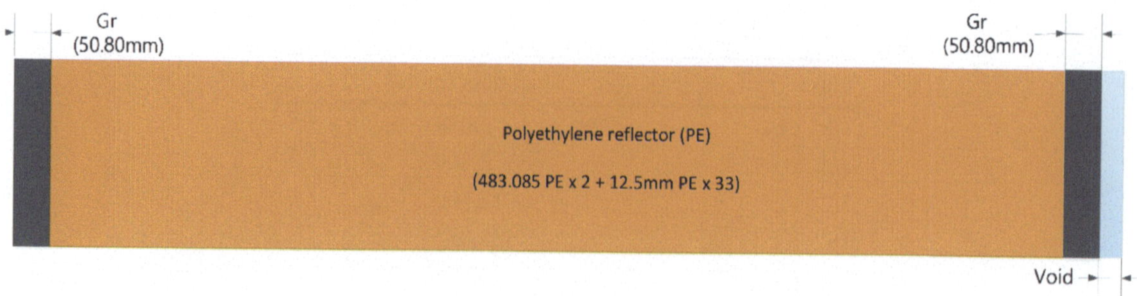

FIG. 2.127. Polyethylene reflector (PE) in the U–Pb benchmark [2.89].

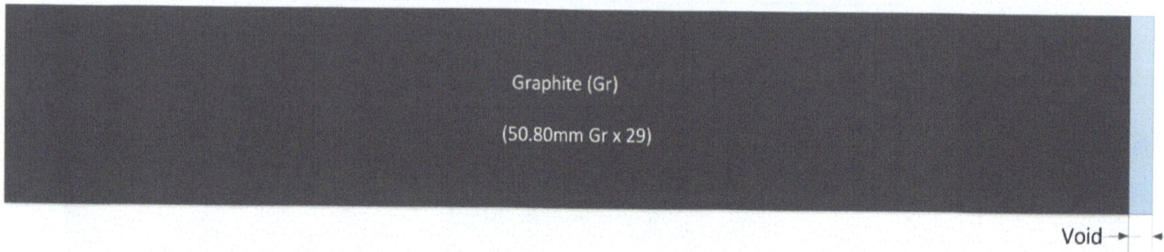

FIG. 2.128. Graphite (Gr) in the U–Pb benchmark [2.89].

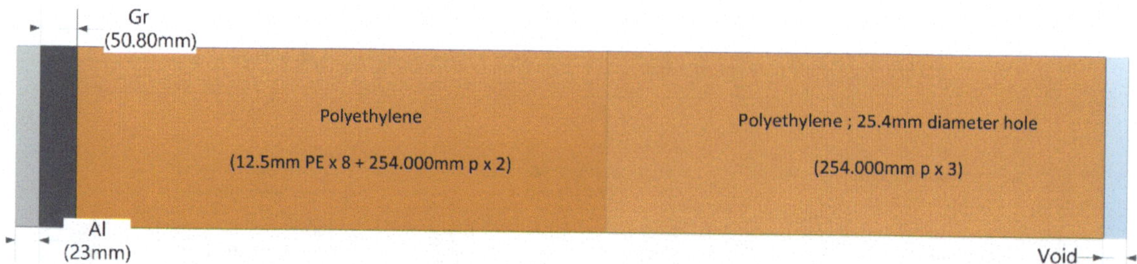

FIG. 2.129. Polyethylene with 25.4 mm diameter hole in the U–Pb benchmark [2.89].

Table 2.71 contains the atomic density data for the fuel, polyethylene reflectors, polyethylene moderator, aluminium sheath, and control and safety rods, and the Pb–Bi target. These parameters were needed for the neutronic calculations.

TABLE 2.71. ATOMIC DENSITIES USED FOR NEUTRONIC CALCULATIONS [2.7]

Item	Material	Isotope	Atomic density ($\times 10^{24}/cm^3$)
HEU fuel plate	1.5875 mm U–Al alloy	^{234}U	1.13659×10^{-5}
		^{235}U	1.50682×10^{-3}
		^{236}U	4.82971×10^{-6}
		^{238}U	9.25879×10^{-5}
		^{27}Al	5.56436×10^{-2}
Polyethylene reflector (PE)	12.500 mm polyethylene	^{1}H	8.06560×10^{-2}
		C	4.03280×10^{-2}
	6.300 mm polyethylene	^{1}H	8.08711×10^{-2}
		C	4.04356×10^{-2}
	3.086 mm polyethylene	^{1}H	8.02167×10^{-2}
		C	4.01084×10^{-2}
Polyethylene moderator (p)	254 mm polyethylene	^{1}H	7.77938×10^{-2}
		C	3.95860×10^{-2}
	483.085 mm polyethylene	^{1}H	7.97990×10^{-2}
		C	4.08960×10^{-2}
Aluminium sheath for the core element	Aluminium	^{27}Al	6.00385×10^{-2}
Lead-Bismuth	44.5% Pb 55.5% Bi	^{204}Pb	1.87461×10^{-4}
		^{206}Pb	3.25860×10^{-3}
		^{207}Pb	3.00266×10^{-3}
		^{208}Pb	7.15378×10^{-3}
		^{209}Bi	1.67670×10^{-2}
Control and safety rods	Absorber	^{10}B	3.87448×10^{-3}
		^{11}B	1.68447×10^{-2}
		^{16}O	3.10787×10^{-2}

2.2.2.   Experiments: Phase I – Kinetic parameters

2.2.2.1. U–Pb zoned core configurations with 14 MeV neutrons

Figure 2.130 shows the base core configuration using 14 MeV pulsed neutrons produced by the D–T reaction in the pulsed neutron generator. A tritium target was located at the bottom centre shown as a grey circle in the figure and the deuterium beam was injected into the target. Four BF$_3$ detectors and four optical fibre detectors were installed in the core. Figure 2.131 shows six subcritical core configurations for which ADS physics experiments were conducted.

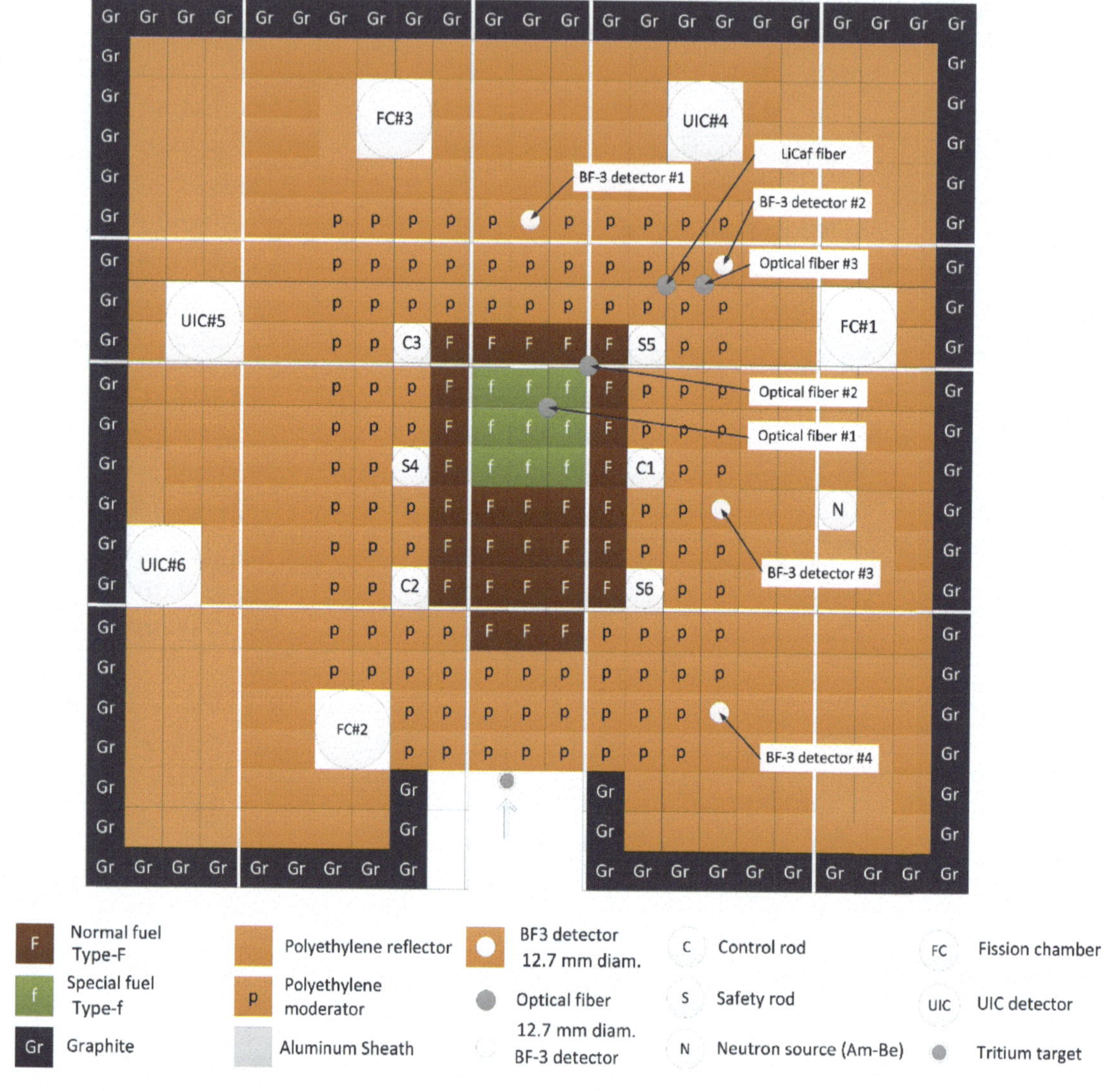

FIG. 2.130. Base configuration with 14 MeV neutrons in the U–Pb benchmark [2.89].

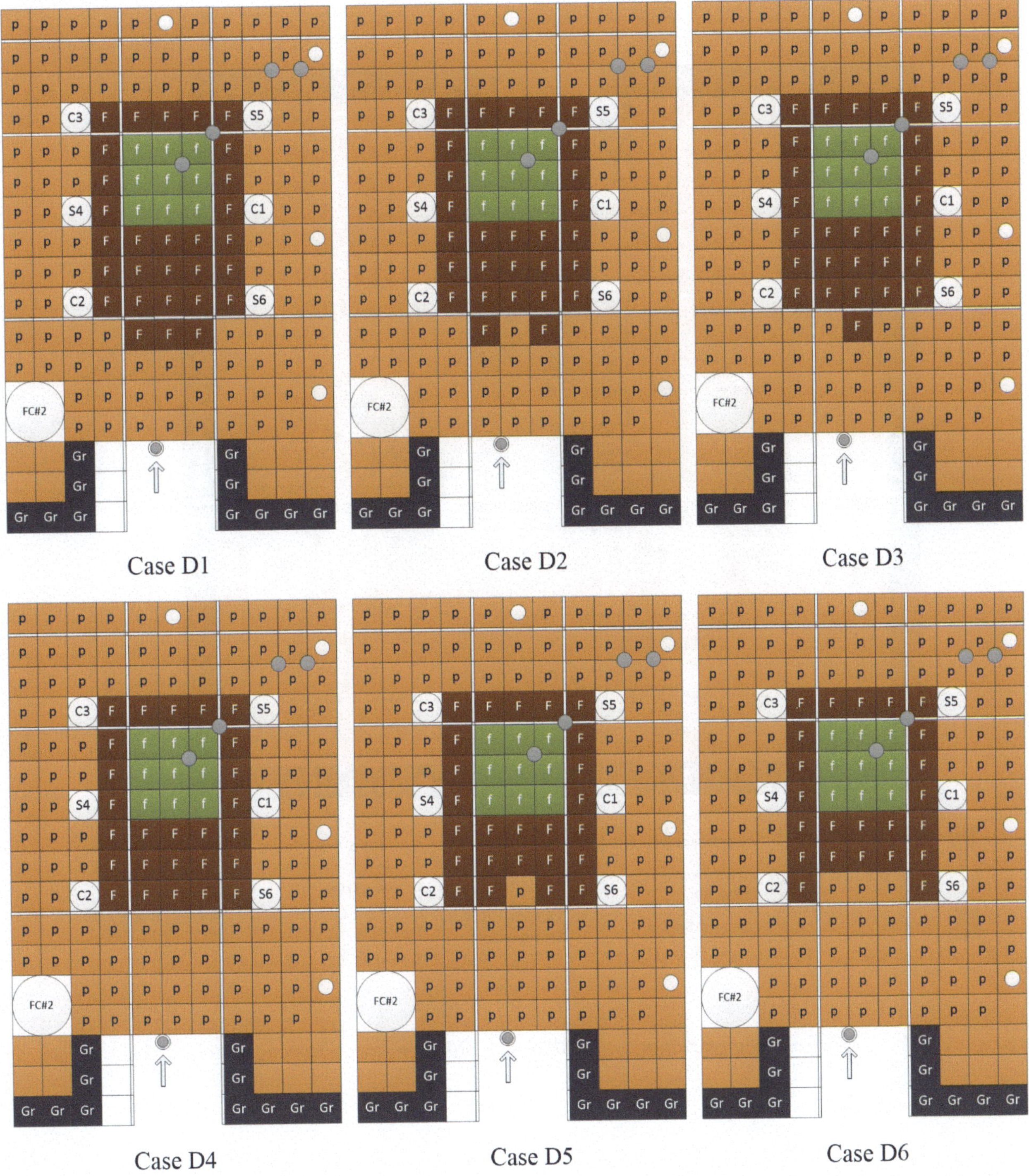

FIG. 2.131. Subcritical core configurations with 14 MeV neutrons in the U–Pb benchmark [2.89].

2.2.2.2. U–Pb zoned core configurations with 100 MeV protons

Figure 2.132 shows the core configuration using spallation neutron source from 100 MeV protons produced at the Pb-Bi target. The target was located at the top centre shown as grey rectangle in the figure. Seven subcritical configurations as shown in Fig. 2.133 were designed by changing the number of fuel assemblies from the base configuration.

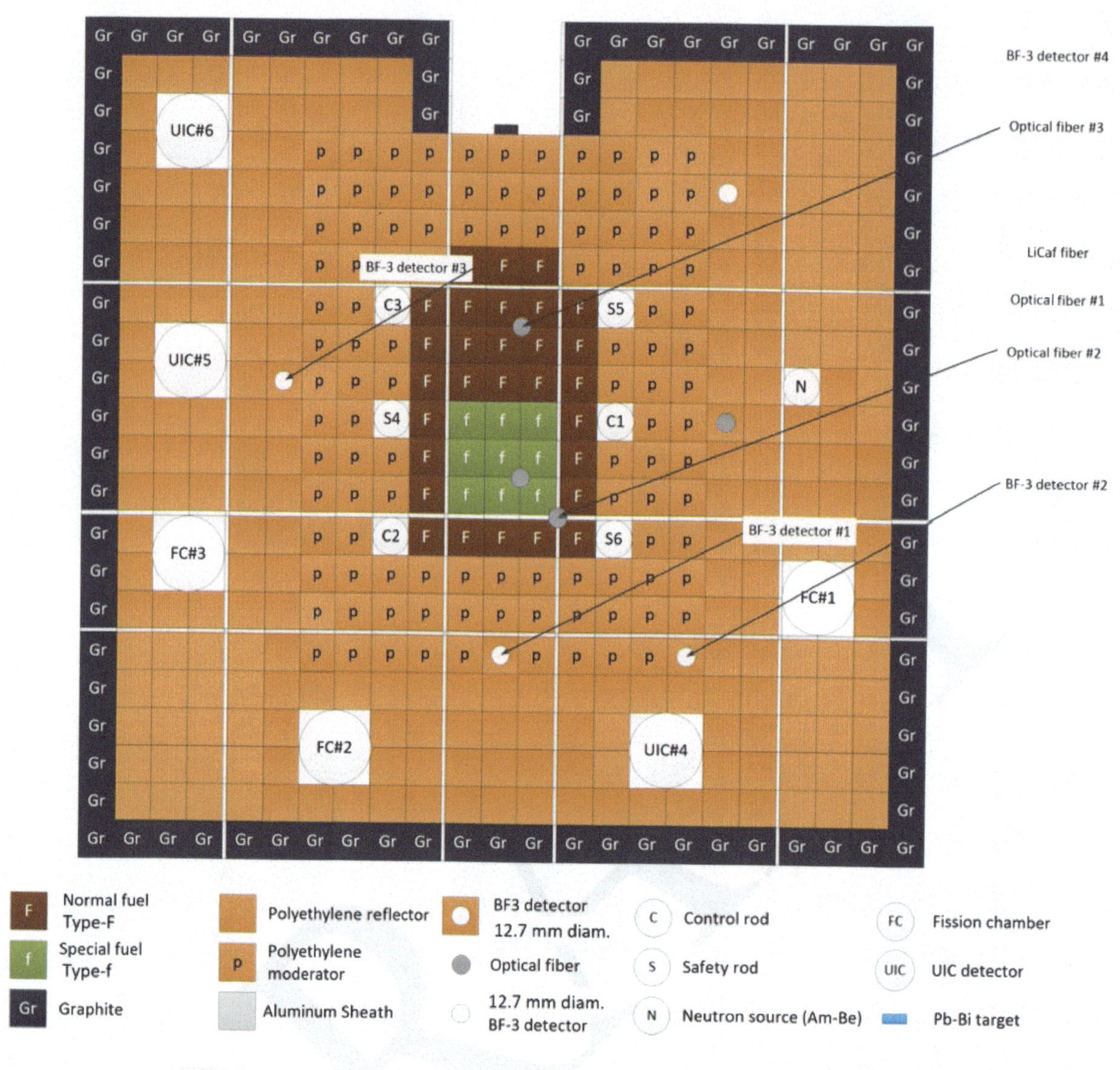

FIG. 2.132. Base configuration with 100 MeV protons in the U–Pb benchmark [2.89].

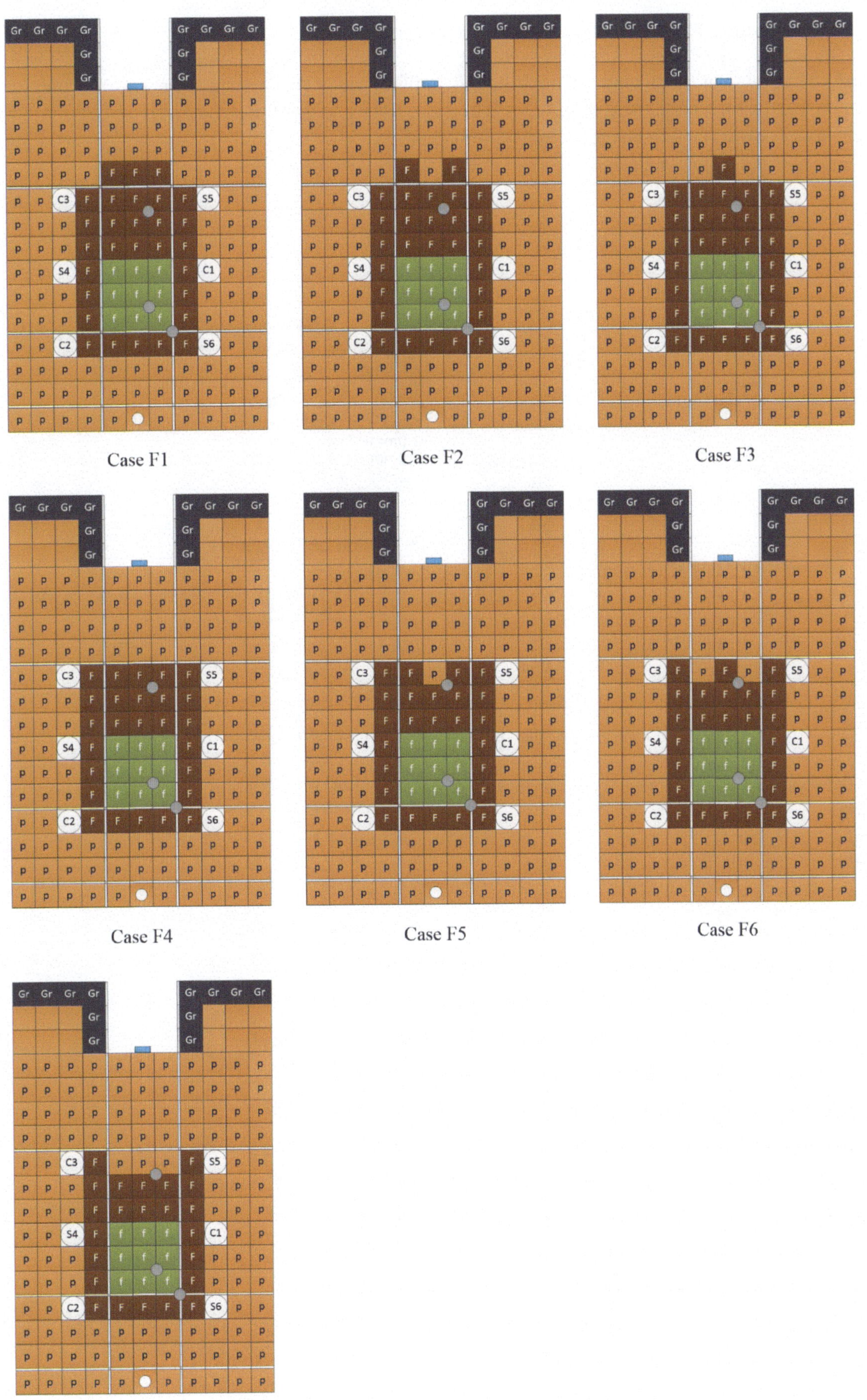

FIG. 2.133. Subcritical core configurations with 100 MeV protons in the U–Pb benchmark [2.89].

2.2.3. Experiments: Phase II – Reaction rate distribution

Figure 2.134 shows an axial view of an installed indium wire. The indium wire was surrounded by an aluminium sheath with a thickness of 0.2 mm. The indium wire with 1.0 mm diameter and 800 mm length is set in the axial centre position along the vertical direction.

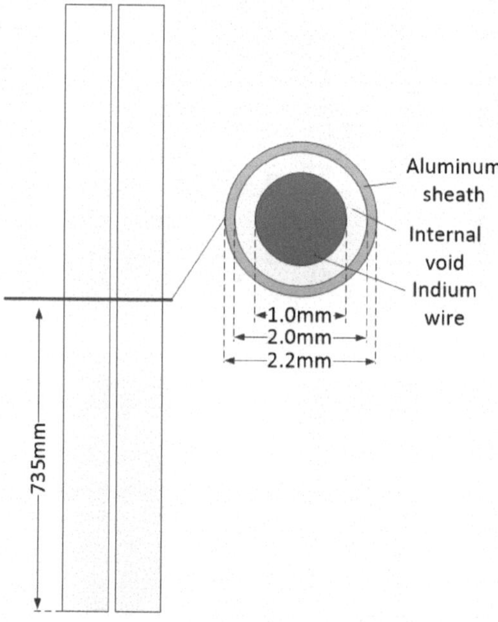

FIG. 2.134. Setting of the indium wire in the U–Pb benchmark [2.89].

Figure 2.135 shows four subcritical configurations of cases F3, F4, F5 and F6 in which the indium wires were installed.

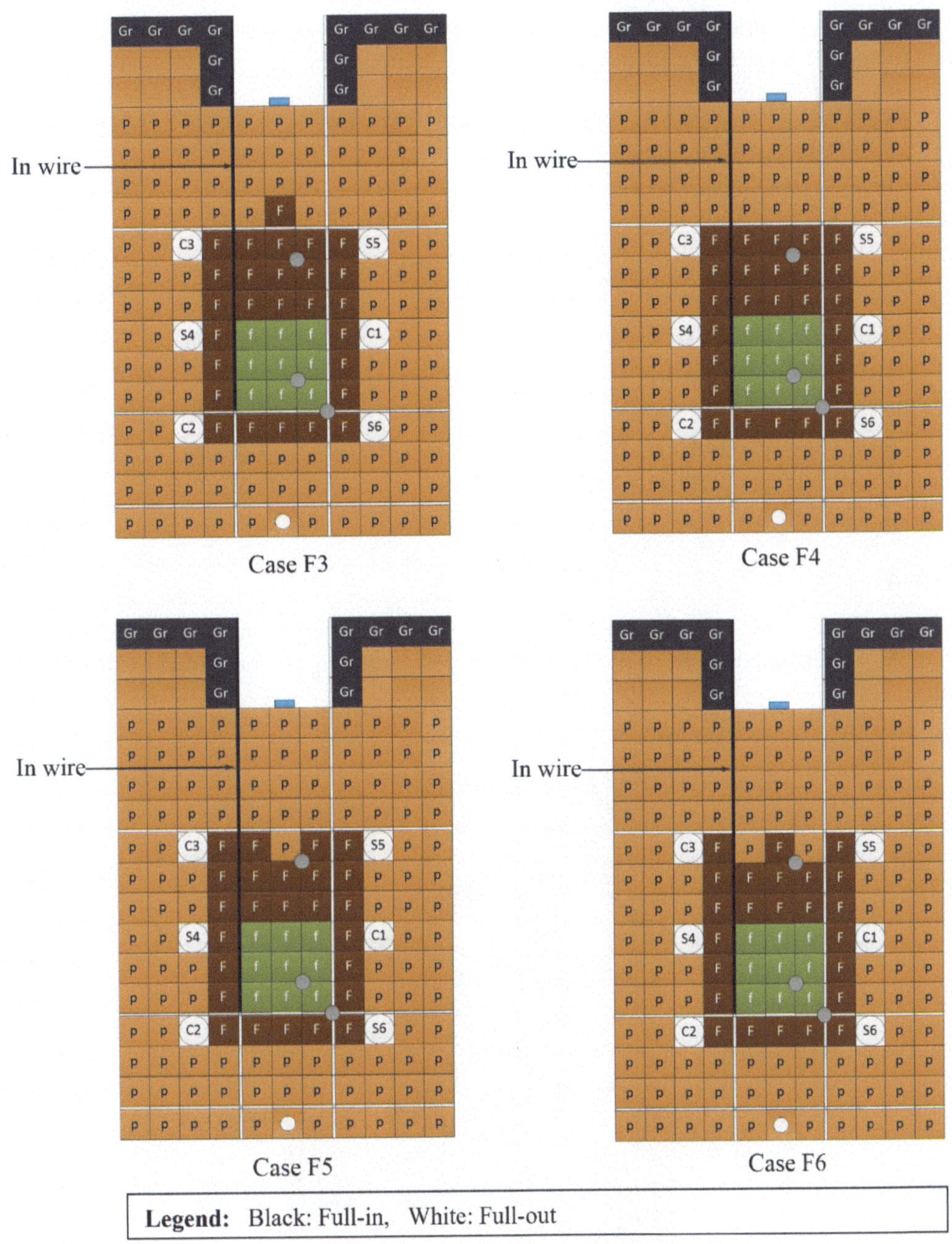

FIG. 2.135. Core configurations of ADS with 100 MeV protons in the U–Pb benchmark [2.89].

2.2.4. Analyses: Phase I – Kinetic parameters (U–Pb zoned cores with 14 MeV neutrons)

2.2.4.1. Modelling

(a) SNU (Republic of Korea)

The effective multiplication factor, the control rod worth, and the excess reactivity were calculated and compared with the experimental results for U–Pb cores with 14 MeV neutrons. The rod positions for the critical state and subcritical systems are given in Table 2.70. McCARD calculation was done with 10^5 histories per cycle, 100 inactive cycles, 500 active cycles using the JENDL-4.0 cross-section libraries.

The prompt neutron decay constants of the six subcritical systems were calculated by the α-iteration method using McCARD. McCARD calculations were conducted with 10^4 histories per cycle and 1000 iterations using

JENDL-4.0 cross-section libraries. The experimental results were measured by the PNS regression and the noise analysis method from detector measurements at different positions.

TABLE 2.72. CONTROL ROD POSITIONS AT THE CRITICAL AND SUBCRITICAL STATE WITH 14 MEV NEUTRONS IN THE U–Pb BENCHMARK [2.89]

Control rod	Critical state (mm)	Subcritical state			
		D1-1 (mm)	D1-4 (mm)	D1-5 (mm)	D2-1 to D5-1 (mm)
C1	766.56	0.00	0.00	0.00	1200.00
C2	1200.00	1200.00	1200.00	0.00	1200.00
C3	1200.00	1200.00	0.00	0.00	1200.00
S4	1200.00	1200.00	1200.00	1200.00	1200.00
S5	1200.00	1200.00	1200.00	1200.00	1200.00
S6	1200.00	1200.00	1200.00	1200.00	1200.00

(b) FDS (China)

Starting from reference core (Fig. 2.130) that was one of subcritical core configurations, additional PNS experiments in ADS were carried out in the KUCA A-core by varying the positions of BF#2 and #3 as shown in Fig. 2.136.

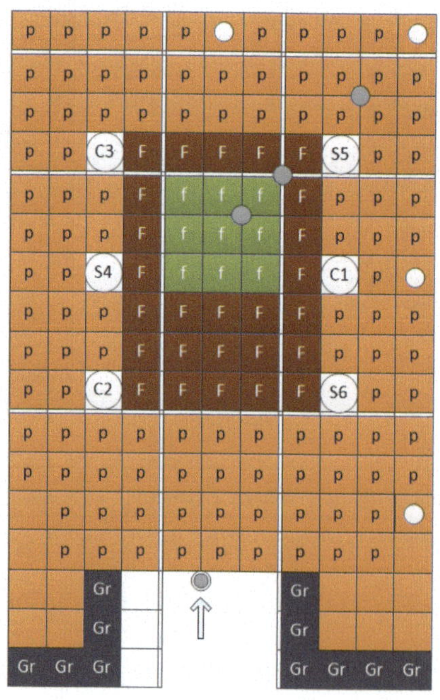

(e) Case 5 (0.96547; 4200)

FIG. 2.136. Subcritical core configurations for PNS method [2.89].

2.2.4.2. Results

(a) Eigenvalue calculations

Table 2.72 shows the McCARD results of the effective neutron multiplication factor, the excess reactivity and the control rod worth. The difference of the effective multiplication factor between McCARD calculation and the experiments at critical system was about 120 pcm. The excess reactivity and the control rod worth calculated by McCARD agreed well with measurements within their 95% confidence intervals.

TABLE 2.72. NEUTRONICS PARAMETERS COMPARISON BETWEEN MCCARD AND EXPERIMENTAL RESULTS WITH 14 MEV NEUTRONS IN THE U–Pb BENCHMARK [2.89]

	McCARD (SD)	Experiment* (SD)
Critical state	0.99881 (0.00012)	1.00000
Excess reactivity (pcm)	99 (18)	80 (2)
Control rod worth (CR1) (pcm)	889 (17)	902 (27)
Control rod worth (CR2) (pcm)	710 (18)	696 (21)
Control rod worth (CR3) (pcm)	229 (18)	232 (7)

* β_{eff} = 853±3 pcm; Λ = (3.24±0.03) ×10⁻⁵ s.

Table 2.73 shows the McCARD and MCNP6.1 results of the effective multiplication factors for the subcritical configurations. The discrepancies between McCARD and MCNP6.1 were attributed to the difference of cross-section libraries, i.e., JENDL-4.0 and ENDF/B-VII.0.

TABLE 2.73. EFFECTIVE MULTIPLICATION FACTORS OF SUBCRITICAL SYSTEMS WITH 14 MEV NEUTRONS IN THE U–PB BENCHMARK [2.89]

Case	McCARD[a] (SD)	MCNP6.1 (SD)
D1-1/2/3	0.99133 (0.00013)	0.99178
D1-4	0.98891 (0.00013)	0.98947
D1-5	0.98019 (0.00014)	0.98225
D2-1/2/3	0.99410 (0.00012)	0.99328
D3-1	0.98197 (0.00013)	0.98004
D4-1	0.96761 (0.00012)	0.96603
D5-1	0.95963 (0.00013)	0.95560
D6-1	0.93106 (0.00013)	0.93000

[a] JENDL-4.0

(b) Prompt neutron decay constant

After injecting a bundle of pulsed neutrons into the fissile nuclear energy system to be measured, the decay characteristic of the prompt and delayed neutrons generated by fission was related to the reactivity of the system. Thus, the reactivity could be reconstructed according to the attenuation characteristic of neutron flux density with time. This is a pulsed neutron source (PNS) method. The area method, which is one of the PNS methods, determined the reactivity by calculating the ratio of the areas under the curves of the prompt neutron and the delayed neutron counts, respectively. The prompt (delayed) neutron flux was integrated over time:

$$I_{p(or\ d)} = \int_0^{\frac{1}{R}} \phi_{p(or\ d)} dt \tag{2.63}$$

To get the reactivity in unit of $, the prompt neutron flux was divided by the delayed neutron flux:

$$\rho_s = \frac{\rho}{\beta} = \frac{I_P}{I_d} = \frac{\int_0^{\frac{1}{R}} \phi_P \, dt}{\int_0^{\frac{1}{R}} \phi_d \, dt} \tag{2.64}$$

Table 2.74 compares the α values. The McCARD estimates agreed better with the PNS results from BF$_3$ #2 and #3 than those from BF$_3$ #1 when the subcriticality became deeper.

TABLE 2.74. ALPHA COMPARISON BETWEEN MCCARD CALCULATION AND EXPERIMENTAL RESULTS WITH 14 MEV NEUTRONS IN THE U–Pb BENCHMARK [2.89]

Case	McCARD α-iteration (SD)	least-squared fitting with PNS method			least-squared fitting with noise method		
		BF$_3$ #1 [Ch#1]	BF$_3$ #2 [Ch#2]	BF$_3$ #3 [Ch#3]	BF$_3$ #1 [Ch#1]	BF$_3$ #2 [Ch#2]	BF$_3$ #3 [Ch#3]
D1-1	566.03 (5.63)	422.96 (12.93)	497.90 (14.80)	486.88 (11.37)	426.81 (28.34)	493.62 (31.92)	494.12 (21.71)
D2-1	433.25 (4.85)	376.67 (8.46)	389.96 (7.21)	370.68 (2.88)	370.33 (4.94)	370.56 (4.21)	369.87 (2.84)
D3-1	756.56 (5.67)	669.23 (5.43)	693.11 (4.60)	684.25 (1.93)	678.46 (4.75)	681.85 (4.28)	686.16 (2.50)
D4-1	1073.43 (6.31)	878.01 (26.55)	979.74 (26.55)	997.15 (13.85)	969.40 (33.08)	970.87 (26.68)	1045.50 (14.28)
D5-1	1221.58 (6.19)	1143.72 (17.67)	1203.76 (15.21)	1188.02 (6.15)	1150.34 (15.00)	1148.70 (13.09)	1186.95 (9.38)
D6-1	1720.22 (6.88)	1493.13 (25.33)	1635.64 (42.06)	1641.15 (8.91)	1604.09 (19.04)	1734.05 (63.78)	1686.35 (13.28)

The pulsed neutrons were injected in the subcritical systems with a time interval of 10 ms. The experiments were carried out in the core configuration with 4200 fuel plates as shown in Fig. 2.136. There were 4 BF#3 detectors recording the neutron number inside the core. As shown in Fig. 2.137, both the BF#1 detector and the BF#3 detector gave reasonable results, with less harmonics effects. Therefore, these two detectors were used for the PNS study.

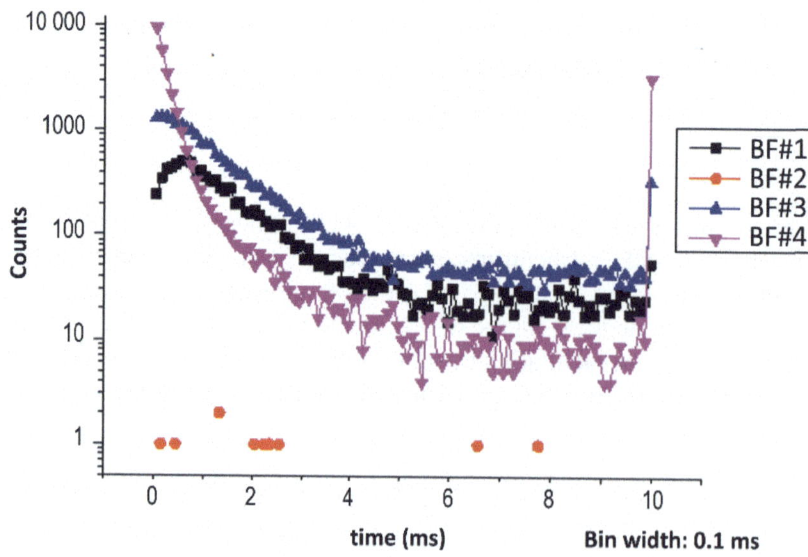

FIG. 2.137. Counts of BF3 detectors in PNS experiments [2.140].

There are three different area methods in PNS: Area method, Extrapolated area method 1, and Extrapolated area method 2. The subcriticality $\rho_\$$ in dollar units was determined by the ratio of two prompt and delayed components in the area method:

$$-\frac{\rho}{\beta_{eff}} = \frac{\int_0^T N_p(t)dt}{\int_0^T N_d(t)dt} \tag{2.65}$$

Where the subscript p means prompt, and d means delayed. N represents the neutron counts of detectors. Subcriticality ρ in dollar units was determined by the ratio of two prompt and delayed components in the extrapolated area method 1:

$$-\frac{\rho}{\beta_{eff}} = e^{\alpha t_w} \frac{\int_{t_w}^T N_p(t)dt}{\int_0^T N_d(t)dt} \tag{2.66}$$

where t_w is the time for decay of prompt neutrons. Subcriticality ρ in dollar units was determined by the ratio of two prompt and delayed components in the extrapolated are method 2:

$$-\frac{\rho}{\beta_{eff}} = \frac{\int_0^T N_p(t)dt}{\int_0^T N_d(t)dt} = \frac{n_{p0}}{\alpha}\frac{R}{n_d} \tag{2.67}$$

where n_{p0} is the number of prompt neutron counts extrapolated to the time $t = 0$ s.

The count distributions were fitted by the above 3 methods. For example, the fit for BF#3 detector is shown in Fig. 2.138.

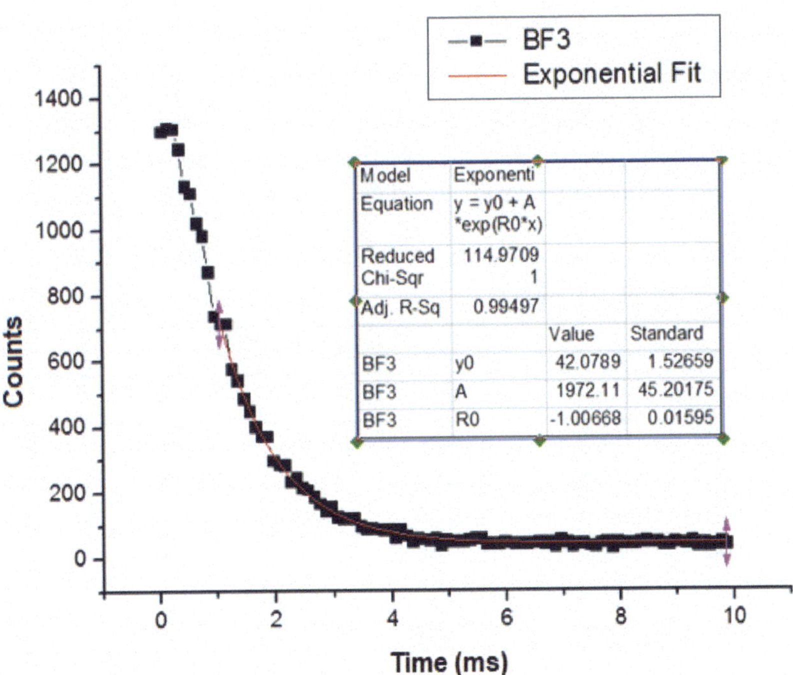

FIG. 2.138. Fit of counts of detector BF#3 in PNS experiments [2.140].

After fitting, the relevant parameters were derived, and the subcriticality was calculated using the above formulas. The results are summarized in Table 2.75.

TABLE 2.75. COMPARISON OF CALCULATED AND MEASURED SUBCRITICALITY USING PNS METHODS [2.140]

	MC simulation	Area method 1	Area method 2	Area method 3
BF#1	–3196 pcm	–3218 ± 279 pcm	–3919 ± 384 pcm	–4062 ± 389 pcm
BF#3	–3196 pcm	–3502 ± 232 pcm	–3799 ± 282 pcm	–3849 ± 229 pcm

Methods 2 and 3 showed less discrepancy between the measured value and theoretical value. The discrepancy was at the level of 600~850 pcm. This conclusion was drawn at subcritical level $k_{eff} \approx 0.97$. However, the discrepancy could be larger in deeper subcritical state. It should be noted that the results of subcriticality for different area methods are dependent on the fitting strategy of detector count distributions.

2.2.5. Analyses: Phase I – Kinetic parameters (U–Pb zoned cores with 100 MeV protons)

2.2.5.1. Modelling

The effective multiplication factor, the control rod worth, and the excess reactivity were calculated and compared with the experimental results for the seven types of subcritical configurations. McCARD calculations were done with 10^5 histories per cycle, 100 inactive cycles and 500 active cycles using the JENDL-4.0 cross-section libraries.

The prompt neutron decay constants of the seven types of subcritical configuration were calculated by the α-iteration method using McCARD. McCARD calculations were conducted with 10^4 histories per cycle and 1000 iterations using JENDL-4.0 cross-section libraries. The experimental results were measured by the PNS regression and the noise analysis method from detector measurements at different positions.

2.2.5.2. Results

(a) Eigenvalue calculations

The control rod positions for the critical and subcritical states were given in Table 2.76. The McCARD calculations were conducted with exchanging the positions of CR2 and CR3 as in Table 2.76 from the benchmark book because they seemed to have switched each other.

TABLE 2.76. CONTROL ROD POSITIONS (IN MM) AT THE CRITICAL AND SUBCRITICAL STATE WITH 100 MEV PROTONS IN THE U–PB BENCHMARK [2.89]

Control rod	Critical state	Subcritical state						
		F1-1	F1-2	F1-3	F1-4	F1-5	F1-6	F2-1 to F7-1
C1	786.14	0.00	0.00	0.00	0.00	0.00	0.00	1200.00
C2	1200.00	0.00	1200.00	1200.00	1200.00	1200.00	1200.00	1200.00
C3	1200.00	0.00	1200.00	0.00	0.00	0.00	0.00	1200.00
S4	1200.00	1200.00	1200.00	1200.00	1200.00	1200.00	0.00	1200.00
S5	1200.00	1200.00	1200.00	1200.00	1200.00	0.00	0.00	1200.00
S6	1200.00	1200.00	1200.00	1200.00	1200.00	1200.00	1200.00	1200.00

Table 2.77 shows the comparison of the McCARD results of the effective multiplication factor, the excess reactivity and the control rod worth with the measurements. The difference of effective multiplication factor between McCARD calculation and the experiments at critical system was about 80 pcm. The excess reactivity and the control rod worth of CR1 calculated by McCARD agreed well with the experimental values within its 95% confidence interval. Note that the experimental values of CR1 and CR2 worth ought to be exchanged.

TABLE 2.77. NEUTRONICS PARAMETERS COMPARISON BETWEEN McCARD AND EXPERIMENTAL RESULTS WITH 100 MEV PROTONS IN THE U–Pb BENCHMARK [2.89]

	McCARD (SD)	Experiment* (SD)
Critical state	0.99927 (0.00013)	1.00000
Excess reactivity (pcm)	59 (18)	37 (1)
Control rod worth (CR1) (pcm)	948 (19)	902 (27)
Control rod worth (CR2) (pcm)	232 (18)	696 (21)
Control rod worth (CR3) (pcm)	709 (18)	232 (7)

* β_{eff} = 853±3 pcm; Λ = (3.24±0.03) ×10^{-5} s.

Table 2.78 shows the McCARD and MCNP6.1 of the effective multiplication factors for the subcritical configurations. Note that the McCARD calculations were conducted with exchanging the positions of CR2 and CR3 as shown in Table 2.75 from the benchmark book. The discrepancies between McCARD and MCNP6.1 were attributed to the difference of cross-section libraries, i.e., JENDL-4.0 and ENDF/B-VII.0.

TABLE 2.78. EFFECTIVE MULTIPLICATION FACTORS OF SUBCRITICAL SYSTEMS WITH 100 MEV PROTONS IN THE U–Pb BENCHMARK [2.89]

Case	McCARD[a] (SD)	MCNP6.1 (SD)
F1-1	0.98174 (0.00012)	0.98208
F1-2	0.99136 (0.00013)	0.99135
F1-3	0.98404 (0.00013)	0.98439
F1-5	0.97722 (0.00013)	0.97744
F1-6	0.96821 (0.00013)	0.96842
F2-1	0.99449 (0.00013)	0.99328
F3-1	0.98188 (0.00013)	0.98004
F4-1	0.96798 (0.00013)	0.96603
F5-1	0.95502 (0.00013)	0.95560
F6-1	0.95408 (0.00013)	0.95047
F7-1	0.92987 (0.00013)	0.92509

[a] JENDL-4.0

(b) Prompt neutron decay constant

Table 2.79 shows the comparison of α values. From the table, one can see that the McCARD estimates agreed better with the PNS results from BF$_3$ #2 and BF$_3$ #3 than those from BF$_3$ #4.

TABLE 2.79. ALPHA COMPARISON BETWEEN MCCARD CALCULATION AND EXPERIMENTAL RESULTS WITH 100 MEV PROTONS IN THE U–Pb BENCHMARK [2.89]

Case	McCARD α-iteration (SD)	Least-squared fitting with PNS method			Least-squared fitting with noise method		
		BF$_3$ #2 (Ch#2)	BF$_3$ #3 (Ch#3)	BF$_3$ #4 (Ch#4)	BF$_3$ #2 (Ch#2)	BF$_3$ #3 (Ch#3)	BF$_3$ #4 (Ch#4)
F1-1	885.65 (6.77)	793.06 (7.12)	807.48 (3.88)	1145.43 (17.58)	791.55 (6.71)	803.10 (3.07)	868.30 (11.13)
F1-2	567.20 (5.45)	483.23 (3.19)	488.39 (1.90)	639.45 (8.43)	481.73 (3.31)	484.30 (1.59)	516.73 (4.89)
F1-3	651.64 (6.13)	544.45 (15.24)	555.53 (9.39)	704.34 (16.18)	548.34 (10.76)	555.50 (4.92)	634.24 (15.92)
F1-5	705.21 (6.09)	611.43 (5.24)	623.65 (2.41)	811.32 (11.35)	612.90 (4.72)	621.46 (2.29)	704.50 (9.52)
F1-6	1018.75 (7.50)	867.68 (17.35)	903.75 (10.95)	1351.25 (29.42)	894.89 (14.37)	911.20 (6.54)	1041.78 (29.28)
F2-1	438.27 (4.85)	368.72 (4.36)	371.42 (3.98)	517.43 (10.05)	375.31 (2.10)	377.47 (1.05)	377.55 (3.46)
F3-1	752.76 (5.73)	690.57 (9.73)	684.02 (4.10)	1255.70 (32.52)	691.99 (5.10)	699.73 (2.81)	706.39 (8.95)
F4-1	1078.83 (6.45)	983.15 (8.05)	1001.46 (4.15)	2410.26 (60.24)	1036.24 (10.26)	1035.20 (5.49)	1046.45 (19.88)
F5-1	1286.36 (6.36)	1110.85 (4.07)	1114.54 (2.81)	2632.74 (26.57)	1143.44 (10.66)	1167.77 (7.55)	2179.82 (535.32)
F6-1	1286.36 (6.36)	1170.63 (4.25)	1178.31 (3.63)	2810.88 (24.95)	1227.65 (4.73)	1236.21 (3.33)	1616.85 (23.53)
F7-1	1733.98 (6.83)	1526.32 (5.96)	1565.34 (4.34)	3620.56 (17.16)	1663.99 (9.55)	1675.08 (7.43)	2546.56 (43.44)

2.2.6. Analyses: Phase II – Reaction rate distribution

2.2.6.1. Modelling

For the reaction rate estimate at the indium wire, McCARD source-mode calculations were conducted 10^8 histories using JENDL-4.0 libraries. The spallation neutron data at the target were obtained from the MCNPX2.6.0 proton simulations.

2.2.6.2. Results

Figures 2.139–2.142 show the comparison between McCARD calculations and experimental results which were measured reaction rates of the ^{115}In$(n, \gamma)^{116m}$In reaction. McCARD results for the ^{115}In$(n, \gamma)^{116m}$In reaction rates were normalized by its area of 1. From the figures, one can see that most of the results agreed well with the measurements.

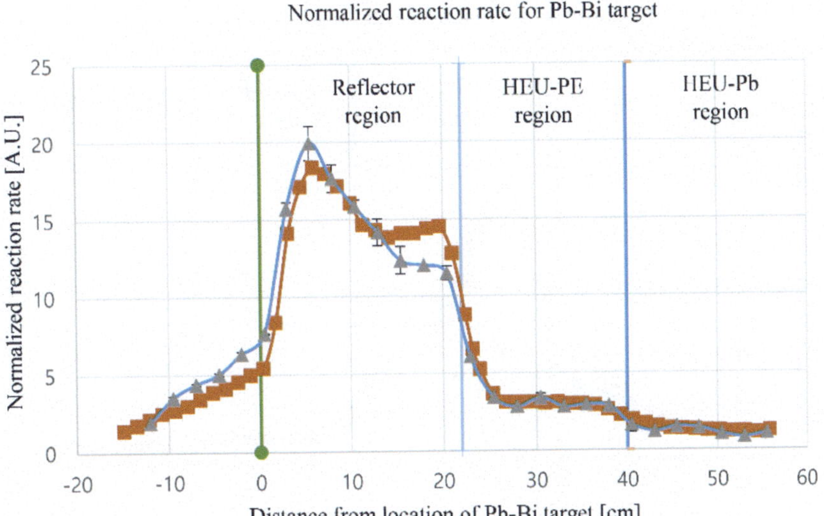

FIG. 2.139. Comparison results of reaction rate distribution for case F3 in the U–Pb benchmark [2.89].

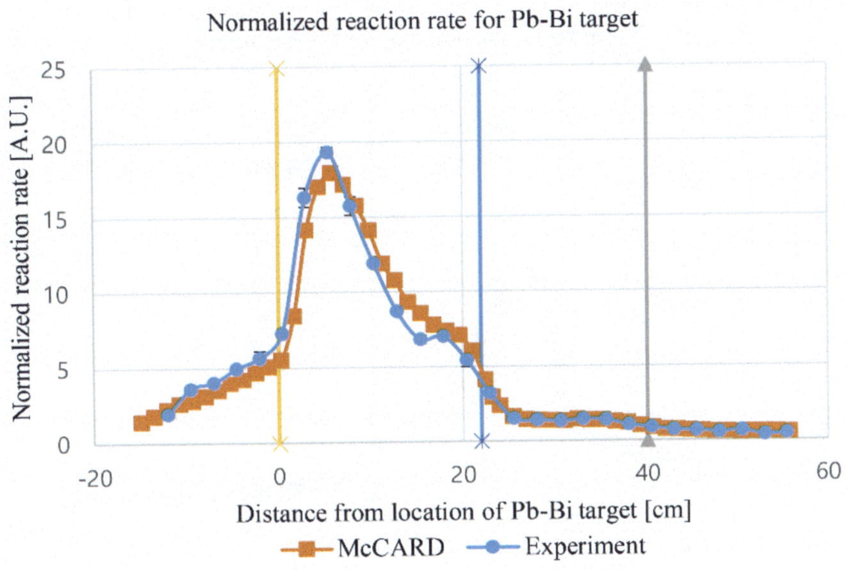

FIG. 2.140. Comparison results of reaction rate distribution for case F4 in the U–Pb benchmark [2.89].

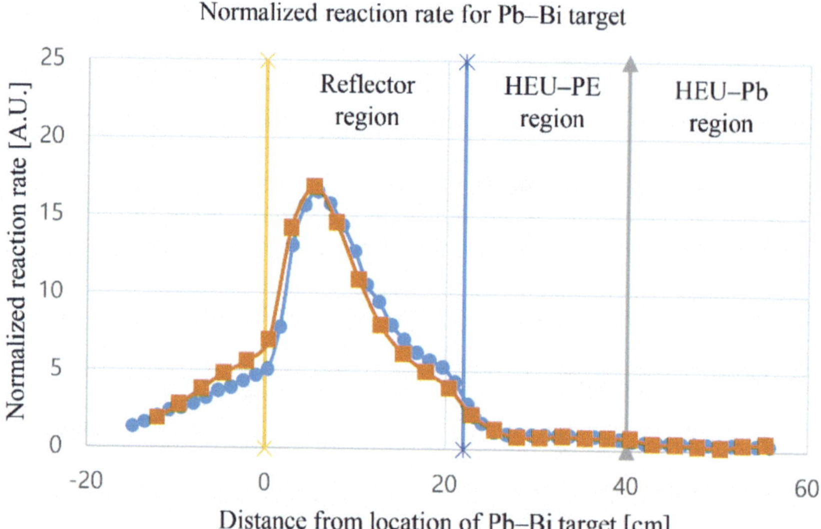

FIG. 2.141. Comparison results of reaction rate distribution for case F5 in the U–Pb benchmark [2.89].

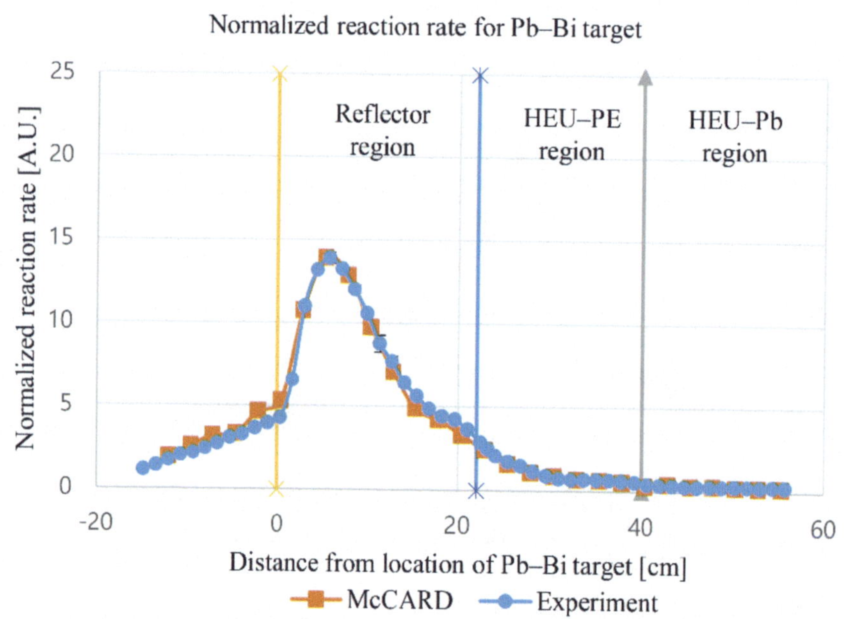

FIG. 2.142. Comparison results of reaction rate distribution for case F6 in the U–Pb benchmark [2.89].

2.3. GIACINT FACILITY

2.3.1. Description of the facility

The GIACINT facility (References [2.154]–[2.157]) consists of an open tank filled with water, fuel rods, detector channels control equipment, and structural supports. The fuel is 19.75% enriched U–Zr–C–N with steel cladding. The fuel rods are arranged in a hexagonal array with 3.2 cm pitch and the fuel array is supported at the bottom by a solid steel plate. Above this steel plate, two aluminium spacer grids and one steel spacer grid at the top hold the fuel array rods. The spacer grids and the steel plate have a hexagonal shape with 88 cm side length. The steel plate and the top steel spacer grid are made of steel type 12X18H10T and have 1.6 cm thickness each. The aluminium spacer grids are made of aluminium alloy AMg61 (1561) and have 1 cm thickness. Twelve aluminium detector channels in Plexiglas tubes are located around the fuel array. The tubes penetrate the spacer grids through holes. The penetration diameters are 1.24 cm and 1.3 cm for the aluminium spacer grids and the steel spacer grids, respectively. The 12 detector channels have 3 CNM-18 neutron detectors, 6 KNK-56M ionization chambers, and 3 KNK-17-1 ionization chambers.

Figures 2.143 and 2.144 show the facility overview and a sketch of the vertical section of the facility, respectively. Experiments have been performed on two critical configurations, as summarized in Table 2.80. The water level reported in Table 2.80 has been measured along the axial direction starting from the bottom surface of the active fuel zone.

TABLE 2.80. CRITICAL CONFIGURATIONS OF CRITICAL ASSEMBLIES

Fuel configuration	Lattice pitch (mm)	Number of fuel rods	Water critical level (mm)	Water temperature (°C)
W-20-2	32	67	494.8 ± 1.0	17.2
W-20-2-confl	32	87	487.9 ± 1.0	19.0

The fuel rod contains cylindrical uranium zirconium carbon nitride tablets with 1.075 cm diameter, as shown in Fig. 2.145. The active and total fuel rod lengths are 50 and 62 cm, respectively; the outer fuel rod radius and steel-clad thickness are 0.6 cm and 0.6 mm, respectively. The gap between the tablets and the steel clad is filled with helium gas at 0.11–0.12 MPa pressure. Figure 2.146 illustrates the fuel rod arrangement in the tank.

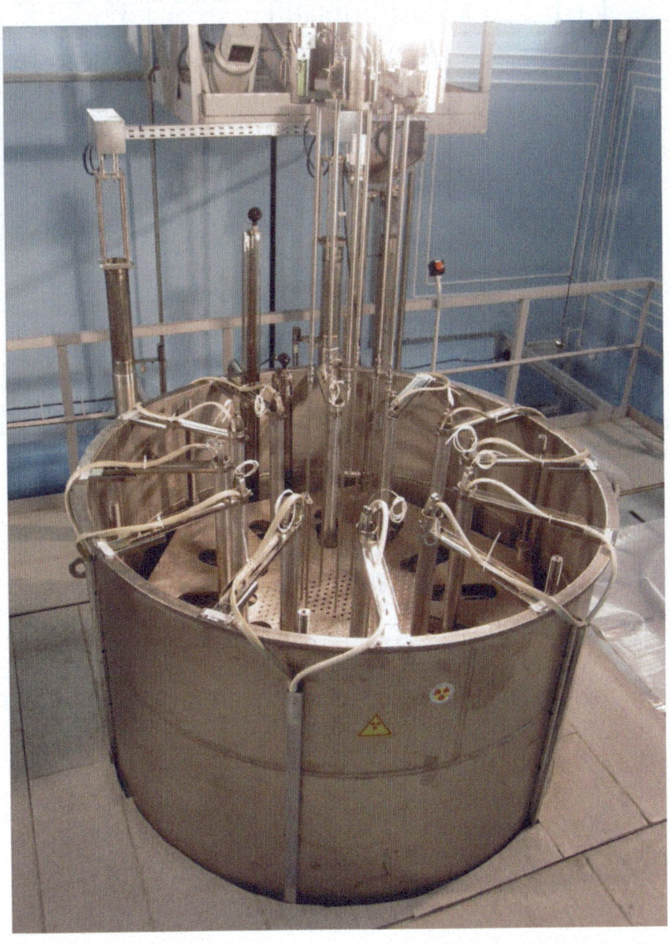

FIG. 2.143. Overview of the GIACINT critical facility.

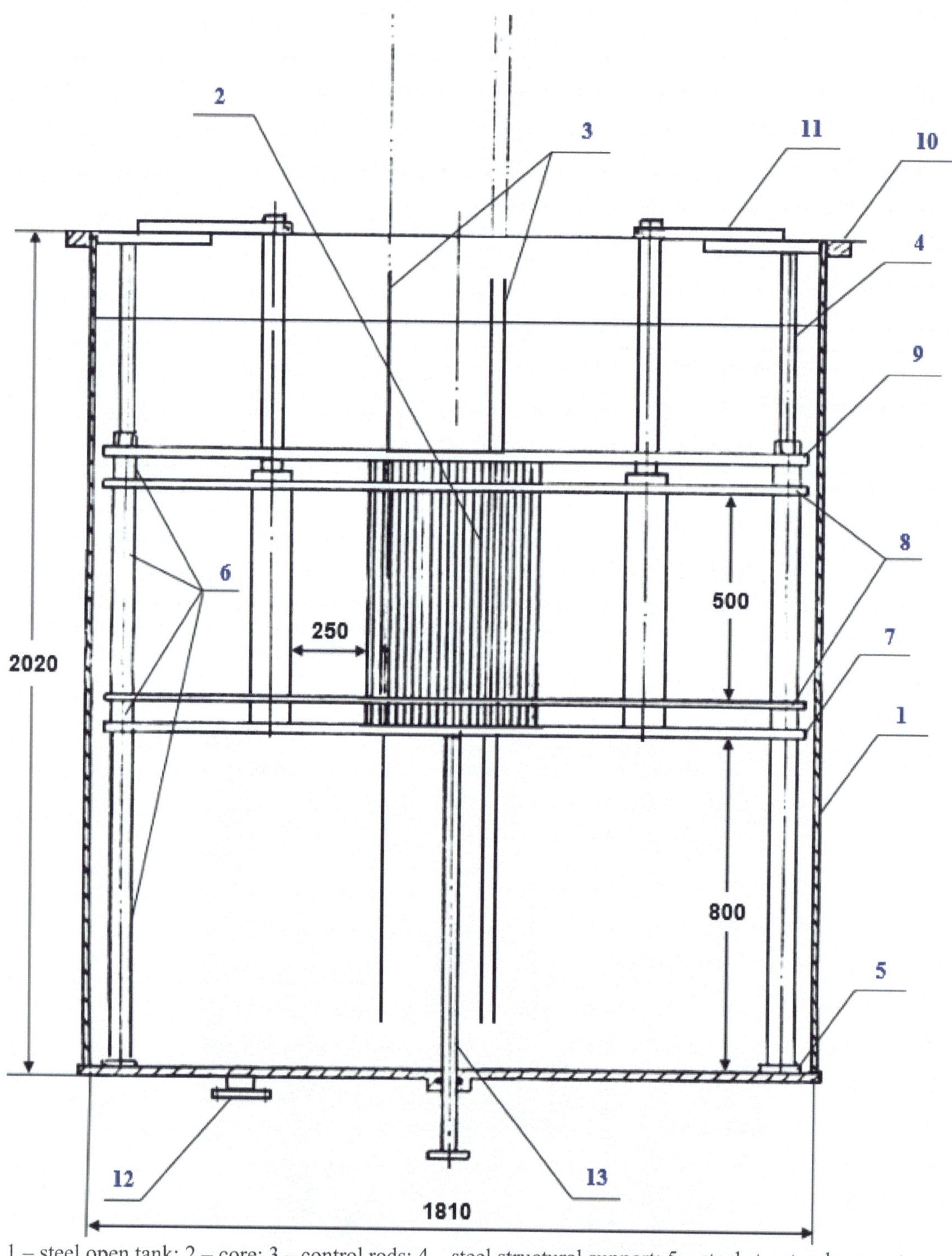

1 – steel open tank; 2 – core; 3 – control rods; 4 – steel structural support; 5 – steel structural support; 6 – steel structural support; 7 – steel support plate; 8 – aluminium spacer grids; 9 – steel spacer grid; 10 – tank flange; 11 – neutron detector equipment; 12 – valve; 13 – tank hole for californium source insertion.

FIG. 2.144. GIACINT critical assembly with a water moderator (vertical section).

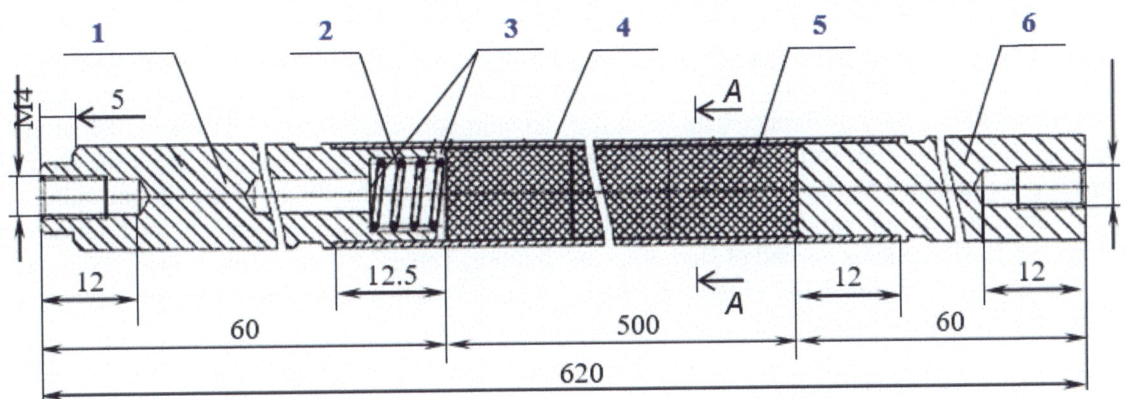

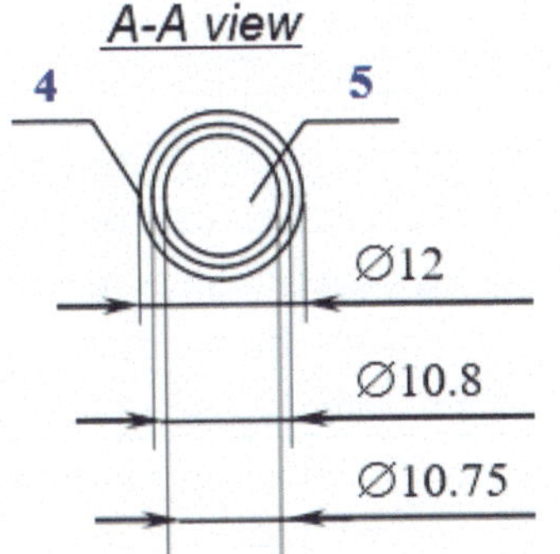

1: lower plug
2: spring
3: gaskets
4: casing
5: fuel core
6: upper plug

FIG. 2.145. Fuel rod cross-section.

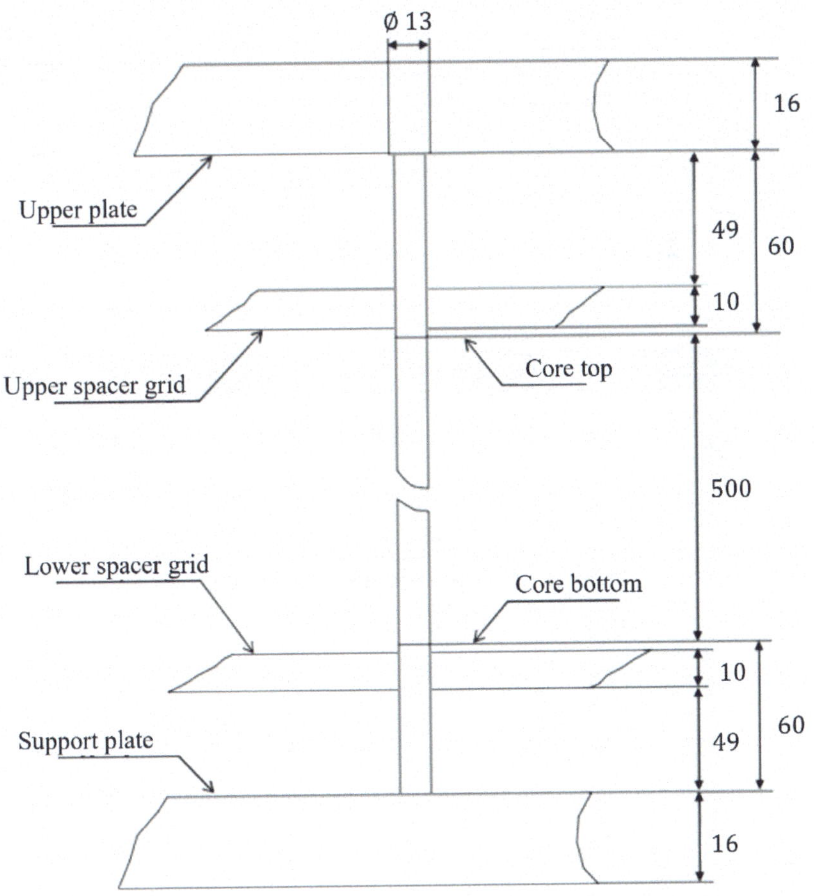

FIG. 2.146. Fuel rod arrangement in the tank; dimensions in mm.

The GIACINT critical assembly is equipped with three pairs of control rod devices: two emergency protection control rod devices, two compensating reactivity control rod devices, and two manual regulation control rod devices. Each control rod device can drive either a single rod or two control rods. The control rods are similar to the fuel rods with the only difference being that they contain boron carbide (B_4C) instead of fuel material. The displacer rods are similar to the fuel rods but contain air instead of fuel material. Both the control and displacer rods are attached to the fuel rods by an 8 cm long solid steel cylinder with 0.6 cm radius. Some control rods are attached to Plexiglas rods instead of fuel rods. The Plexiglas rods are like the fuel rods except that they are made of Plexiglas instead of steel, fuel and helium. Table 2.81 shows the control rod devices for each fuel configuration. Figure 2.147 illustrates a fuel rod connected to a control rod. Figures 2.148 and 2.149 illustrate the fuel configurations and the control rods arrangements.

TABLE 2.81. CONTROL ROD DEVICES

Fuel configuration	Control rod device [a]	Number of control rods
W-20-2	EP-1	one rod (fuel rod + control rod)
	EP-2	one rod (fuel rod + control rod)
	CR-1	one rod (fuel rod + control rod)
	CR-2	two rods (Plexiglas rod + control rod)
	MR-1	two rods (Plexiglas rod + control rod)
	MR-1	two rods (Plexiglas rod + control rod)
W-20-2-conf1	EP-1	one rod (fuel rod + control rod)
	EP-2	one rod (fuel rod + control rod)
	CR-1	one rod (fuel rod + control rod)
	CR-2	two rods (fuel rod + control rod)
	MR-1	two rods (fuel rod + displacer rod)
	MR-2	two rods (fuel rod + displacer rod)

[a] EP = emergency protection, CR = compensating reactivity, MR = manual regulation

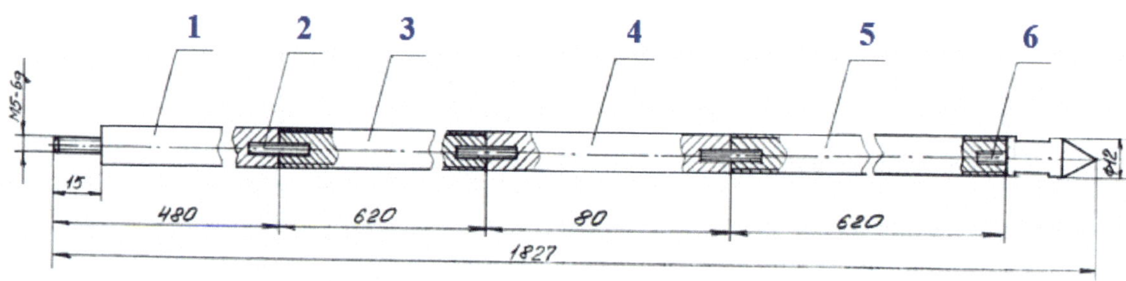

FIG. 2.147. Fuel rod and control rod.

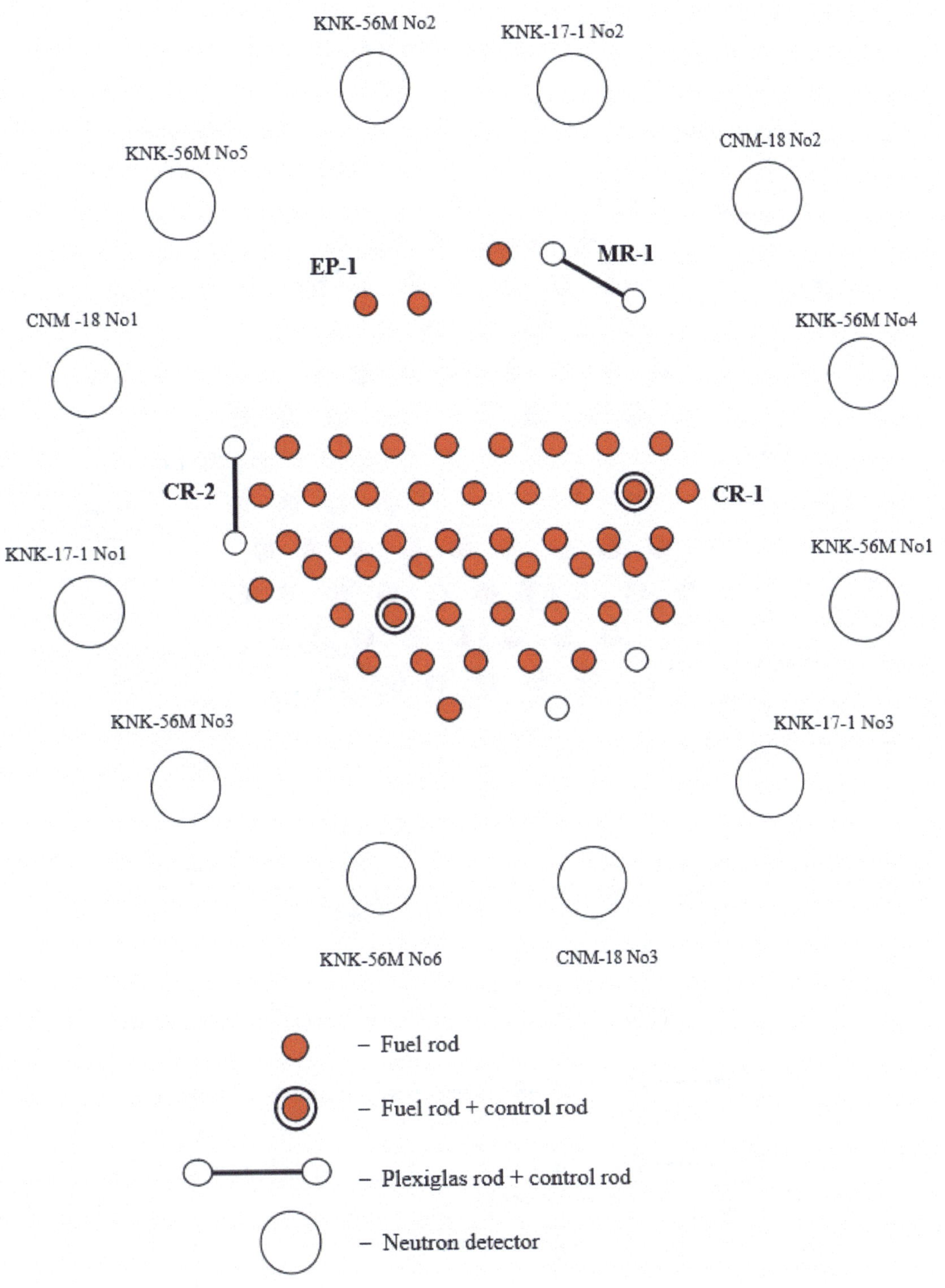

FIG. 2.148. Fuel rod loading for the critical assembly W-20-2.

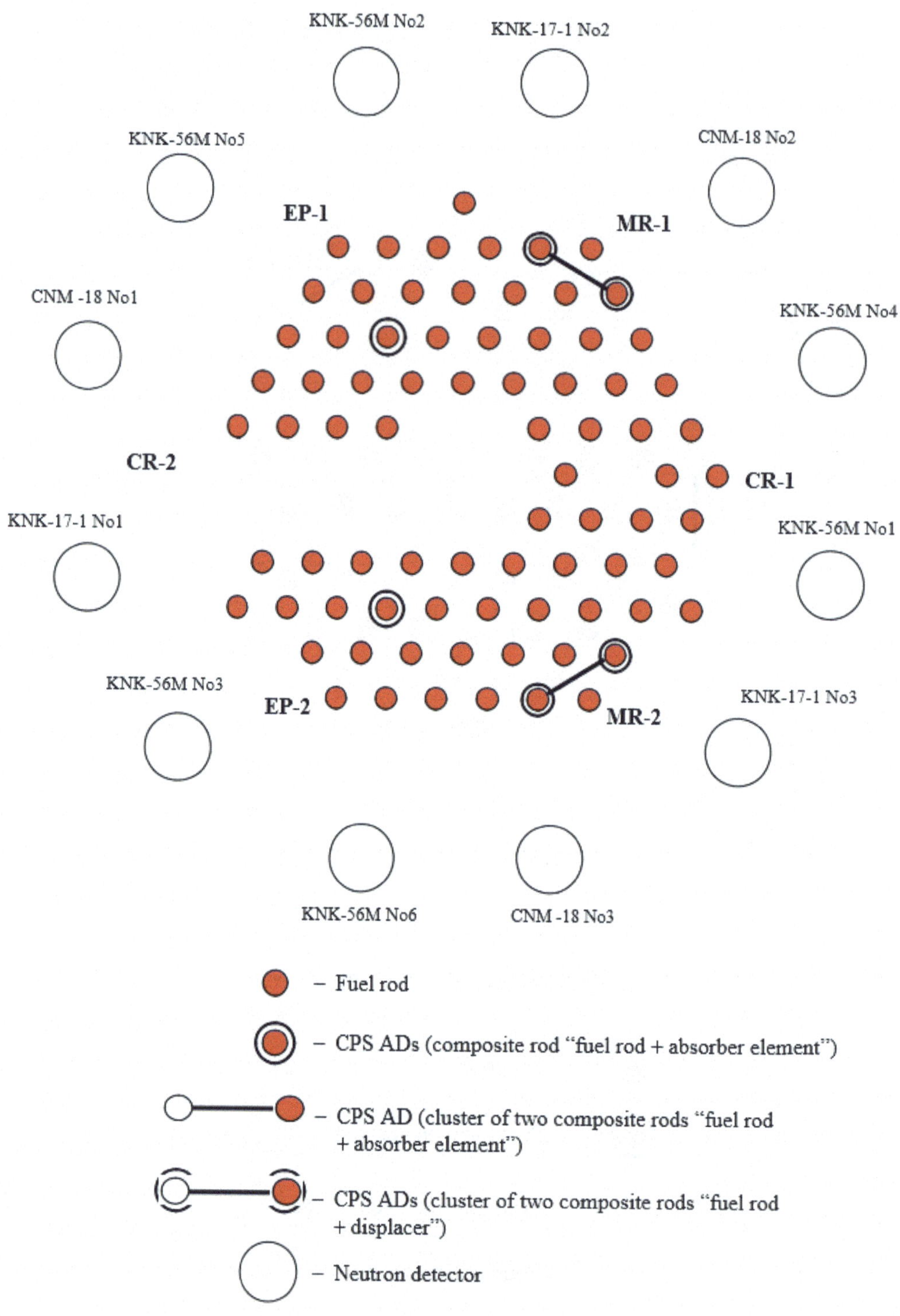

FIG. 2.149. Fuel rod loading for the critical assembly W-20-2-conf1.

2.3.2. Materials specifications

The fuel material consists of uranium zirconium carbon nitride (chemical composition $U_{0.9}Zr_{0.1}C_{0.5}N_{0.5}$). In each fuel rod, the weights of the fuel material, uranium, and ^{235}U are (540±15) g, (489±14.5) g, and (96.7±2.5) g, respectively. Tables 2.82 and 2.83 give the composition of the U–Zr–C–N fuel. The lower part of the fuel rod includes: a lower plug (weight 39 g, 06X18H10T steel), a spring (weight 0.44 g made from the wire II-P-0.8), a gasket (weight 0.13 g, 12X18H10T steel), seven gaskets (weight 0.017 g, 08X18H10T steel), and a pin (weight 0.15 g, 10X17H13M3T steel). The upper part of the fuel rod includes an upper plug (steel 06X18H10T). The fuel rod clad is made of 06X18H10T steel. The steel compositions are listed in Table 2.84; all steels have a density of 7.9 g/cm^3.

TABLE 2.82. MASS FRACTION (%) OF THE MAJOR ELEMENTS COMPOSING THE U-ZR-C-N FUEL

U	Zr	C	N
91 ± 1	3.6 ± 0.3	$2.5^{+0.2}_{-0.5}$	$2.5^{+0.7}_{-0.3}$

TABLE 2.83. UPPER LIMIT OF THE MASS FRACTION (%) OF IMPURITIES IN THE U-ZR-C-N FUEL

W	O	B	Ni	Cr	Fe	Mo	Cu	Mn	Si	Hf	Co
0.5	0.1	3E-4	3E-2	3E-2	3E-2	3E-2	5E-3	5E-3	1.5E-2	2E-3	1E-2

TABLE 2.84. MASS FRACTIONS (%) OF THE STEELS

Element	Steel 12X18H10T	Steel 08X18H10T	Steel 6X18H10T	Steel 10X17H13M3T
C	≤0.12	≤0.08	≤0.06	≤0.10
Si	≤0.8	≤0.8	≤0.8	≤0.8
P	≤0.035	<0.035	<0.035	<0.035
S	≤0.02	<0.02	<0.02	<0.02
Ti	5·C – 0.7	5·C – 0.7	5·C – 0.6	5·C – 0.7
Cr	17.0 – 19.0	17.0 – 19.0	17.0 – 19.0	16.0 – 18.0
Mn	≤2.0	≤2.0	1.0 – 2.0	≤2.0
Ni	9.0 – 11.0	9.0 – 11.0	9.0 – 11.0	12.0 – 14.0
Mo	–	–	–	3.0 – 4.0
Fe	Balance	Balance	Balance	Balance

All metal in the control and displacement rods consists of 12X18H10T stainless steel. Boron carbide has a density of 1.38 g/cm^3. The Plexiglas density is 1.19 g/cm^3. The tank, steel support plate and spacer grid are made of 12X18H10T stainless steel. The aluminium spacer grids are made of AMg61 (1561) aluminium alloy with 2.68 g/cm^3 density, shown in Table 2.85.

TABLE 2.85. MASS FRACTION OF THE AMG61 (1561) ALUMINIUM ALLOY

Element	Mass fraction (%)
Be	0.0001 – 0.003
Mg	5.5 – 6.5
Si	<0.4
Mn	0.7 – 1.1
Fe	<0.4
Cu	<0.1
Zn	<0.2
Zr	0.02 – 0.12
Al	Balance

2.3.3. Computational models

The SERPENT computer program [2.158] was used to model the facility using combinatorial geometry. In this model, the water temperature was considered and the water density modified accordingly. As discussed in Refs [2.159] and [2.160], a moving geometry transient was simulated by SERPENT using two separate calculations. In the first calculation, SERPENT rans in criticality mode and wrote two binary files. The first binary file stored prompt and delayed neutrons and the second binary file stored delayed neutron precursors; all stored data were calculated at the start of the transient before any geometrical change. In all criticality calculations, the k_{eff} value was ~1.00000 ± 0.00002; this value referred to the start of the transient. In order to set the criticality condition at the beginning of the transient, the ^{235}U enrichment was tuned. Without the ^{235}U enrichment tuning, the k_{eff} calculated by SERPENT was about 200 pcm above the critical value. In the second calculation (transient), SERPENT ran in dynamic-source mode, reads, and updated all data in the two binary files from the first calculation. In the transient calculation, the transient time was divided into several time bins (e.g.,70). At the beginning of each time bin, the prompt and delayed neutrons, and the delayed neutron precursors data were read from the two binary files and the data were updated according to the neutron random walk and the natural decay of the delayed neutron precursors. At the end of each time bin, the two binary files were updated for use at the start of the next time bin. In addition, at the end of each time bin, Monte Carlo variance reduction techniques, splitting and Russian roulette, were used to maintain the neutron population and the delayed neutron precursors approximately constant over time. In the SERPENT methodology, delayed neutrons were emitted from the precursors if they contributed to the time bin under consideration.

In addition, MCNP [2.161] models were developed to perform the transient experiments. In the MCNP models, geometrical transients used a single fixed-source simulation [2.162]. First, the uranium fuel enrichment was tuned to get k_{eff} equal to ~1.00000 ± 0.00002, as in the SERPENT methodology. The MCNP methodology did not require any criticality simulation to write data files for starting the transient simulation. The fixed-source simulation began 200 seconds before the start of the geometrical change by injecting a neutron pulse into the system. The neutron pulse occurred 200 seconds before the geometrical change, and it had a constant spatial distribution over the fuel zone, an isotropic angular distribution, and fission energy spectrum. When the geometrical change started, the prompt and delayed neutron populations had reached an equilibrium. The total simulation length was set by the *cut* card of MCNP, and it was equal to the experimental transient time plus 200 seconds.

Two different approaches were implemented in MCNP to simulate geometrical change transients. In the first approach, the geometry was modified at the end of each neutron track. The second approach accounted for the geometry change during the neutron random walk sampling. Simulations of different transients showed that there was no difference between the results from these two approaches for the considered transients. However, the second approach was required only when the transient speed was close to the thermal neutron speed.

The main differences in the geometrical change transients modelling with SERPENT and MCNP were the following:

- MCNP did not track the delayed neutron precursors and directly tracked delayed neutrons; the latter were sampled from nuclear data libraries. SERPENT tracked the delayed neutron precursors to sample delayed neutrons;
- MCNP did not divide the transient time into time bins and time bins were only used in the tallies. SERPENT divided the transient time into time bins during the neutron transport simulation and used different time bins for the tallies;
- MCNP needed one fixed-source calculation for the transient simulation. SERPENT needed one criticality calculation and one fixed-source calculation for the transient simulation;
- MCNP could either update the geometrical change at the end of each neutron track (simple method) or sample the particle random walk considering the geometrical change (within the time of the particle random walk) and update the geometrical change at the end of each particle random walk (improved method). SERPENT updated the geometrical changes only at the end of each time bin.

Figure 2.150 shows the SERPENT and MCNP model of the GIACINT facility. The two Monte Carlo codes shared the same geometry and materials definitions.

Legend:
1 – fuel rods (uranium fuel, stainless steel clad, and helium gap); 2 – stainless steel tank; 3 – channel for californium source (air); 4 – stainless steel support rods; 5 – water moderator; 6 – assembly top zone (air); 7 – stainless steel bottom support plate; 8 – aluminium fuel rods bottom grid; 9 – aluminium fuel rods top grid; 10 – stainless steel top support plate; 11 – detectors channels (air, aluminium, and Plexiglas).

FIG. 2.150. Overview of the GIACINT experimental facility (top-right plot) and its Monte Carlo computational model (top-left plot). The bottom-left and bottom-right plots show the computational model without the steel tank and water.

2.3.4. Experiments and results

Two sets of experiments were carried out. In one set of experiments, some control rod was inserted into the core. For slow transients (few tens of seconds), the control rods insertion occurred at constant speed, driven by an electrical motor. For fast transients (less than a second), the control rods insertion was driven first by a mechanical spring injecting the control rods into the core and then by gravity. For fast transients, the control rod movement did not occur at constant speed because of the mechanical spring, gravity accelerations and the water buoyancy force resisting the control rod movement. In the other set of experiments, water was drained by a guillotine valve and the water level decreased at constant speed.

In the fast transient experiments, ionization chambers measured the reaction rate, which was assumed to be proportional to the neutron flux and the generated power. The inverse kinetics method was applied to measure the reactivity using the reactivity meter TsVR-10.

Tables 2.86 and 2.87 summarize all experiments performed at the GIACINT facility; Figs 2.151–2.154 show some of the experimental results. A few experiments were modelled with the MCNP and SERPENT computer programs. Figures 2.155 and 2.156 show the comparison between experimental and Monte Carlo simulation results. The agreement between the two Monte Carlo programs was very good. Experimental results slightly differ from the computational results. This difference can be explained by uncertainties in the material compositions, geometry specifications, and experiment start and end times.

TABLE 2.86. EXPERIMENTS WITH THE CONFIGURATION W-20-2

Exp. №	Position of control rods at time 0 (mm)	Water level (mm)	Description of the experiment	Reactivity from KNK-56M (№1+№2+№3) ($)	Reactivity from KNK-17-1 ($)	Water temp. (°C)
1	EP-1 = EP-2 = 0 CR-1 = 0; CR-2 = 501.4; MR-1 = MR-2 = 0	706.9 ± 1.0	CR-1 is thrown in the core (from level 0 mm to the level 700 mm) for 0.85 s	1.89 ± 0.06	KNK-17-1 №1: 1.79 ± 0.05	16.9 ± 0.1
2	EP-1 = EP-2 = 0; CR-1 = 0; CR-2 = 505.4; MR-1 = MR-2 = 0	706.9 ± 1.0	CR-1 is thrown in the core (from level 0 mm to the level 700 mm) for 0.8 s	1.89 ± 0.06	KNK-17-1 №1: 1.79 ± 0.05	17.2 ± 0.1
3	EP-1 = EP-2 = 0; CR-1 = 0; CR-2 = 505.4; MR-1 = MR-2 = 0	706.9 ± 1.0	CR-1 is lowered from level 0 mm to level 700 mm for 47.60 s	1.94 ± 0.06	KNK-17-1 №1: 1.79 ± 0.05	17.2 ± 0.1
4	EP-1 = EP-2 = 0; CR-1 = 0; CR-2 = 505.4; MR-1 = MR-2 = 0	706.9 ± 1.0	CR-1 is thrown in the core (from level 0 mm to the level 700 mm) for 0.86 s	1.89 ± 0.06	KNK-17-1 №2: 1.87 ± 0.06	17.2 ± 0.1
5	EP-1 = EP-2 = 0; CR-1 = 0; CR-2 = 505.7; MR-1 = MR-2 = 0	706.9 ± 1.0	CR-1 is thrown in the core (from level 0 mm to the level 700 mm) for 0.79 s	1.89 ± 0.06	KNK-17-1 №3: 1.87 ± 0.06	17.2 ± 0.1
6	EP-1 = EP-2 = 0; CR-1 = 0; CR-2 = 505.7; MR-1 = MR-2 = 0	706.9 ± 1.0	CR-1 is lowered from level 0 mm to level 700 mm for 27.77 s	1.92 ± 0.06	KNK-17-1 №3: 1.87 ± 0.06	17.2 ± 0.1
7	EP-1 = EP-2 = 0; CR-1 = CR-2 = 0; MR-1 = MR-2 = 0	494.8 ± 1.0	Water moderator drains 0 to level 436.5 mm for 105 s	1.78 ± 0.5	KNK-17-1 №3: 1.55 ± 0.05	17.2 ± 0.1

[a] EP = emergency protection, CR = compensating reactivity, MR = manual regulation

TABLE 2.87. EXPERIMENTS WITH CONFIGURATION W-20-2-CONF1

Exp. №	Position of control rods at time 0 (mm)	Water level (mm)	Description of the experiment	Reactivity from KNK-56M (№1+№2+№3) ($)	Reactivity from KNK-17-1 ($)	Water temp. (°C)
1	EP-1 = EP-2 = 0; CR-1 = CR-2 = 0; MR-1 = 0; MR-2 = 311.4	528.9 ± 1.0	MR-1 is lowered to the level 312.4 mm (reactivity margin)	0.38 ± 0.01	—	19.4 ± 0.1
2	EP-1 = EP-2 = 0; CR-1 = 0; CR-2 = 295.5; MR-1 = MR-2 = 0	528.9 ± 1.0	CR-1 is thrown in the core (from level 0 mm to level 700 mm)	1.95 ± 0.06	KNK-17-1 No1: 1.85 ± 0.06	19.4 ± 0.1
3	EP-1 = EP-2 = 0; CR-1 = 0; CR-2 = 295.3; MR-1 = MR-2 = 0	528.9 ± 1.0	CR-1 is thrown in the core (from level 0 mm to level 700 mm) for 0.82 s	1.95 ± 0.06	KNK-17-1 No3: 1.96 ± 0.06	19.4 ± 0.1
4	EP-1 = EP-2 = 0; CR-1 = 0; CR-2 = 295.2; MR-1 = MR-2 = 0	528.9 ± 1.0	CR-1 is thrown in the core (from level 0 mm to level 700 mm)	1.95 ± 0.06	KNK-17-1 No2: 2.04 ± 0.06	19.4 ± 0.1
5	EP-1 = EP-2 = 0; CR-1 = 0; CR-2 = 295.0; MR-1 = MR-2 = 0	528.9 ± 1.0	MR-2 is thrown in the core (from level 0 mm to level 700 mm) for 1.15 s	0.965 ± 0.03	KNK-17-1 No3: 1.04 ± 0.03	19.4 ± 0.1

[a] EP = emergency protection, CR = compensating reactivity, MR = manual regulation

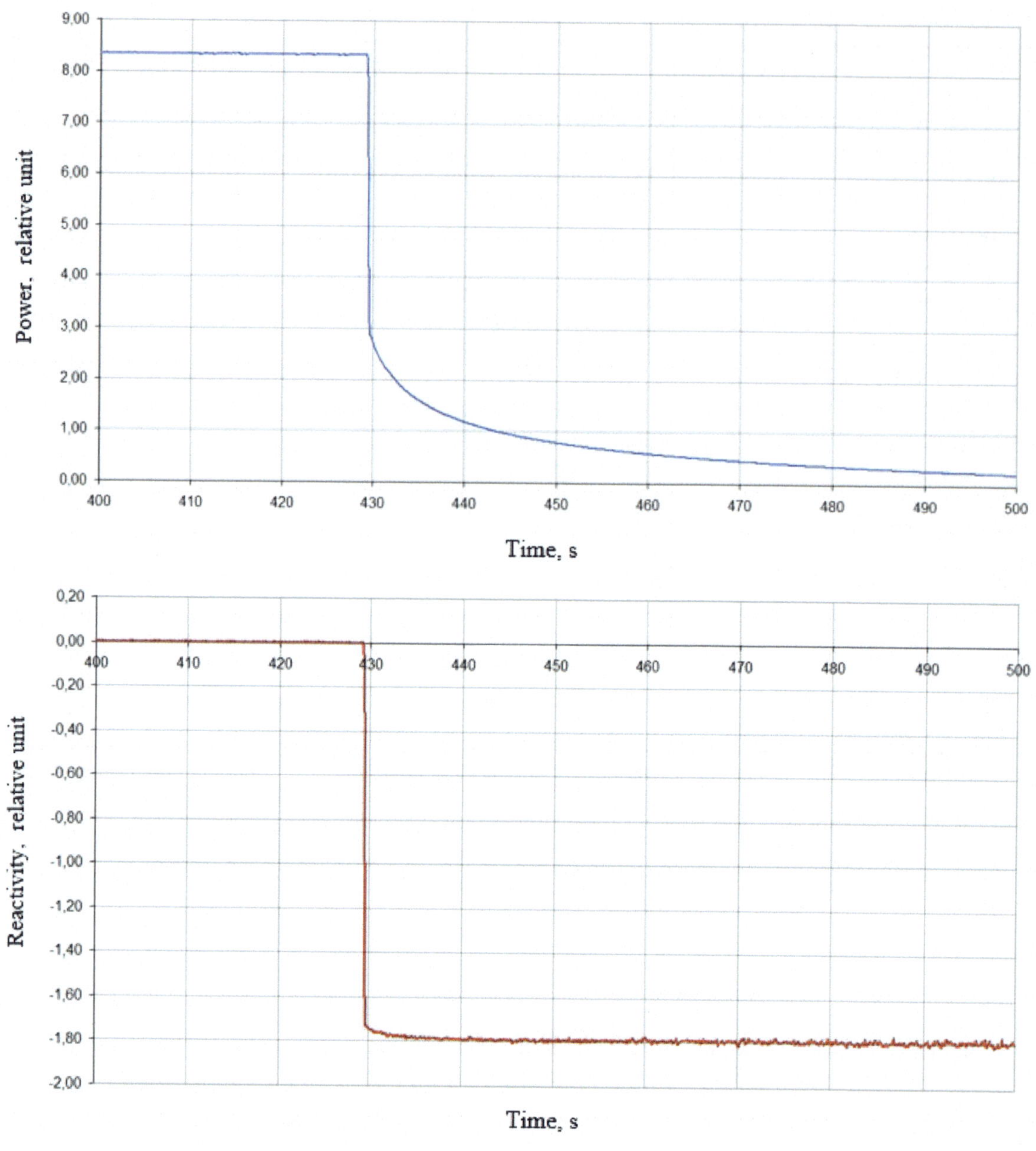

FIG. 2.151. Power and reactivity as a function of time for experiment 2 of Table 2.86.

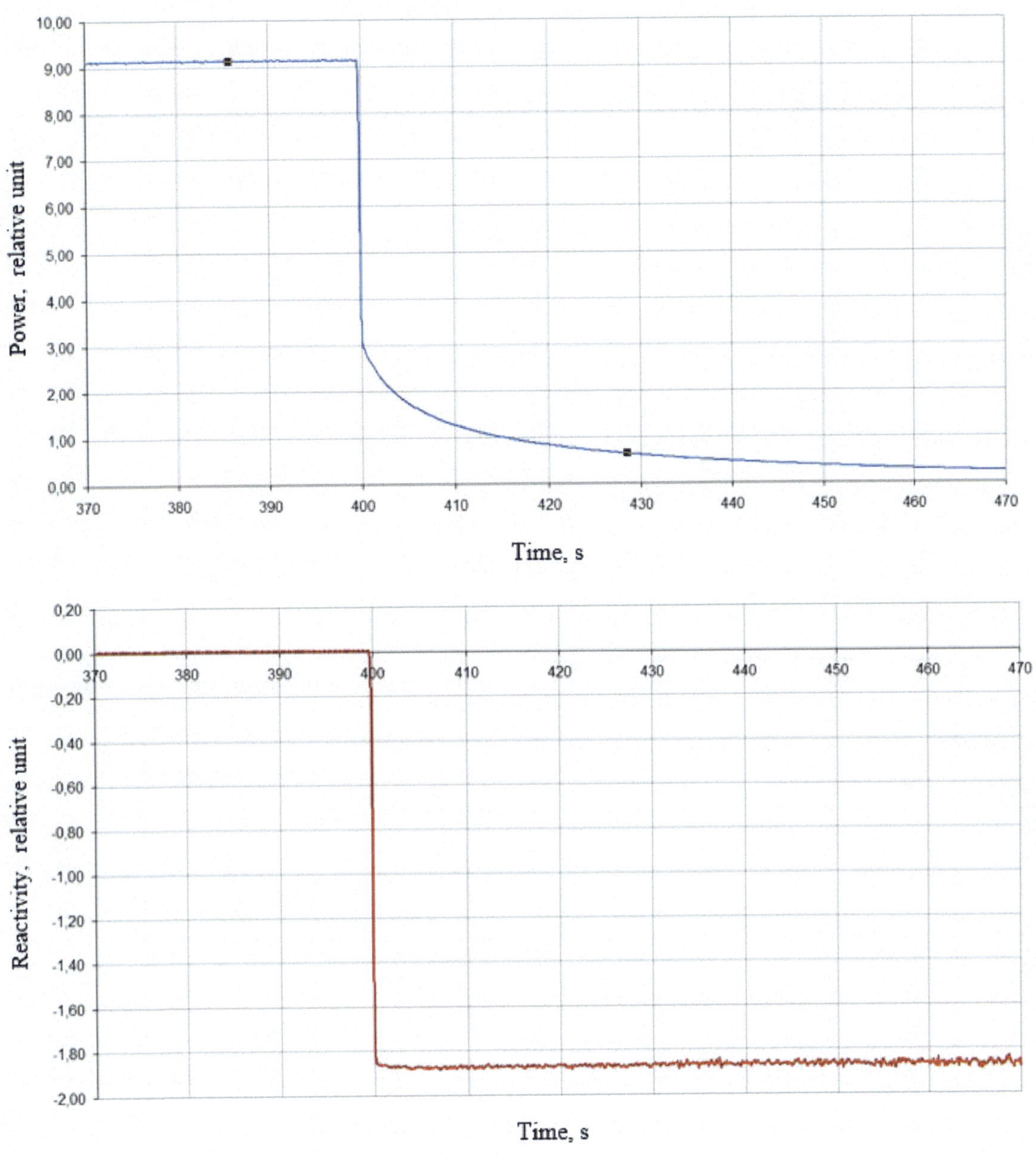

FIG. 2.152. Power and reactivity as a function of time for experiment 4 of Table 2.86.

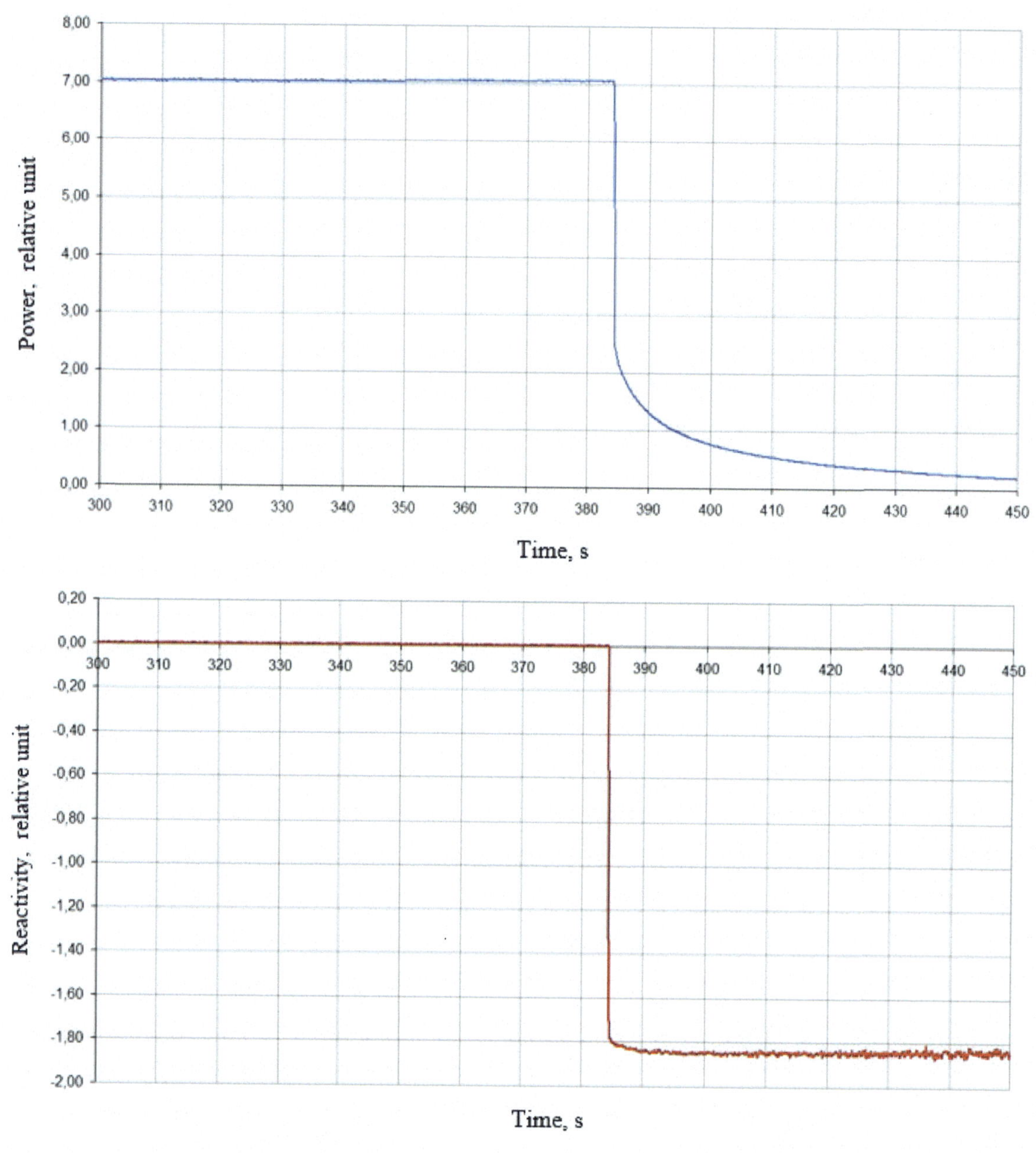

FIG. 2.153. Power and reactivity as a function of time for experiment 2 of Table 2.87.

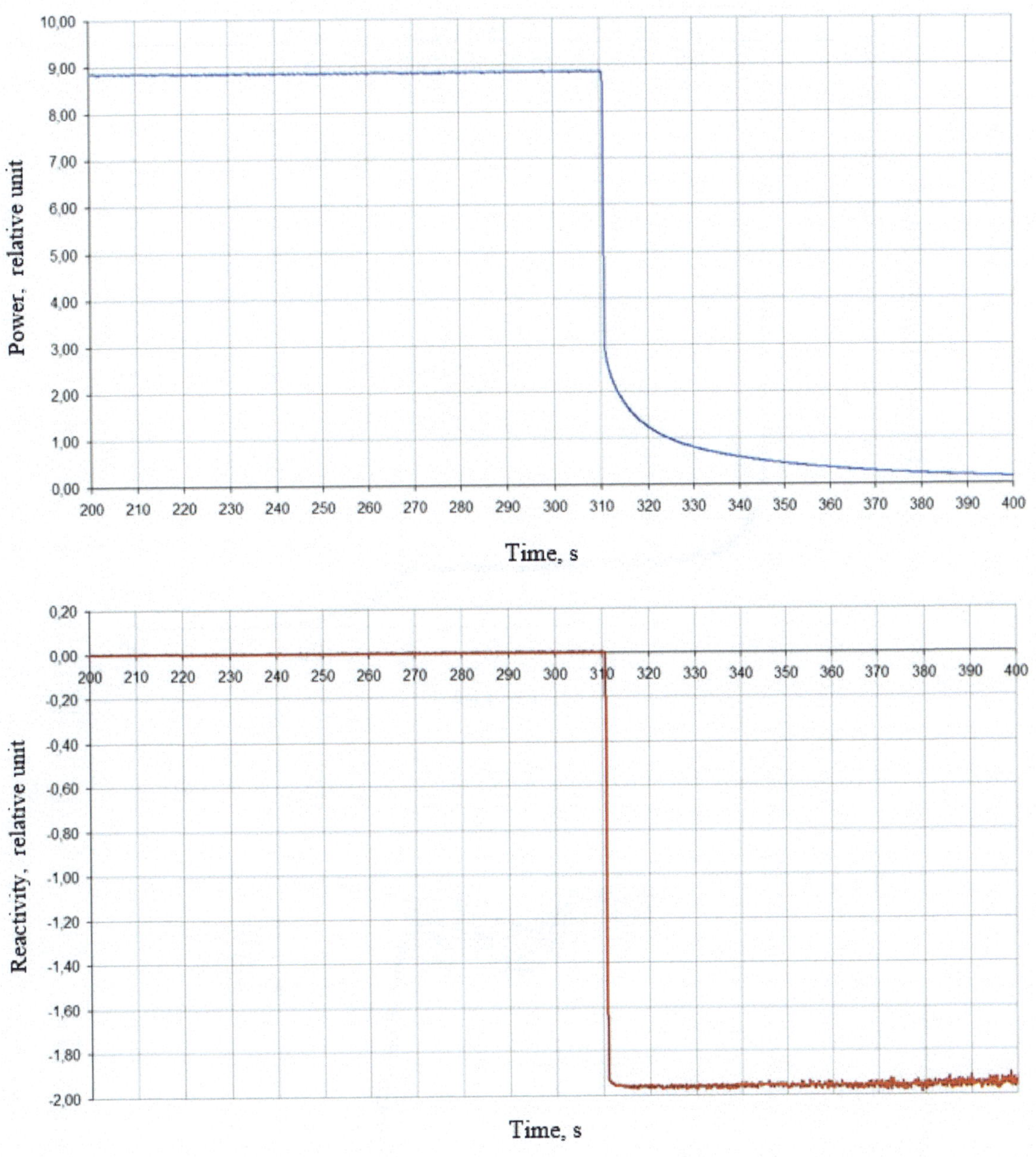

FIG. 2.154. Power and reactivity as a function of time for experiment 3 of Table 2.87.

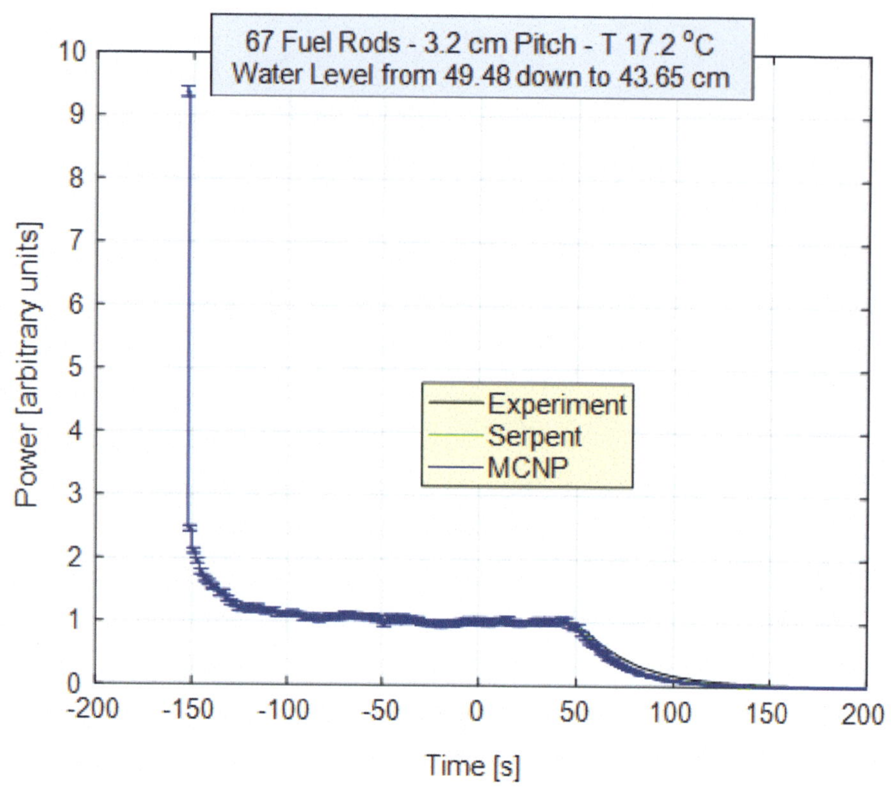

FIG. 2.155. Fission rate as a function of time for water level change from 49 down to 43.6 cm.

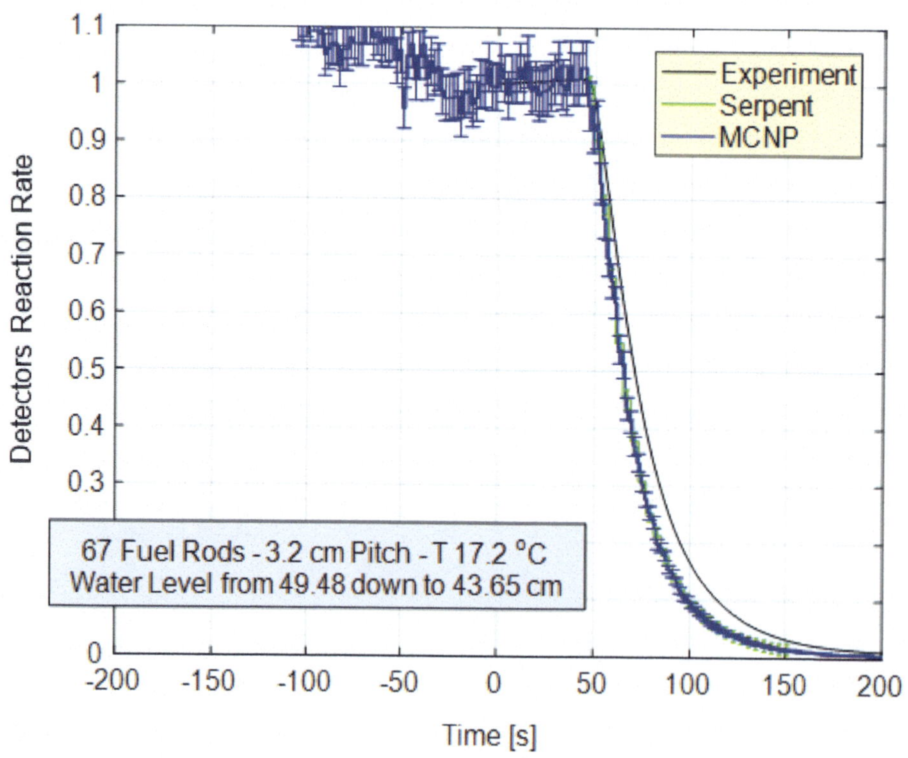

FIG. 2.156. Zoom of the fission rate as a function of time for water level change from 49 down to 43.6 cm.

2.3.5. Concluding remarks

The geometry changes transient experiments at the GIACINT facility were performed and simulated successfully using validated MCNP and SERPENT computer programs. These two computer programs used different methodologies to analyse these transients driven by geometry changes. SERPENT tracked delayed neutron precursors, divided the transient time into time bins, and updated the geometry at the end of each time bin. MCNP tracked delayed neutrons, did not divide the transient time into time bins, and updated the geometry change at the end of each neutron random walk.

2.4. QUINTA REACTION RATE MEASUREMENTS

Minor actinides (such as Np, Am, Cm) significantly contribute to the radiotoxicity and heat generation of spent nuclear fuel. They also generate significant amounts of helium due to the decay process.

2.4.1. Experiments

2.4.1.1. Experimental assembly

The QUINTA assembly (References [2.163]–[2.166]), located at the Joint Institute for Nuclear Research (JINR) in Dubna, Russian Federation, has been used for the application of a fast neutron spectrum to the nuclear incineration of actinides, i.e., their transmutation into stable or short-lived products (Figs 2.157 and 2.158). QUINTA is a target-blanket subcritical assembly, which consists of a total of 512 kg of natural uranium, surrounded by a lead shielding. It is composed of five sections, each being 114 mm long and separated by a 17 mm air gap, which allows the placement of samples mounted onto special plates (Fig. 2.159). The uranium target is composed of cylindrical rods with 36 mm in diameter, 104 mm in length and 1.72 kg in mass. The uranium rods are cladded in a 1 mm thick aluminium casing. Excluding the first section, 61 rods are arranged in a hexagonal lattice with a pitch size of 36 mm and enclosed in a hexagonal aluminium container with a wall thickness of 5 mm (Fig. 2.160). The first section contains 4 rods. The 7 central rods are removed to create the beam window. The beam window is 80 mm in diameter and serves to reduce the loss of backward emitted and/or scattered neutrons. The front and back of each section are then bounded by additional aluminium plates (dimension 350 mm × 350 mm × 5 mm). The deuteron beam is misaligned from the assembly axis by an 2° angle (Fig. 2.161).

The entire uranium target-blanket reaches 638 mm in length and a mass of 538 kg. The five sections are mounted onto a single slab of aluminium with thickness 25 mm and surrounded by lead bricks (100 mm thick) with a total mass of 1780 kg. This serves as a neutron reflector and, to some extent, as a biological shielding for γ-rays. The top section of the lead is further supported by 16 mm of aluminium (Fig. 2.157). The front of the lead castle has a square window (dimension 150 mm × 150 mm), and one side of the lead castle has a rectangle window for placing the actinide samples as shown in Fig. 2.158. Sample plates can be inserted and removed quickly into the air gaps between the sections, as well as on the front and back of the target, by the presence of slots and lids located on the roof of the lead castle. The sample plates are labelled 1–6, starting from the direction of the incident beam.

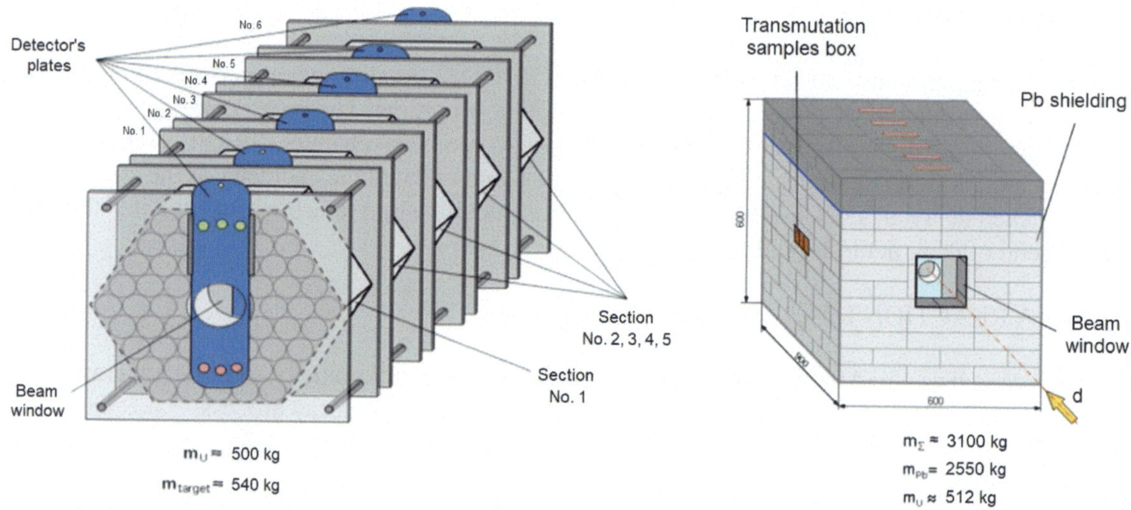

FIG. 2.157. Scheme of the QUINTA assembly. Left side: view on the uranium target with supporting structures and plastics used for sample placement (detector's plates). Right side: view on the lead shielding enfolding the target with marked the transmutation samples box (window) for the actinides sample location in the shielding.

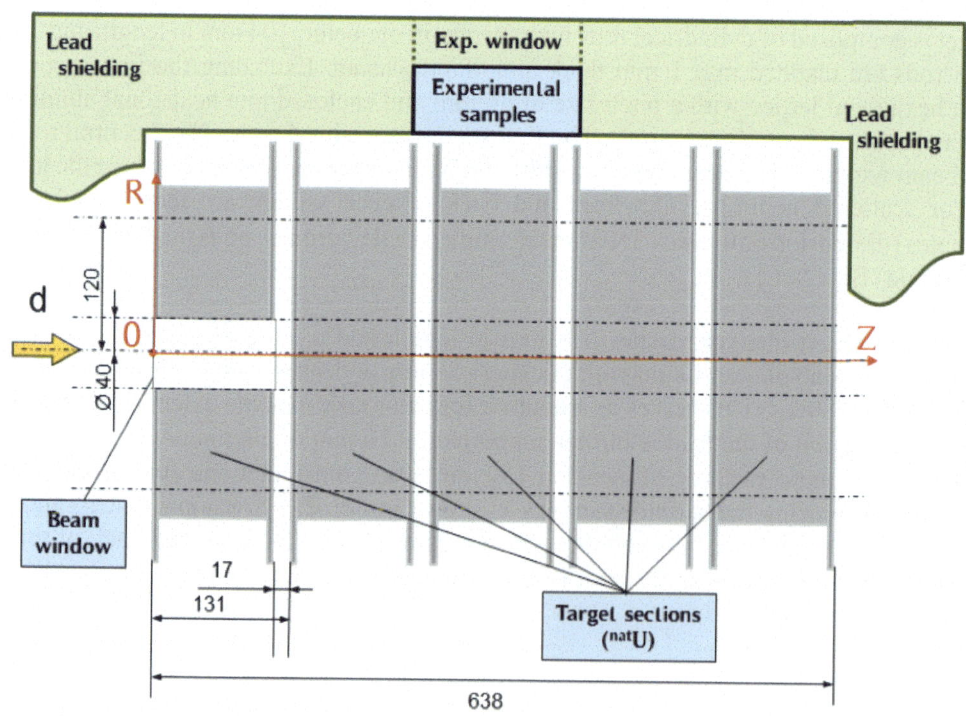

FIG. 2.158. Top view of actinide experimental samples position in the QUINTA assembly (R = 180 mm, Z = 319 mm).

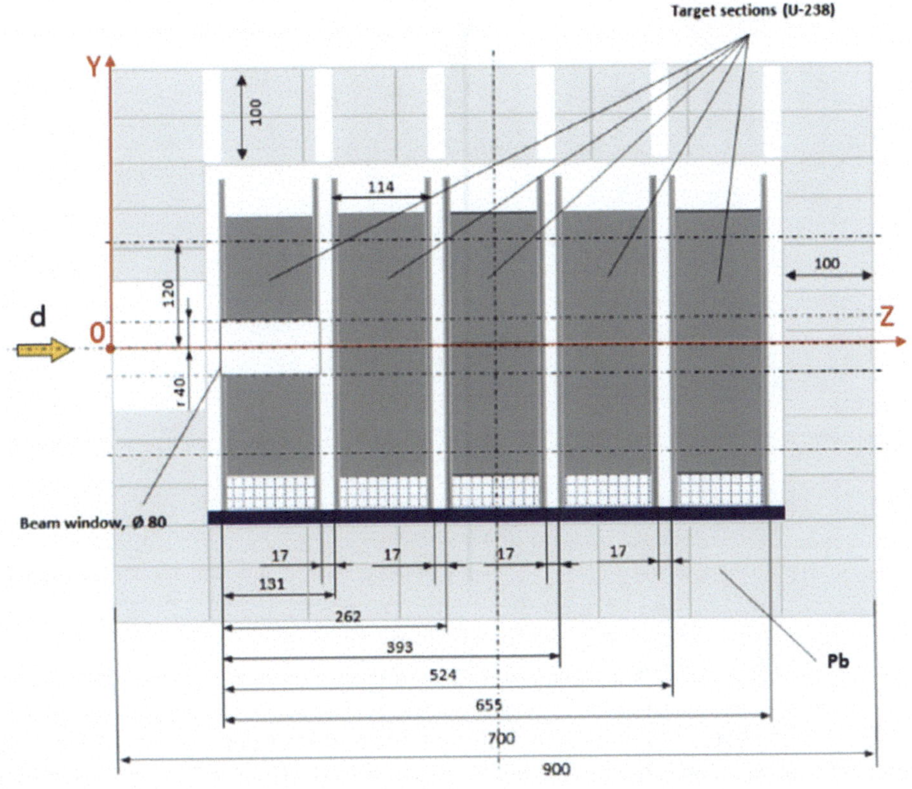

FIG. 2.159. Dimensions of the QUINTA assembly (mm) with the five U-target cylinders.

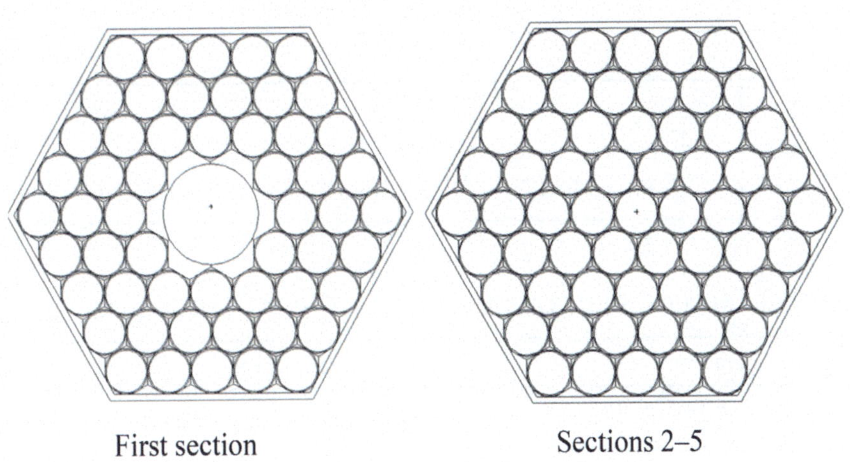

FIG. 2.160. Uranium rods cladding and packaging.

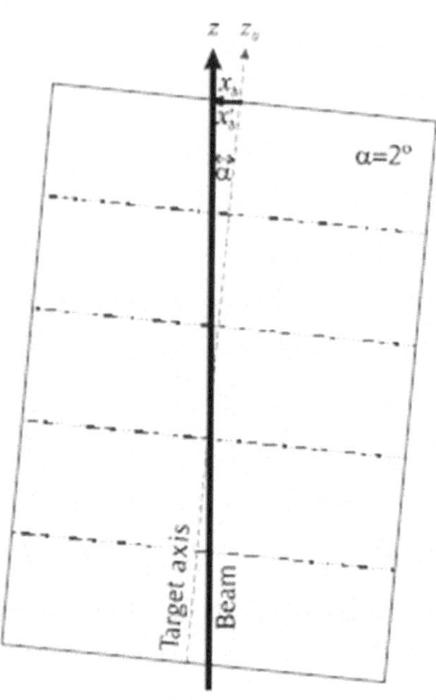

FIG. 2.161. Position of the deuteron beam direction with respect to the target assembly.

The QUINTA target was irradiated with a pulsed deuteron beam of energies 2, 4, and 8 GeV, and carbon ions of 24 GeV extracted from the Nuclotron accelerator, also located at the JINR, as reported in Table 2.88. Additional measurements were done with protons provided by the Phazotron accelerator available in Dubna with the parameters specified in Table 2.89. The ion fluence in both tables refers to the total number of particles impinging the QUINTA target during the irradiation time.

TABLE 2.88. DEUTERON BEAM CHARACTERISTICS FROM NUCLOTRON ACCELERATOR AT JINR

Incident ion	Deuteron (D+)			Carbon (C+)
Ion energy (GeV)	2	4	8	24
Irradiation time (h)	6.27	9.35	16.7	22.8
Integral ion fluence (number of particles)	$3.02(30) \times 10^{13}$	$2.73(27) \times 10^{13}$	$0.91(9) \times 10^{13}$	1.75×10^{11}

TABLE 2.89. PROTON BEAM PROVIDED FOR QUINTA BY THE PHAZOTRON ACCELERATOR

Incident ion	Proton (p+)
Ion energy (GeV)	0,66
Irradiation time (h)	5.71
Integral ion fluence (number P+)	$8.64(30) \times 10^{14}$

2.4.1.2. Irradiation of ^{237}Np samples

Neptunium-237 samples were placed inside QUINTA. The assembly was irradiated by a pulsed deuteron beam (the energy was 2 and 4 GeV, and irradiation times were 6–10 h). After the irradiation, gamma-rays of ^{92}Sr, ^{97}Zr, ^{132}I, ^{133}I, ^{135}I, and ^{238}Np from the samples were measured for evaluating the fission/capture rates of ^{237}Np. The average neutron energy and fluence were estimated to be about 0.4 MeV and in the order of 10^{12} n/cm², respectively.

Gamma spectrum analysis of the neutron irradiated actinide samples was performed in terms of fission product amount generated due to fast neutron induced actinide fissions, and isotopes amount generated due to neutron capture by the actinides.

After irradiation, the actinide samples were removed and transported away to be analysed with gamma spectrometry using a high-purity germanium (HPGe) n-type coaxial detector. Measurements began ~1.5 hours after irradiation had stopped, continuing for up to 6 days afterwards. The spectra collection times ranged from 15 minutes to just over 3 hours. All spectra were analysed using the DEIMOS32 program [2.167] to determine the absolute intensity and half width of the line (full width at half maximum – FWHM).

2.4.1.3. Irradiation of natural uranium samples

Investigations of nuclear-physical characteristics of neutron fields, generated in a massive uranium target irradiated by protons with an energy of 0.66 GeV, were performed. 23 natural uranium samples, spatially arranged in the subcritical assembly QUINTA (Fig. 2.157), were irradiated with spallation neutrons. The experimental data from gamma-ray spectroscopy was processed to obtain the number of neutron induced fissions and neutron captures in the detector foils, and to evaluate the neutron flux and fluence in the assembly by spectral indexes analysis.

2.4.2. Analyses on actinide transmutation rate

The neutronic parameters in the QUINTA facility were obtained by the experimental determination of the number of neutron induced fission and capture events in actinide samples. In particular, ^{237}Np was used in most of the analyses.

The number of capture and fission yields was estimated by detection of the gamma spectra emitted either by the decay of the nuclei generated by the capture process, or the fission products. The total yield of observed isotopes per gram of activated material (N_{yield}) was calculated with Eq. (2.68):

$$N_{yield} = \frac{S_y}{m \cdot \varepsilon_p \cdot I_y \cdot \phi \cdot COI} \cdot \frac{\lambda\, t_{ir}}{(1 - e^{-\lambda \cdot t_{ir}})} \cdot \frac{1}{(1 - e^{-\lambda\, t_{real}})} \cdot \frac{t_{real}}{t_{live}} \cdot e^{\lambda t} \qquad (2.68)$$

Where

- γ – gamma line index;
- S_γ – gamma peak area;
- m – activation sample mass (g);
- ε_p – gamma spectrometer efficiency;
- I_γ – correction for gamma line intensity (%);
- ϕ – deuteron fluence;

- λ – isotope decay constant (s^{-1});
- COI – correction for gamma quanta coincidence;
- t_+ – cooling time (s);
- t_{ir} – irradiation time (s);
- t_{real} – real time of measurement (s);
- t_{live} – live time of measurement (s).

The procedure of evaluation of actinide fission rate per deuteron and per gram ($I_{f\gamma}$) consisted of calculating the number of produced nuclei N_{yield} (Eq. (2.69)) for each isotope and then dividing it by the isotope production efficiency per fission (γ_f):

$$I_{f\gamma} = \frac{N_{yield}}{\gamma_f} \qquad (2.69)$$

Where

- $If\gamma$ – actinide fission rate per deuteron and per gram;
- f – reaction index (f = fission).

The actinide fission rate as well as the gamma line indexes were extracted from decay tables and nuclei databases for fission product generation.

The number of fission events (N_{yf}) for the actinide isotope (usually ^{237}Np) in the QUINTA facility was provided, knowing $I_{f\gamma}$, the actinide mass and the deuteron beam fluence. For the case of capture events (N_{yc}), the procedure included the spectral analysis of the gamma decay of the produced isotope to evaluate N_{yc}.

The relation between the number of fission and capture events and the integrated neutron flux is described by Eq. (2.70):

$$N_y = V_p \, \overline{\phi} \, N \, \overline{\sigma} \, t \tag{2.70}$$

Where

- $\overline{\phi}$ – average neutron flux in the place of actinide sample location (n·cm^{-2}·s);
- N – number of actinide isotopes per volume unit (cm^{-3});
- V_p – actinide sample volume (cm^3);
- $\overline{\sigma}$ – average microscopic cross-section for the reactions (n, f) or (n, γ), (barns);
- t – irradiation time (s).

The average neutron energy (E_n) was obtained from the comparison of the measured fission to capture ratio (α_m) with the spectral indexes available in nuclear databases. Spectral indexes depend on the neutron energy $\alpha(E_n)$ as follows (Eq. (2.71)):

$$\alpha_m = \frac{N_{yf}}{N_{yc}} = \frac{\overline{\sigma_f}}{\overline{\sigma}_c} = \frac{\sigma_f(E_n)}{\sigma_c(E_n)} = \alpha(E_n) \tag{2.71}$$

For the case of ^{237}Np, Fig. 2.162 shows the cross-sections for fission and neutron captures, and the corresponding spectral indexes ($\alpha(E_n)$) are listed in Table 2.90 for a range of expected neutron energies in the QUINTA experiments. The average neutron energy (E_n) is obtained when $\alpha_m = \alpha(E_n)$.

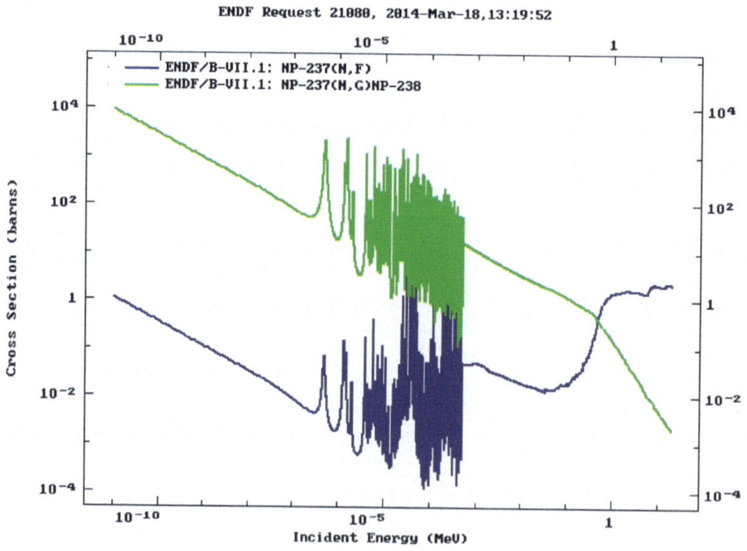

FIG. 2.162. Cross-sections of ^{237}Np(n, γ)^{238}Np and ^{237}Np(n, f) reactions [2.168].

E_n (MeV)	σ(n, f) (barn)	σ(n, γ) (barn)	$\alpha(E_n) = \sigma_f/\sigma_c$
0.39	0.169	0.564	0.299
0.40	0.183	0.548	0.333
0.41	0.211	0.533	0.395
0.42	0.239	0.519	0.461
0.43	0.268	0.504	0.531
0.44	0.297	0.491	0.605

The number of neutron-induced-fissioned (N_{yf}) and neutron-captured (N_{yc}) actinide isotopes in the actinide sample of volume V_p can be expressed by Eqs (2.72a) and (2.72b), respectively:

$$N_{yf} = V_p \, \overline{\phi} \, N \, \overline{\sigma}_f \, t \tag{2.72a}$$

$$N_{yc} = V_p \, \overline{\phi} \, N \, \overline{\sigma}_c \, t \tag{2.72b}$$

For the neutron energy that was estimated, the average cross-section is available. The neutron flux ($\overline{\phi}$) in the location of the actinide sample is then determined by Eqs (2.73a) and (2.73b):

$$\overline{\phi} = \frac{N_{yf}}{V_p N \, \overline{\sigma}_f \, t} \tag{2.73a}$$

$$\overline{\phi} = \frac{N_{yc}}{V_p N \, \overline{\sigma}_c \, t} \tag{2.73b}$$

Two different equations for fissioned (N_{yf}) and captured (N_{yc}) actinide isotopes should give the same value of the average neutron flux, which is a proof for correct measurements.

It is useful to look closely at the ratios $\alpha(E_d)) = \sigma_f(E_d)/ \sigma_c(E_d)$ of the fission and capture cross-sections of the different minor actinide isotopes (see Fig. 2.162 for ^{237}Np and Fig. 2.163 for ^{241}Am). The fission/absorption ratios are consistently higher for the fast spectrum. This means that, in a fast neutron spectrum, actinides are preferentially fissioned, instead of being transmuted into higher actinides.

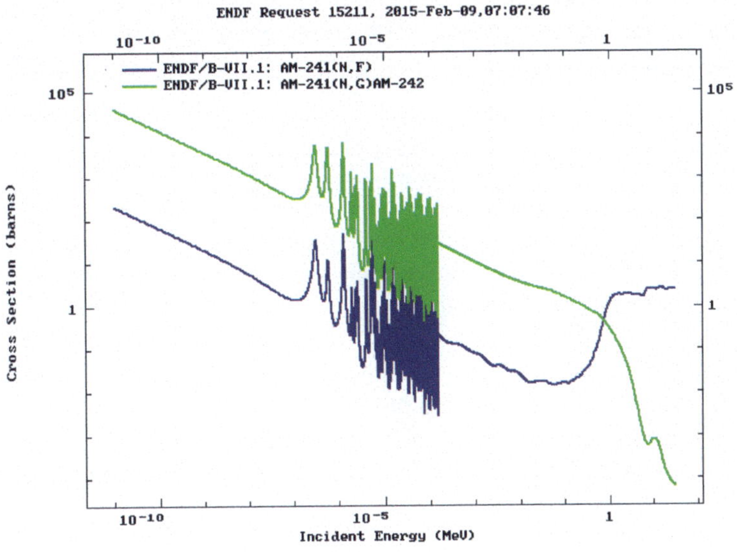

FIG. 2.163. Cross-sections of 241Am(n, γ)242Am and 241Am(n, f) reactions [2.168].

Apart from the ^{237}Np irradiation tests, spectral analyses of the gamma-ray results of the irradiation of natural uranium samples were performed. Metallic natural uranium is composed of ^{238}U (99.2752%), ^{235}U (0.7202%), and a very small amount of ^{234}U. The number of nuclei per volume of ^{238}U and ^{234}U are comparable in a high energy neutron spectrum.

The number of induced fission events (N_{yfs}) is then given by Eq. (2.74):

$$N_{yfs} = N_{yf8} + N_{yf5} = V_p \overline{\phi} N_8 \overline{\sigma} f_8 t + V_p \overline{\phi} N_5 \overline{\sigma} f_5 t = V_p \overline{\phi} N_8 t (\overline{\sigma} f_8 + \frac{N_5}{N_8} \overline{\sigma} f_5) \qquad (2.74)$$

Where

- $\overline{\phi}$ – average neutron flux in the place of actinide sample location (n·cm^{-2}·s);
- N – number of ^{238}U (N_8) and ^{235}U (N_5) uranium isotopes per volume unit (cm^{-3});
- Vp – uranium sample volume (cm^3);
- $\overline{\sigma} f$ – average microscopic cross-section for the fission reaction for ^{238}U or ^{235}U (barns);
- t – irradiation time (s).

The number of captures in the sample is measured by the amount of ^{239}Pu that is generated by direct neutron capture by ^{238}U, in the well-known chain:

$$^{238}U(n,\gamma) \rightarrow {}^{239}U \frac{\beta}{23.5\ min} \rightarrow {}^{239}Np \frac{\beta}{2.36\ day} \rightarrow {}^{239}Pu$$

The total number of measured neutron captures is then given by Eq. (2.75):

$$N_{yc\,8} = V_p \overline{\phi} N_8 \overline{\sigma}_{c8} t \qquad (2.75)$$

Where

- $\overline{\phi}$ – average neutron flux in the place of actinide sample location (n·cm^{-2}·s);
- N_8 – number of ^{238}U uranium isotopes per volume unit (cm^{-3});
- V_p – uranium sample volume (cm^3);
- $\overline{\sigma}_c$ – average microscopic cross-section for the neutron capture reaction of ^{238}U (barns);
- t – irradiation time (s).

The spectral neutron energy (E_d) was obtained to achieve the balance between the measured spectral index ($\alpha_{m85}(E_d)$) of the natural uranium ($N_5/N_8 = 0.7202/99.2752 = 0.00725$) irradiated sample, computing the energy weighted fission and capture the uranium cross-section from the available database according to Eq. (2.76):

$$\alpha_{m85}(E_d) = \frac{N_{yfs}}{N_{yc8}} = \frac{\overline{\sigma}_{f8}}{\overline{\sigma}_{c8}} + \frac{N_5}{N_8} \frac{\overline{\sigma}_{f5}}{\overline{\sigma}_{c8}} \qquad (2.76)$$

Additional evaluation of the neutron fluence was done by a silicon detector. Fast neutrons with energies of E_n > 100 keV induced damage in the volume of the silicon semiconductor by formation of vacancies and interstitials (so-called Frenkel defects) in the crystal lattice.

Owing to this phenomenon, the silicon detector changes its electrical parameters (characteristics) under neutron irradiation. The radiation defects lead, among others, to reverse current growth.

Reverse dark current of a silicon detector grows linearly [2.169] with increasing fast neutron fluence (Eq. (2.77)):

$$\Delta I = \alpha \cdot V \cdot \Phi \qquad\qquad (2.77)$$

Where

- $\Delta I - (I_{irrad} - I_{nonirrad})$, increment detector current (A);
- V – volume of the detector (cm^3);
- Φ – neutron fluence (cm^{-2}), equivalent for silicon damage fluence of 1 MeV fast neutrons;
- α – 2×10^{-17} A/cm, reverse current damage coefficient.

The α value depends on the annealing time and temperature. Neutron radiation induced damage of silicon detectors is presented in Fig. 2.164, extracted from Ref. [2.170]. The damage cross-section as function of energy starts with the threshold mode for the energy 170 keV.

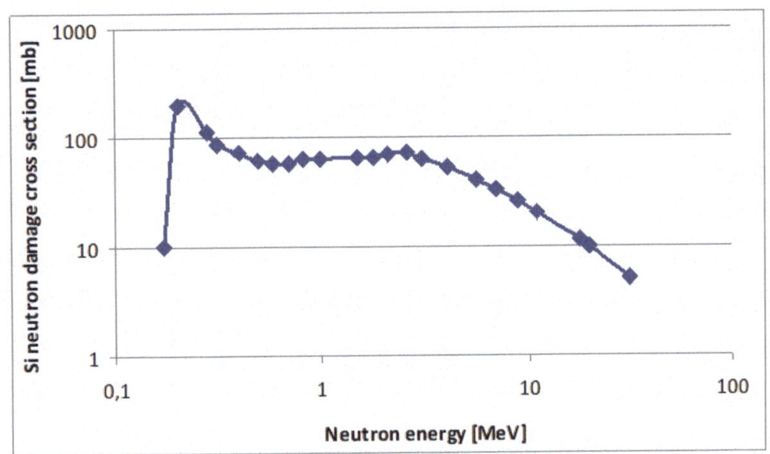

FIG. 2.164. Neutron damage cross-section of silicon detectors as function of neutron energy [2.170].

In the neutron energy range from 0.4 MeV to 4 MeV, it can be roughly assumed that the damage cross-section of silicon detectors is constant with an accuracy of 25%. The measured neutron fluence with this detector is the equivalent to neutron energies around 1 MeV.

2.4.2.1. Gamma spectrum analysis of irradiated ^{237}Np samples and evaluation of capture and fission rates

The neutron fluence and fission and capture rates of ^{237}Np were evaluated on QUINTA for a set of particle beams.

From the gamma-ray spectrum of the irradiated ^{237}Np, the fission products in Table 2.91 were selected (^{92}Sr, ^{97}Zr, ^{132}I, ^{133}I, and ^{135}I) to evaluate its fission rate per gram, normalized by impinging deuteron applying Eqs (2.70) and (2.76). For ^{135}I, two different gamma peaks were used: 1131.51 keV and 1260.41 keV.

The resulting normalized fission rate I_{fy} is presented in Figs 2.165 and 2.166 for deuteron beams of 2 and 4 GeV, respectively. Irrespective of the which fission fragments were used in the calculation, the fission rate I_{fy} should be the same.

Isotope	$T_{1/2}$ (h)	γ line (keV)	I_γ (%)	γ_f (%) cumulate
^{92}Sr	2.66	1383.93	90	4.01
^{97}Zr	16.744	507.64	5.03	5.35
^{132}I	2.295	772.6	75.6	4.39
^{133}I	20.87	529.87	87	4.45
^{135}I	6.57	1260.41	28.7	4.16
		1131.51	22.6	

Source of γ_f: https://www.nndc.bnl.gov/

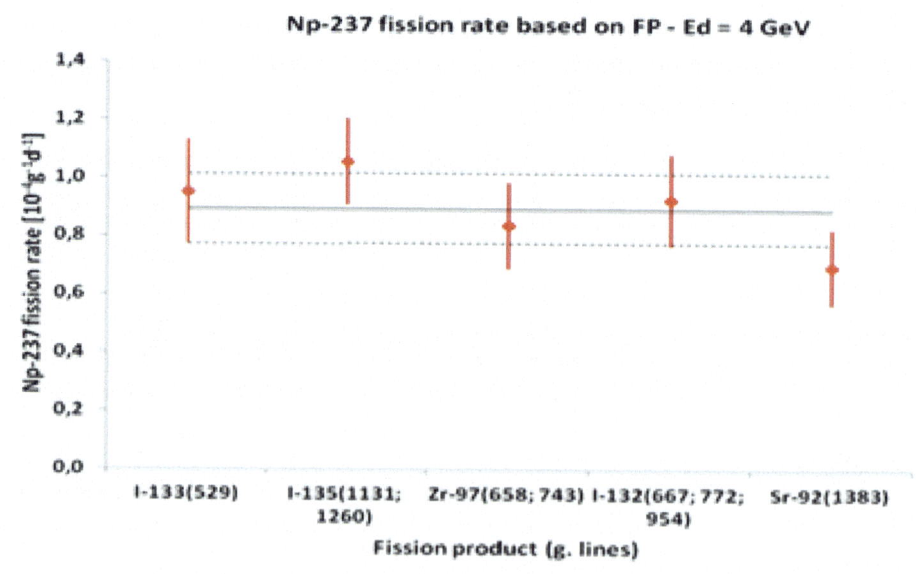

FIG. 2.165. Fission rate for different fission products of a ^{237}Np sample (mass 0.987 g) for deuteron beam of 2 GeV energy.

FIG. 2.166. Fission rate for different fission products of a ^{237}Np sample (mass 1.115 g) for deuteron beam of 4 GeV energy.

The neutron capture rate in ^{237}Np was determined as well from gamma spectrum analysis by the measurement of the ^{238}Np isotope decay, as follows:

$$^{237}\text{Np}(n, \gamma)^{238}\text{Np} \ (\beta^-, \ T_{1/2} = 2.117 \text{ days})$$

The gamma lines of the ^{238}Np decay scheme are specified in Table 2.92.

TABLE 2.92. GAMMA LINE DATA OF ^{238}Np GENERATED IN THE ^{237}Np(n, γ)^{238}Np REACTION

Isotope	γ line (keV)	I_γ (%)
	984.45	27.8
	1028.54	20.38
^{238}Np	1025.87	9.65
	923.98	2.869
	962.77	0.702

The resulting normalized capture rate experimental estimations of ^{237}Np are presented in Figs 2.167 and 2.168 for deuteron beams of 2 and 4 GeV, respectively. In both cases, fission rate and capture rate of ^{238}Np were disregarded, as they were negligible compared to the respective rates of ^{237}Np during the transmutation analysis.

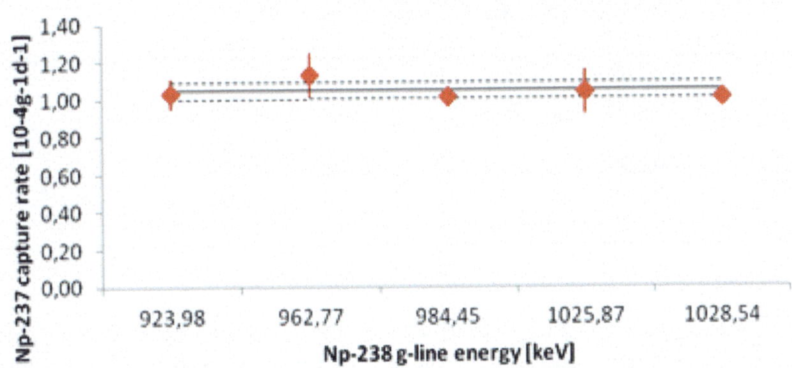

FIG. 2.167. Capture rate of a ^{237}Np sample (mass 0.987 g) for different gamma lines of ^{238}Np for deuteron beam of 2 GeV energy.

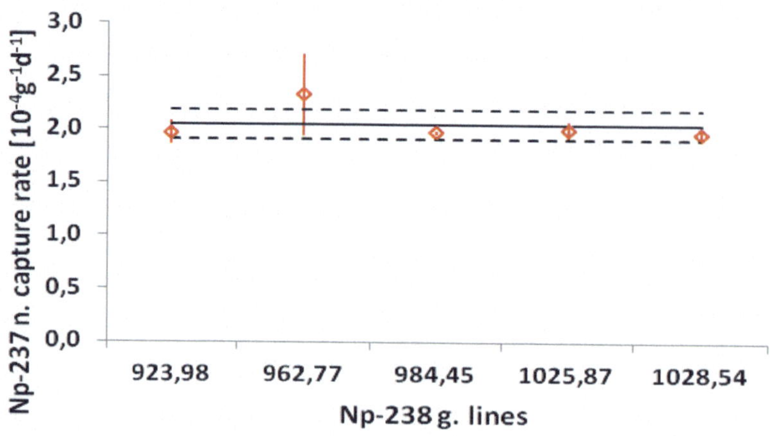

FIG. 2.168. Capture rate of a ^{237}Np sample (mass 1.115 g) for different gamma lines of ^{238}Np for deuteron beam of 4 GeV energy.

From the measured spectral index, defined as the ratio between the fission and capture rates, the neutron energy was obtained by matching with the spectral indexes ($\sigma(n, f)/\sigma(n, \gamma)$) calculated with the cross-section data from the ENDF/B-VII.1 database [2.168]. The results are listed in Table 2.93.

TABLE 2.93. FISSION/ABSORPTION RATIOS AS FUNCTION OF NEUTRON ENERGY FOR ^{237}Np

Neutron energy (MeV)	Fission cross-section, $\sigma(n, f)$ (barn)	Capture cross-section, $\sigma(n, \gamma)$ (barn)	$\sigma(n, f) / \sigma(n, \gamma)$
0.39	0.169	0.564	0.299
0.40	0.183	0.548	0.333
0.41	0.211	0.533	0.395
0.42	0.239	0.519	0.461
0.43	0.268	0.504	0.531
0.44	0.297	0.491	0.605

The total number of fission events (N_{yf}) and capture events (N_{yc}) per sample were then calculated from the neutron flux, the average cross-section at the neutron energy that was estimated, the sample volume, and the irradiation time.

A summary of the results is compiled in Table 2.94 for the ^{237}Np sample and particle beam energies of 2, 4 and 8 GeV/deuteron, respectively.

TABLE 2.94. COMPILATION OF FISSION RATES AND CAPTURE RATES FOR ^{237}NP AT THREE DIFFERENT DEUTERON BEAM ENERGIES [2.171], [2.172]

Deuteron beam energy	2 GeV	4 GeV	8 GeV
Integral deuteron fluence (d)	$3.02(10) \times 10^{13}$	$2.73(10) \times 10^{13}$	$0.91(4) \times 10^{13}$
Mass of Np-237 sample (g)	0.987	1.115	1.085
Number of Np-237 fissions per gram sample (N_{yf})	$(1.12 \pm 0.39) \times 10^9$	$(1.69 \pm 0.70) \times 10^9$	$(1.03 \pm 0.54) \times 10^9$
Number of neutron captured Np-237 atoms per gram sample (N_{yc})	$(2.13 \pm 0.27) \times 10^9$	$(4.73 \pm 0.61) \times 10^9$	$(1.99 \pm 0.27) \times 10^9$
Fission rate per gram sample and deuteron	$(3.72 \pm 1.30) \times 10^{-5}$	$(6.20 \pm 2.56) \times 10^{-5}$	$(1.13 \pm 0.60) \times 10^{-4}$
Capture rate per gram sample and deuteron	$(7.064 \pm 0.897) \times 10^{-5}$	$(1.73 \pm 0.22) \times 10^{-4}$	$(2.19 \pm 0.30) \times 10^{-4}$
Fission/capture rate (α_m)	0.53 ± 0.20	0.36 ± 0.15	0.52 ± 0.28
Fission rate per gram sample, deuteron and its energy	$(1.86 \pm 0.65) \times 10^{-5}$	$(1.55 \pm 0.63) \times 10^{-5}$	–
Capture rate per gram sample, deuteron and its energy	$(3.53 \pm 0.45) \times 10^{-5}$	$(4.33 \pm 0.55) \times 10^{-5}$	–

2.4.2.2. Actinide incineration effectiveness by distribution of fission fragments from fast neutron induced fissions

Table 2.95 presents the fission and capture rates for ^{237}Np with reference to different particles and energies, including protons, deuterons and C^{6+} nuclei. The first two ratios give the values per gram of ^{237}Np and per charged particle. The next two rows give these values normalized in addition by the number of nucleons in the particle. In the following two rows, the results normalized in addition by the energy of particles in GeV are shown. The last row presents the fission-to-capture ratios in these cases, which are relevant for the assessment of the incineration effectiveness of the considered transmutation options.

TABLE 2.95. EXPERIMENTAL FISSION AND CAPTURE RATES FOR ^{237}NP [2.173]

Particle energy/particle	0.66 GeV/p	2 GeV/d	4 GeV/d	8 GeV/d	24 GeV/C^{6+}
Fission (10^{-5} g^{-1} p^{-1})	1.67(35)	5.56(87)	8.95(12)	13.3(24)	51.1(135)
Capture (10^{-5} g^{-1} p^{-1})	2.40(42)	10.10(43)	20.40(14)	22.6(14)	86.1(84)
Fission (10^{-5} g^{-1} p^{-1} nucleon^{-1})	1.67(35)	2.78(43)	4.47(60)	6.65(12)	4.26(11)
Capture (10^{-5} g^{-1} p^{-1} nucleon^{-1})	2.40(42)	5.22(22)	10.21(71)	11.31(7)	7.17(7)
Fission (10^{-5} g^{-1} p^{-1} nucleon^{-1} GeV^{-1})	2.53(53)	1.39(22)	1.12(15)	0.83(15)	0.18(5)
Capture (g^{-1} p^{-1} nucleon^{-1} GeV^{-1})	3.63(63)	2.61(11)	2.55(18)	1.41(9)	0.30(3)
Fission (10^{-5} g^{-1} p^{-1} proton^{-1} GeV^{-1})	2.53(53)	2.78(43)	2.24(30)	1.66(30)	0.35(9)
Capture (10^{-5} g^{-1} p^{-1} proton^{-1} GeV^{-1})	3.63(63)	5.22(22)	5.11(35)	2.83(18)	0.60(6)
Ratio fission/capture	0.70(19)	0.53(9)	0.44(7)	0.59(11)	0.59(17)

2.4.2.3. Neutron fluence comparison with fission/capture estimation and silicon detectors

The measurements of the neutron fluence with silicon detectors in position 5 (see Fig. 2.169) are reported and compared in Table 2.96 with the measurements obtained by the fission/capture method. The silicon detectors estimated 0.73, 1.1, and 0.6 (10^{12} n/cm^2), respectively, for deuteron beams (E_d) of 2, 4, and 8 GeV. The values of neutron fluencies measured by actinide detectors were about 2.5 times higher than that measured by silicon detectors. This difference arose from the attenuation of the neutron fluence by a 10 cm lead reflector when measuring with the silicon detectors, and the fact that the silicon detector effectively measures the fast neutrons of energies higher than 170 keV. The neutron fluence was also measured with the 0.66 GeV proton beam provided by the Phazotron accelerator. The results are presented in Table 2.96.

Figure 2.170 shows the measurement in five points depicted in Fig. 2.169. As expected, a clear correlation of the neutron fluence with the deuteron beam energy was found, as neutron yield ought to be proportional to the particle energy.

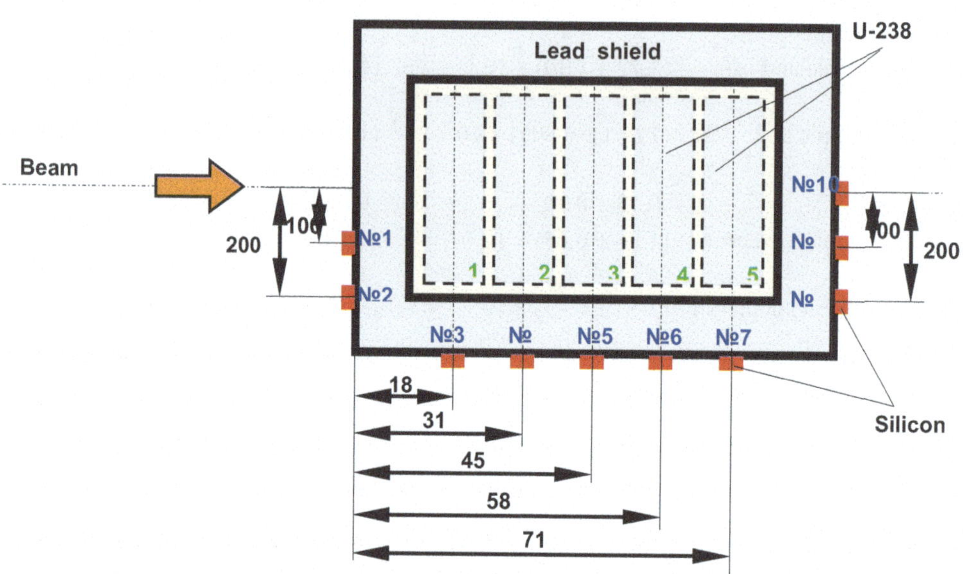

FIG. 2.169. Location of silicon detectors along the perimeter of the QUINTA assembly.

TABLE 2.96. COMPARISON OF THE RESULTS OF NEUTRON FLUENCE MEASUREMENTS

Particle beam energy (d: deuteron; p: proton)	0.66 GeV/p	2 GeV/d	4 GeV/d	8 GeV/d
Number of fissions in the sample (10^{12} n/cm^2)	1.67	1.64	3.63	3.13
Number of captures in the sample (10^{12} n/cm^2)	1.64	1.66	3.39	1.55
Silicon detector neutron fluence (10^{12} n/cm^2)	1.11	0.73	1.1	0.6

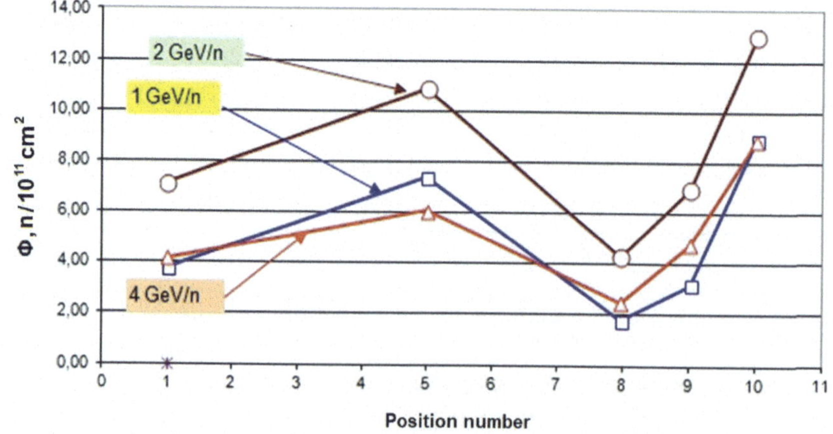

FIG. 2.170. Neutron fluence measured using the silicon detector in five points at different deuteron energies.

180

2.5. BRAHMMA FACILITY

2.5.1. Introduction

India is undertaking an R&D programme for ADS development, motivated in particular by the interest in the transmutation of thorium to ensure long term sustainability of nuclear energy production. Development of high-power accelerators and suitable targets and coupling these targets with the subcritical reactor are the focus areas of these efforts. On the reactor front, the R&D effort is mainly directed towards the development of accurate simulation tools, nuclear data and experiments oriented towards understanding the physics of accelerator driven subcritical reactors (which is quite different from that of critical reactors) and methods for determining the degree of subcriticality.

Unlike critical reactors, there are many physics issues which are critical to understanding and safely operating a subcritical reactor using external neutrons. One such key issue is the accurate measurement of the effective neutron multiplication factor k_{eff}, which determines the power in case of a subcritical reactor. Accurate monitoring of the subcritical reactivity value is one of the central operational and safety issues of a future ADS. The reactivity variations should under no circumstances lead to either prompt criticality or delayed criticality. However, conventional methods of subcriticality monitoring are no longer valid in ADS. Unlike critical reactors, the subcritical system has higher modes besides the fundamental mode. The total subcritical flux can be approximated by a series of eigenmodes with the fundamental mode being the predominant one. Any experimental measurement of the system parameters will be influenced by the higher harmonics and, hence, necessary corrections need to be applied to extract the fundamental mode. For subcritical systems, the development of new online methods for monitoring of subcriticality level is therefore required. Moreover, the accuracy of the existing methods, such as the Sjöstrand area ratio method or noise methods, strongly depend on the detector position, even when the system behaviour is point kinetic. Therefore, the development of accurate reactivity measurement techniques is quite important. This may also find application during fuel loading of critical reactors or management of a radioactive waste assay containing minor actinides.

In India, physics related R&D efforts include basic research activities in the area of subcritical reactor physics, development of new computer codes and nuclear data for analysis of ADS, conceptual design studies of ADS for thorium utilization, and an experimental programme for understanding the physics of ADS and for developing methods for measuring and monitoring the subcriticality of ADS. An initial effort has been also to understand the physics of subcritical reactors using surrogate neutron sources such as D–D or D–T neutron generators.

As part of the ADS programme in India, a subcritical assembly named BRAHMMA, driven by an indigenously developed deuteron accelerator, has been in operation at BARC since 2013. This facility was conceived for the purpose of investigating the static and dynamic neutronic properties of accelerator driven subcritical systems. It provides an opportunity to study various physics issues related to accelerator driven subcritical systems on a low power scale. The uniqueness of the system is its modular design where subcriticality level can be varied by using different fuel configurations, such as natural or slightly enriched uranium. The other unique feature is the use of beryllium oxide as reflector which has resulted in a compact system. This Section provides a description of the BRAHMMA facility with respect to the experimental programme at as well as analytical investigations aimed at validating codes and computational procedures.

BRAHMMA [2.174], [2.175] is a thermal subcritical reactor core that can be coupled to a D–D or D–T neutron generator. The subcritical assembly is a polyethylene moderated natural uranium system with a beryllium oxide reflector. The assembly uses nearly 2.8 tonnes of natural uranium fuel in the form of 160 fuel rods. The k_{eff} value of the subcritical system is 0.890. The modular subcritical core consists of metallic natural uranium as fuel with Al cladding and high-density polyethylene as moderator in a 13×13 square lattice configuration followed by beryllium oxide as reflector in the radial direction. The pitch of the fuel–moderator lattice is 48 mm. The whole assembly is finally surrounded by borated polyethylene to isolate the system from scattered neutrons. The central 3×3 positions of the lattice (dimension 144 mm × 144 mm) are empty and would serve as the cavity for inserting the neutron source. It is coupled to a D-T or D-D neutron generator producing neutrons of energy 14.1 MeV or 2.45 MeV, respectively. The neutron producing target is located at the centre of the subcritical assembly.

There are three axial experimental channels (EC1, EC2 and EC3) of 10.0 mm diameter each and four radial experimental channels of 7.2 mm diameter each, which cover the core region (EC4, EC5, EC6 and EC7) as well as the reflector region (EC4R, EC5R, EC6R and EC7R). The relative positions of the experimental channels are such that their influence on each other is minimized (Fig. 2.171). The description of experimental channels is provided in Ref. [2.174].

FIG. 2.171. BRAHMMA subcritical system coupled to the neutron source.

2.5.2. Experiments

Extensive measurements were carried out for reactivity measurements using PNS techniques, axial and radial flux profiles, and noise measurements [2.175]–[2.179]. A brief list of the experiments is given below:

- PNS techniques
- Area ratio;
- Slope fit;
- Source jerk;
- Noise techniques
- Feynman-alpha;
- Rossi-alpha;
- Spatial correction for modal effects;
- Development of miniature neutron detectors;
- Development of techniques for fundamental mode extraction by suppression of higher harmonics in deep subcritical systems;
- Measurement of source multiplication;
- Neutron flux profiles.

2.5.3. Results for pulsed neutron source measurements

Since the axial and radial experiment channels are narrow, special miniature ^{3}He detectors have been developed [2.177] for PNS measurements. The miniature ^{3}He detector has a diameter of 6.2 mm, an active length of 70 mm, and a sensitivity of 0.4 cps/nv^2. For measurements in lattice locations, ^{3}He detectors with 25 mm diameter, 100 mm active length, and 10 cps/nv sensitivity have been used [2.176].

PNS techniques [2.178] are one of the simplest methods for reactivity measurements of a subcritical system. In PNS techniques, the detector response to a short neutron pulse is measured. Following the neutron pulse, the neutron population first rises to a peak and then dies away quickly due to the subcritical nature of the system. After a short period of time, the detector response decays much more slowly to represent the emission of delayed neutrons in the reactor. Three techniques – area ratio method, slope fit method, and source jerk method – have been employed to measure the reactivity ρ and the prompt neutron decay constant α.

For area ratio and slope fit experiments, the external pulsed D–T neutron source is operated with a pulse period of 10 ms (100 Hz). The neutron flux in the subcritical assembly is recorded by the detectors from a number of repetitive neutron pulses (typically 50 000 pulses) once the delayed neutron level has reached an equilibrium value. Figures 2.172(a) and (b) show the time response for detectors located in the core and reflector regions, respectively. Table 2.97 lists the prompt decay constant α and the effective neutron multiplication factor k_{eff} measured using the slope fit method.

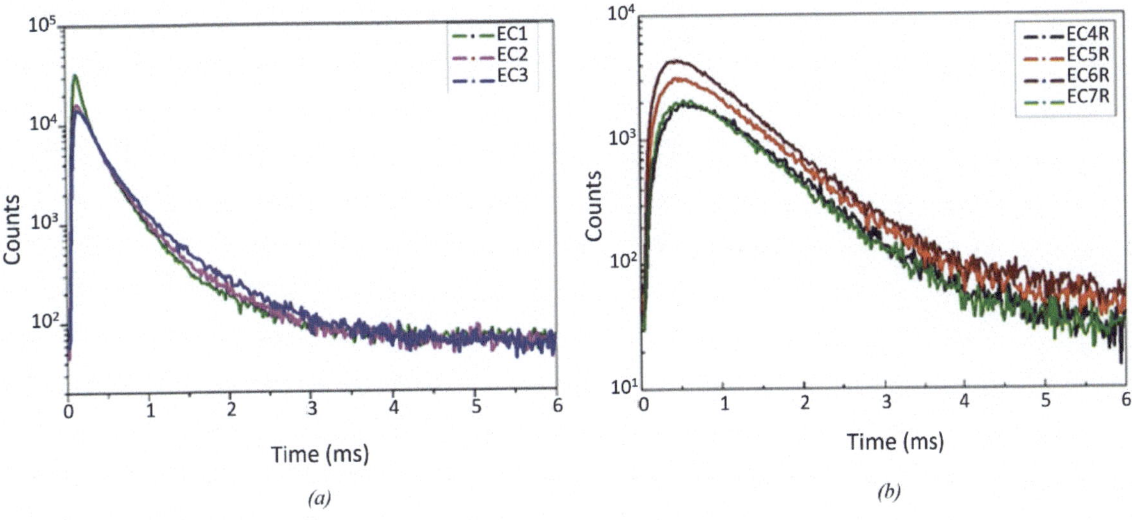

FIG. 2.172. Counts vs time for detectors located in core (a) and reflector (b).

Reactivity predictions are affected by spatial effects that need to be considered by introducing correction factors. More details about spatial effects in reactivity measurements are provided in Section 5.4.

---

² The symbol nv is commonly used to designate a flux of thermal neutrons equal to one thermal neutron per square centimetre per second.

TABLE 2.97. MEASURED REACTIVITY VALUES FOR AXIAL AND RADIAL EXPERIMENTAL CHANNELS USING THE AREA RATIO METHOD AND THE SOURCE JERK METHOD

Experimental channels	Spatial correction factor	Area ratio method		Source jerk method	
		ρ^{meas} (\$)	ρ^{corr} (\$)	ρ^{meas} (\$)	ρ^{corr} (\$)
EC1	0.62 ± 0.02	-28.03	-17.38 ± 0.56	-25.87	-16.04 ± 0.51
EC2	0.92 ± 0.03	-19.04	-17.52 ± 0.57	-17.39	-16.00 ± 0.52
EC3	0.99 ± 0.04	-16.61	-16.44 ± 0.66	-15.69	-15.53 ± 0.62
EC4R	1.28 ± 0.05	-12.10	-15.49 ± 0.60	-11.24	-14.39 ± 0.56
EC5R	1.00 ± 0.04	-16.25	-16.25 ± 0.65	-13.31	-13.31 ± 0.53
EC6R	1.11 ± 0.04	-14.99	-16.64 ± 0.59	-15.43	-17.13 ± 0.61
EC7R	1.38 ± 0.05	-12.59	-17.37 ± 0.62	-13.96	-19.26 ± 0.69

2.6.   REFERENCES TO CHAPTER 2

[2.1]   GAMDINI, A., SALVATORES, M., The Physics of Subcritical Multiplying Systems, J. Nucl. Sci. Technol. **39[6]** (2002) 673–686.

[2.2]   KOBAYASHI, K., NISHIHARA, K., Definition of Subcriticality Using the Importance Function for the Production of Fission Neutrons, Nucl. Sci. Eng. **136** (2000) 272.

[2.3]   SOULE, R., et al., Neutronic Studies in Support of Accelerator-Driven Systems: the MUSE Experiments in the MASURCA Facility, Nucl. Sci. Eng. **148** (2004) 124.

[2.4]   PYEON, C., et al., Static and Kinetic Experiments on Accelerator-Driven System with 14MeV Neutrons in Kyoto University Critical Assembly, J. Nucl. Sci. Technol. **45** (2008) 1171.

[2.5]   PYEON, C., et al., First Injection of Spallation Neutrons Generated by High-Energy Protons into the Kyoto University Critical Assembly, J. Nucl. Sci. Technol. **46** (2009) 1091.

[2.6]   PYEON, C., "Experimental benchmarks of neutronics on solid Pb-Bi in accelerator-driven system with 100 MeV protons at Kyoto University Critical Assembly, KURRI-TR-447, Research Reactor Institute, Kyoto University, Kyoto (2017).

[2.7]   PYEON, C., YAMANAKA, M., "Experimental benchmarks of neutron characteristics on uranium-lead zoned core in accelerator-driven system at Kyoto University Critical Assembly, KURNS-EKR-1, Institute for Integrated for Radiation and Nuclear Science, Kyoto University, Kyoto (2018).

[2.8]   CHABOD, S., et al., Neutron Flux Characterization of MEGAPIE Target, Nucl. Instrum. Methods A **562** (2006) 618.

[2.9]   MICHEL-SENDIS, F., et al., Neutronics of the MEGAPIE Liquid Pb-Bi Spallation Target: Inner Flux Measures and Simulations, Nucl. Technol. **168** (2008) 322.

[2.10]   TSUJIMOTO, K., et al., Neutronics Design for Lead-Bismuth Cooled Accelerator-Driven System for Transmutation of Minor Actinide, J. Nucl. Sci. Technol. **41** (2004) 21.

[2.11]   NISHIHARA, K., et al., Neutronics Design of Accelerator-Driven System for Power Flattening and Beam Current Reduction, J. Nucl. Sci. Technol. **45** (2008) 812.

[2.12]   SUGAWARA, T., et al., Analytical Validation of Uncertainty in Reactor Physics Parameters for Nuclear Transmutation Systems, J. Nucl. Sci. Technol. **47** (2010) 521.

[2.13]   ABDERRAHIM, H., D'HONDT, P., MYRRAHA, A European Experimental ADS for R&D Applications Status at Mid-2005 and Prospective towards Implementation, J. Nucl. Sci. Technol. **44** (2007) 491.

[2.14]   MANSANI, L., et al., The European Lead-Cooled EFIT Plant: An Industrial-Scale Accelerator-Driven System for Minor Actinide Transmutation, Nucl. Technol. **180** (2012) 241.

[2.15]   RIMPAULT, G., et al., The Issue of Accelerator Beam Trips for Efficient ADS Operation, Nucl. Technol. **184** (2013) 249.

[2.16]   PYEON, C., et al., Reaction Rate Analysis of Nuclear Spallation Reactions Generated by 150, 190 and 235 MeV Protons, J. Nucl. Sci. Technol. **47** (2010) 1090.

[2.17]   PYEON, C., et al., Preliminary Study on the Thorium-Loaded Accelerator-Driven System with 100 MeV Protons at the Kyoto University Critical Assembly, Ann. Nucl. Energy **38** (2011) 2298.

[2.18]   LIM, J., et al., Subcritical Multiplication Parameters of the Accelerator-Driven System with 100 MeV Protons at the Kyoto University Critical Assembly, Sci. Tech. Nucl. Inst. **2012** ID: 395878 (2012).

[2.19]   PYEON, C., et al., et al., Experimental Analyses of Spallation Neutrons Generated by 100 MeV Protons at the Kyoto University Critical Assembly, Nucl. Eng. Technol. **45** (2013) 81.

[2.20]   SAKON, A., et al., Measurement of Large Negative Reactivity of an Accelerator-Driven System in the Kyoto University Critical Assembly, J. Nucl. Sci. Technol. **51** (2014) 116.

[2.21]   TAKAHASHI, Y., et al., Conceptual Design of Multi-Target for the Accelerator-Driven System Experiments with 100 MeV Protons, Ann. Nucl. Energy **54** (2013) 162.

[2.22]   YAGI, T., et al., Application of Wavelength Shifting Fiber to Subcriticality Measurements, Appl. Radiat. Isot. **72** (2013) 11.

[2.23]   PYEON, C., et al., Mockup Experiments on the Thorium-Loaded Accelerator-Driven System in the Kyoto University Critical Assembly, Nucl. Sci. Eng. **177** (2014) 156.

[2.24]   J.B. Lagrange, et al., "Straight Scaling FFAG Beam Line, Nucl. Instrum. Methods A **691** (2013) 55.

[2.25]   YAMAKAWA, E., et al., "Serpentine Acceleration in Zero-Chromatic FFAG Accelerators, Nucl. Instrum. Methods A **716** (2013) 46.

[2.26]   SHAHBUNDER, H., et al., Experimental Analysis for Neutron Multiplication by using Reaction Rate Distribution in Accelerator-Driven System, Ann. Nucl. Energy **37** (2010) 592.

[2.27]   PYEON, C., et al., "Neutron Characteristics of Solid Targets in Accelerator-Driven System with 100 MeV protons at Kyoto University Critical Assembly, Nucl. Technol. **192** (2015) 181.

[2.28]   HENDRICKS, J., et al., MCNPX User's Manual, Version 2.5.0, LA-UR-05-2675, Los Alamos National Laboratory, Los Alamos (2005).

[2.29]   ZIEGLER, J., BIERSACK, J., ZIEGLER, M., The Stopping and Range of Ions in Matter, Pergamon Press, New York (1985).

[2.30]   TAKADA, H., et al., Validation of JENDL High-Energy File through Analyses of Spallation Experiments at Incident Proton Energies from 0.5 to 2.83 GeV, J. Nucl. Sci. Technol. **46** (2009) 589.

[2.31]   PYEON, C., et al., Experimental Benchmarks on Kinetic Parameters in Accelerator-Driven System with 100 MeV Protons at Kyoto University Critical Assembly, Ann. Nucl. Energy 105 (2017) 346.

[2.32]   PYEON, C., et al., "Validation of Pb Nuclear Data by Monte Carlo Analyses of Sample Reactivity Experiments at Kyoto University Critical Assembly, J. Nucl. Sci. Technol. **53** (2016) 602.

[2.33]   SJÖSTRAND, N., Measurement on a Subcritical Reactor using a Pulsed Neutron Source, Arkiv för Fysik **13** (1956) 233.

[2.34]   KITAMURA, Y., et al., General Formulae for the Feynman-α Method with the Bunching Technique, Ann. Nucl. Energy **27** (2000) 1199.

[2.35]   KITAMURA, Y., et al., Calculations of the Pulsed Feynman- and Rossi-alpha Formulae with Delayed Neutrons, Ann. Nucl. Energy **32** (2005) 671.

[2.36]   PERSSON, C., et al., Pulsed Neutron Source Measurements in the Subcritical ADS Experiment YALINA-Booster, Ann. Nucl. Energy **35** (2008) 2357.

[2.37]   UYTTENHOVE, W., et al., The Neutronic Design of a Critical Lead Reflected Zero-Power Reference Core for On-Line Subcriticality Measurements in Accelerator-Driven Systems, Ann. Nucl. Energy **38** (2011) 1519.

[2.38]   UYTTENHOVE, W., et al., Methodology for Modal Analysis at Pulsed Neutron Source Experiments in Accelerator-Driven Systems, Ann. Nucl. Energy **72** (2014) 286.

[2.39]   GOORLEY, J., et al., Initial MCNP6 Release Overview – MCNP6 Version 1.0, LA-UR-13-22934, Los Alamos National Laboratory, Los Alamos, NV (2013).

[2.40]   SHIBATA, K., et al., JENDL-4.0: A New Library for Nuclear Science and Engineering, J. Nucl. Sci. Technol. **48** (2011).

[2.41]   GOZANI, T., A Modified Procedure for the Evaluation of Pulsed Source Experiments in Subcritical Reactors, Nukleonik 4 (1962) 348.

[2.42]   PYEON, C., et al., Reaction Rate Analyses for an Accelerator-Driven System with 14 MeV Neutrons in Kyoto University Critical Assembly, J. Nucl. Sci. Technol. **46** (2009) 965.

[2.43] PYEON, C., et al., Accuracy of Reaction Rates in the Accelerator-Driven System with 14 MeV Neutrons at the Kyoto University Critical Assembly, Ann. Nucl. Energy **40** (2012) 229.

[2.44] DULLA, S., et al., Analysis of KUCA Measurements by the Reactivity Monitoring MAρTA Method, Ann. Nucl. Energy **101** (2017) 397.

[2.45] YAMANAKA, M., et al., Effective delayed neutron fraction by rossi-α method in accelerator-driven system experiments with 100 MeV Protons at Kyoto University critical assembly. J. Nucl. Sci. Technol. **54** (2017) 293.

[2.46] RUBBIA, C., A High Gain Energy Amplifier Operated with Fast Neutrons, Proc. Int. Conf. Accelerator-Driven Transmutation Technol. And Appl., Las Vegas (1994).

[2.47] IWAMOTO, H., et al., On-Line Subcriticality Measurement using A Pulsed Spallation Neutron Source, J. Nucl. Sci. Technol. **54** (2017) 432.

[2.48] KEEPIN, G., Physics of Nuclear Kinetics, AddisonWesley, Reading, Massachusetts (1965).

[2.49] PYEON, C.H., et al., "Reaction rate analyses of accelerator-driven system experiments with 100 MeV protons at Kyoto University Critical Assembly" J. Nucl. Sci. Technol. **55** [2] (2018) 190

[2.50] KOBAYASHI, K., et al., JENDL Dosimetry File 99 (JENDL/D-99), JAERI Report 1344 (2002).

[2.51] MIHALCZO, J., PARE, V., Theory of Correlation Measurement in Time and Frequency Domains with ^{252}Cf, Ann. Nucl. Energy **2** (1975) 97.

[2.52] SAKURI, T., et al., Experimental Cores of Benchmark Experiments on Effective Delayed Neutron Fraction β_{eff} at FCA, Prog. Nucl. Energy **35** (1999) 131.

[2.53] DOULIN, V., et al., The β_{eff} Measurement Results on FCA-XIX Cores, Prog. Nucl. Energy **35** (1999) 163.

[2.54] SPRIGGS, G., et al., Rossi-α Measurements in a Fast Critical Assembly, Prog. Nucl. Energy **35** (1999) 169.

[2.55] YAMANE, Y., et al., Effective delayed neutron fraction measurements in FCA-XIX cores by using modified Bennet method, Prog. Nucl. Energy **35** (1999) 183.

[2.56] SAKURAI, T., OKAJIMA, S., Measurement of Effective Delayed Neutron Fraction β_{eff} by ^{252}Cf Source Method for Benchmark Experiment of β_{eff} at FCA, Prog. Nucl. Energy **35** (1999) 195.

[2.57] SAKURAI, T., et al., Measurement of Effective Delayed Neutron Fraction β_{eff} by Covariance-to Mean Method for Benchmark Experiment of β_{eff} at FCA, Prog. Nucl. Energy **35** (1999) 203.

[2.58] SAKURAI, T., OKAJIMA, S., Analysis of Benchmark Experiment of Effective Delayed Neutron Fraction β_{eff} at FCA, Prog. Nucl. Energy **35** (1999) 209.

[2.59] SPRIGGS, G., A Measurement of the Effective Delayed Neutron Fraction of the Westinghouse Idaho Nuclear Company Slab Tank Assembly using Rossi-α Techniques, Nucl. Sci. Eng. **115** (1993) 76.

[2.60] SPRIGGS, G., Two Rossi-α Techniques for Measuring the Effective Delayed Neutron Fraction, Nucl. Sci. Eng. **113** (1993) 161.

[2.61] ORNDOFF, J., Prompt neutron periods of metal critical assemblies, Nucl. Sci. Eng. **2** (1957) 450.

[2.62] YAMANAKA, M., et al., Measurement of Effective Delayed Neutron Fraction with External Neutron Source at Kyoto University Critical Assembly, Proc. The Reactor Physics Asia Conference 2015 (RPHA15), Jeju, Korea (2015).

[2.63] DEGWEKER, S., RANA, Y., Reactor Noise in Accelerator Driven Systems – II, Ann. Nucl. Energy **34** (2007) 463.

[2.64] CHADWICK, M., et al., ENDF/V-II.0: Next Generation Evaluated Nuclear Data Library for Nuclear Science and Technology, Nucl. Data Sheet **107** (2006) 2931.

[2.65] SHIM, H., et al., McCARD: Monte Carlo code for Advanced Reactor Design and Analysis, Nucl. Eng. Technol. **44**(2) (2012) 161.

[2.66] SHIM, H., et al., Monte Carlo Alpha Iteration Algorithm for a Subcritical System Analysis, Sci. Technol, Nucl. Ins. 2015, Article ID 859242 (2015) 7.

[2.67] GABRIELLI, F., et al., Deterministic Analyses of the Phase-1 Kinetics Experiments in the KUKA Subcritical A-Core Configurations, Proc. of PHYSOR 2016, Sun Valley (2016).

[2.68] DORIATH, J., et al., ERANOS1: The Advanced European System of Codes for Reactor Physics Analysis, Proc. of Int. Conf. on Mathematical Methods and Supercomputing for Nuclear Application, Karlsruhe, Germany (1993).

[2.69] ALCOUFFE, R., et al., PARTISN: A Time-Dependent, Parallel Neutral Particle Transport Code System, LA-UR-08-07258, Los Alamos National Laboratory, Los Alamos, NV (2008).

[2.70] RINEISKI, A., DORIATH, J., Time dependent neutron transport with variational nodal method, Proc. Int. Conf. on Mathematical Methods and Supercomputing for Nuclear Application, Saratoga Springs, NY, 1997 2 (1997) 1661.

[2.71] GABRIELLI, F., et al., Reactor Transient Analyses with KIN3D/PARTISN, Proc. of Int. Conf. M&C 2013, Sun Valley, ID, USA, (2013).

[2.72] CARRICO, C., et al., Three-dimensional Variational Nodal Transport Methods for Cartesian, Triangular and Hexagonal Criticality Calculations, Nucl. Sci. Eng. **111** (1992) 168.

[2.73] DILBER, I., LEWIS, E., Variational Nodal Methods for Neutron Transport, Nucl. Sci. Eng. **91** (1985) 132.

[2.74] PELOWITZ, D., et al., MCNPX User's Manual, Version 2.7.0, LA-CP-11-00438, Los Alamos National Laboratory, Los Alamos, NV (2011).

[2.75] RIMPAULT, G., Algorithmic Features of the ECCO Cell Code for Treating Heterogeneous Reactor Subassemblies, International conference on mathematics and computations, reactor physics, and environmental analyses; Portland, OR (1995).

[2.76] TALAMO, A., et al., Advances in the computation of the Sjöstrand, Rossi, and Feynman distributions, Prog. Nucl. Energy **101/C** (2017) 299–311.

[2.77] TALAMO, A., et al., Calculation of the cross and auto power spectral densities for low neutron counting from pulse mode detectors, Ann. Nucl. Energy **131** (2019) 138–147.

[2.78] TALAMO, A., et al., Calculation of the prompt neutron decay constant for the KUCA facility driven by a stationary or pulsed external neutron source, J. Nucl. Sci. Technol. **57** (2020) 145–156.

[2.79] TALAMO, A., et al., Dead-Time and spatial corrections for the KUCA subcritical assembly experiments, Proc. of PHYSOR 2018, Cancun, Mexico (2018).

[2.80] PELOWITZ, D., et al., MCNP6 User's Manual, Code Version 6.1.1beta, LA-CP-14-00745, Los Alamos National Laboratory, Los Alamos, NM (2014).

[2.81] OECD NUCLEAR ENERGY AGENCY, The JEFF-3.1 Nuclear Data Library, JEFF Report 21, NEA No. 6190, Paris (2006).

[2.82] KONING, A., et al., Status of the JEFF nuclear data library, J. Korean Phys. Soc. **59/2** (2011) 1057-1062.

[2.83] KAHLER, A., et al., The NJOY Nuclear Data Processing System, Version 2012.

[2.84] MACFARLANE, R., MUIR, D., The NJOY Nuclear Data Processing System Version 91, LA-12740-M, Los Alamos National Laboratory, Los Alamos, NM (1994).

[2.85] RUGGIERI, J., et al., ERANOS-2.1: The International code system for GEN-IV fast reactor analysis, Proc. Int. Conf. ICAPP'06, Reno, NV (2006).

[2.86]   RAHLFS, S., Validation Physique du Nouveau Code de Cellule Européen ECCO pour le Calcul des Coefficients de Réactivité des Réacteurs REP et RNR. Ph.D. Dissertation. Université de Provence, Marseille, France (1995).

[2.87]   ASKEW, J., et al., A General Description of Lattice Code WIMSD, Journal of the British Nuclear Energy Society **5** (4) (1966) 564.

[2.88]   LATRHOP, K., BRINKLEY, F., Theory and use of the general geometry TWOTRAN Program, LA-4432, Los Alamos Scientific Laboratory, Los Alamos, NM (1970).

[2.89]   Monte Carlo Lab., Nuclear Engineering Department, Seoul National University Contribution to IAEA CRP, unpublished data

[2.90]   SINGH, K., et al., Iterative Schemes for obtaining dominant alpha-modes of the neutron diffusion equation, Annals of Nuclear Energy **36** (2009) 1086–1092.

[2.91]   SJÖSTRAND, N., Measurement on a Subcritical Reactor using a Pulse Neutron Source, Arkiv för Fysik **11** (1956) 13.

[2.92]   BELL, G., GLASSTONE, S., Nuclear Reactor Theory, Van Nostrand Reinhold Company, New York (1970).

[2.93]   GABRIELLI, F., et al., MUSE-4 Analyses by means of spatial kinetics capabilities in the ERANOS code, Proc. Eighth Intl Topical Meeting on Nuclear Applications and Utilization of Accelerators (AccApp'07), ANS, ISBN: 0-89448-054-5, Pocatello, ID, USA, (2007) CD-ROM.

[2.94]   ALIBERTI, G., GOHAR, Y., YALINA Analytical Benchmark Analyses Using the Deterministic ERANOS Code System, ANL-09/23, Argonne National Laboratory, Lemont, IL (2009).

[2.95]   DULLA, S., et al., Interpretation of Experimental Measurements on the SC-1 Configuration of the VENUS-F Core, Proc. of PHYSOR 2014, Kyoto, Japan, (2014).

[2.96]   PYEON, C., et al., Neutronic Characteristics of Solid Targets in Accelerator-Driven System at KYOTO University Critical Assembly, Proc. of PHYSOR 2014, Kyoto, Japan, (2014).

[2.97]   TALAMO, A., GOHAR, Y., Neutron Detector Signal Processing to Calculate the Effective Neutron Multiplication Factor of Subcritical Assemblies, ANL-16/14, Argonne National Laboratory, Lemont, IL (2016).

[2.98]   DUDERSTADT, J., HAMILTON, L., Nuclear Reactor Analysis. John Wiley & Sons, ISBN 978-0471223634 (1976) 216–218.

[2.99]   TALAMO, A., et al., MCNPX, MONK, and ERANOS analyses of the YALINA Booster subcritical assembly. Nucl. Eng. Des. **241/5** (2011) 1606–1615.

[2.100] TALAMO, A., et al., Monte Carlo and deterministic neutronics analyses of YALINA thermal facility and comparison with experimental results. Nucl. Technol. **184/2** (2013) 131–147.

[2.101] PÁZSIT, I., PÁL, L., Neutron fluctuations: A treatise on the physics of branching processes, Elsevier Science. ISBN-10: 0080450644, Oxford (2007).

[2.102] O'KELLY, D., Operation and Reactivity Measurements of an Accelerator Driven Subcritical TRIGA Reactor, Ph.D. Dissertation, University of Texas at Austin (2008).

[2.103] TALAMO, A., et al., Correction factor for the experimental prompt neutron decay constant, Ann. Nucl. Energy **62** (2013) 421–428.

[2.104] KURAMOTO, R., et al., Rossi-α experiment in the IPEN/MB-01 research reactor, Braz. J. Phys. **35** (2005).

[2.105] TALAMO, A., et al., Monte Carlo Modelling and Analyses of YALINA-booster Subcritical Assembly Part II: Pulsed Neutron Source, ANL-08/33, Argonne National Laboratory, Lemont, IL (2008).

[2.106] TALAMO, A., et al., Pulse superimposition calculational methodology for estimating the subcriticality level of nuclear fuel assemblies, Nucl. Instrum. Methods Phys. Res. A **606/3** (2009) 661–668.

[2.107] TALAMO, A., GOHAR, Y., Numerical Application of the Sjöstrand Method without the Pulses Superimposition Methodology, ANS Winter Meeting, Washington, DC, USA (2015).

[2.108] TALAMO, A., et al., Monte Carlo and deterministic calculation of the Bell and Glasstone spatial correction factor. Comput. Phys. Commun. **183/9** (2012) 1904–1910.

[2.109] PACILIO, N., Reactor-Noise Analysis in the Time Domain. AEC Critical Review Series (1969).

[2.110] UHRIG, R., Random Noise Techniques in Nuclear Reactor Systems ISBN-10: 047106694X, J. Wiley & Sons (1970).

[2.111] THIE, J., Power Reactor Noise, ISBN-10: 0894480251, American Nuclear Society (1983).

[2.112] TALAMO, A., et al., Calculation of the Cross Power Spectral Density for Pulse Mode Detectors, American Nuclear Society Summer Meeting, Philadelphia, PA, USA (2018).

[2.113] SAKON, A., et al., Power spectral analysis for a thermal subcritical reactor system driven by a pulsed 14 MeV neutron source, J. Nucl. Sci. Technol. **50 (5)** (2013) 481–492.

[2.114] WANG, C. et al., Improved neutron-gamma discrimination for a 3He neutron detector using subspace learning methods, Nucl. Instrum. Methods Phys. Res. A **853** (2017) 27–35.

[2.115] KELLEY, R., et al., Pulse shape discrimination in helium-4 scintillation detectors, Nucl. Instrum. Methods Phys. Res. A **830** (2016) 44–52.

[2.116] TALAMO, A., GOHAR, Y., Application of the backward extrapolation method to pulsed neutron sources, Nucl. Instrum. Methods Phys. Res. A **877** (2018) 16–23.

[2.117] ELTER, Zs. et al., Performance of higher order campbell methods, part II: calibration and experimental application, Nucl. Instrum. Methods Phys. Res. A **835** (1) (2016) 86–93.

[2.118] YAMAMOTO, T., Frequency domain Monte Carlo simulation method for cross power spectral density driven by periodically pulsed spallation neutron source using complex-valued weight Monte Carlo, Ann. Nucl. Energy **63** (2014) 711–720.

[2.119] AGUILLAR, A., et al., Design specification for the European Spallation Source neutron generating target element, Nucl. Instrum. Methods Phys. Res. A **856** (2017) 99–108.

[2.120] VALENTINE, T., MIHALCZO, J., MCNP-DSP: a neutron and gamma ray Monte Carlo calculation of source-driven noise-measured parameters, Ann. Nucl. Energy **23** (16) (1996) 1271–1287.

[2.121] KITAMURA, Y., et al., Reactor noise experiments by using acquisition system for time series data of pulse train, J. Nucl. Sci. Technol. **36** (8) (1999) 653–660.

[2.122] PYEON, C., et al., Perspectives of research and development of accelerator-driven system in Kyoto University Research Reactor Institute, Prog Nucl. Energy. **82** (2015) 22–27.

[2.123] PYEON, C., Neutronics on solid Pb-Bi in accelerator driven system with 100 MeV Protons at Kyoto University Critical Assembly, Research Reactor Institute, Kyoto University (2015).

[2.124] YAMAMOTO, T., Applicability of non-analog Monte Carlo technique to reactor noise simulation, Ann. Nucl. Energy **38** (2–3) (2011) 647–655.

[2.125] HUMBERT, P., Simulation and Analysis of List Mode Measurements on SILENE Reactor, J. Comput. Theor. Transport. **47** (4–6) (2018)350–363.

[2.126] NAGAYA, Y., et al., MVP/GMVP Version 3: general purpose monte carlo codes for neutron and photon transport calculations based on continuous energy and multigroup methods, JAEA-Data/Code 2016-18. Japan Atomic Energy Agency, Japan (2017).

[2.127] HOH, S., Development of methods for the determination of reactivity from flux measurements in nuclear reactors [Ph.D. Dissertation]. Politecnico di Torino, Italy (2017).

[2.128] ARKANI, M., RAISALI, G., Measurement of Dead Time by Time Interval Distribution Method, Nuclear Instruments and Methods in Physics Research Section A, **774** (2015) 151–158.

[2.129] GILAD, E., et al., Dead Time Corrections Using the Backward Extrapolation Method, Nuclear Instruments and Methods in Physics Research A, **864** (2017) 53–60.

[2.130] TALAMO, A., et al., Impact of the Neutron Detector Choice on Bell and Glasstone Spatial Correction Factor for Subcriticality Measurement, Nuclear Instruments and Methods in Physics Research Section A, **668** (2012) 71–82.

[2.131] SAVITZKY, A., GOLAY, J., Smoothing and differentiation of data by simplified least squares procedures, Analytical Chemistry **36** (1964) 1627–1639.

[2.132] CHARTRAND, R., Numerical differentiation of noisy, nonsmooth data, ISRN Applied Mathematics 2011, doi:10.5402/2011/164564 (2011).

[2.133] DULLA, S., et al, A method for on-line reactivity monitoring in nuclear reactors, Annals of Nuclear Energy **65** (2014) 433–440.

[2.134] SIMMONS, B., KING, J., A Pulsed Neutron Technique for Reactivity Determination, Nucl. Sci. Eng., 3, **595** (1958).

[2.135] MCDONNEL, F., HARRIS, M., Pulsed-Source Experiments in a Reflected Coupled-Core Reactor – I. Reactivity Measurements, J. Nucl. Energy **26** (1972) 113.

[2.136] STUMBUR, E., et al., Combined Pulsed Method of Measuring High Reactivities for Reactors with Reflectors, Atomic Energy **36** (1974) 224.

[2.137] SUZAKI, T., Subcriticality Determination of Low-Enriched UO$_2$, Lattice in Water by Exponential Experiment, J. Nucl. Sci. Technol., **28** (1991) 1067.

[2.138] PERSSON, C., et al., Analysis of Reactivity Determination Methods in the Subcritical Experiment Yalina, Nuclear Instruments and Methods in Physics Research Section A: Accelerators, Spectrometers, Detectors and Associated Equipment, **554** (2005) 374-383.

[2.139] JANG, S., SHIM, H., Determination of an Effective Detector Position for Pulsed-Neutron-Source Alpha Measurement by Time-Dependent Monte Carlo Neutron Transport Simulations, Science and Technology of Nuclear Installations, 2018, Article ID 2350458, https://doi.org/10.1155/2018/2350458 (2018).

[2.140] 'INEST-, China, Contribution to IAEA CRP, unpublished data.

[2.141] HENDRICKS, J., et al., MCNPX 2.6.0 Extensions, LA-UR-08-2216, Los Alamos National Laboratory, Los Alamos, NM (2008).

[2.142] WU, Y., et al., CAD-Based Monte Carlo Program for Integrated Simulation of Nuclear System SuperMC, Annals of Nuclear Energy (2014).

[2.143] WU, Y. FDS Team, CAD-Based Interface Programs for Fusion Neutron Transport Simulation. Fusion Engineering and Design **84** (7-11) (2009) 1987–1992.

[2.144] SIMMONS, B., KING, J., A Pulsed Neutron Technique for Reactivity Determination, Nucl. Sci. Eng., 3, **595** (1958).

[2.145] LI, Y., et al., Benchmarking of MCAM4.0 with the ITER 3D Model. Fusion Engineering and Design **82** (2007) 2861–2866.

[2.146] CULLEN, D., et al., Static and Dynamic criticality: are they different? UCRL-TR-201506, Lawrence Livermore National Laboratory, Livermore, CA (2003).

[2.147] TRUCHET, G., et al., Application of the modified neutron source multiplication method to the prototype FBR Monju. Annals of Nuclear Energy **51** (0) (2013) 94–106.

[2.148] TSUJI, M., et al, Subcriticality measurement by neutron source multiplication method with a fundamental mode extraction, Journal of nuclear science and technology **40** (3) (2003) 158–169.

[2.149] NAING, W., et al., The effect of neutron source distribution on subcriticality measurement of pressurized water reactors using the modified neutron source multiplication method [J], J. Nucl. Sci. Technol. **40** (11) (2003) 951–958.

[2.150] NAING, W., et al., Subcriticality measurement of pressurized water reactors by the modified neutron source multiplication method. Journal of Nuclear Science and Technology **40** (12) (2003) 983–988.

[2.151] YAMANAKA, M., et al., Monte Carlo Approach of Effective Delayed Neutron Fraction by k-Ratio Method with External Neutron Source, Nucl. Sci. Eng., **184** (2016) 551.

[2.152] PYEON, C., et al., Preliminary Experiments for Accelerator Driven Subcritical Reactor with Pulsed Neutron Generator in Kyoto University Critical Assembly, J. Nucl. Sci. Technol. **44** (2007) 1368.

[2.153] CHIBA, G., et al., On Effective Delayed Neutron Fraction Calculations with Iterated Fission Probability, J. Nucl. Sci. Technol., **48** (2011) 1163.

[2.154] SIKORIN, S.N., Testing of the prototype of the local system of the diagnostic complex for reactivity measurement in the Giacint critical facility, Report SSI JIPNR - Sosny 1243, Minsk, Belarus (2012).

[2.155] TALAMO, A., et al., Criticality Experiments and Analyses of Uranium Zirconium Carbon Nitride LEU Fuel, Proc. of PHYSOR 2014, Kyoto, Japan (2014).

[2.156] TALAMO, A., et al., Serpent Stereolithography Modeling of the GIACINT Critical Assembly Experimental Facility, Proc. of PHYSOR 2016, Sun Valley, ID, USA (2016).

[2.157] TALAMO, A., et al., Serpent Validation and Optimization with Mesh Adaptive Search on Stereolithography Geometry Models, Annals of Nuclear Energy **115** (2018) 619–632.

[2.158] LEPPÄNEN, J., et al., The Serpent Monte Carlo code: Status, Development and Applications in 2013, Annals of Nuclear Energy **82** (2015) 142–150.

[2.159] VALTAVIRTA, V., HESSAN, M., LEPPÄNEN, J., Delayed Neutron Emission Model for Time Dependent Simulations with the Serpent 2 Monte Carlo Code – First Results, Proc. of PHYSOR 2016, Sun Valley, ID, USA (2016).

[2.160] TALAMO, A., et al., The Precursor-Based Delayed Neutrons Emission Model Applied to a Water Level Transient, ANS Winter Meeting, Washington, DC, USA (2017).

[2.161] PELOWITZ, D.B., et al., MCNP6 User's Manual, Code Version 6.1.1beta, Los Alamos National Laboratory, LA-CP-14-00745 (2014).

[2.162] CAO, Y., GOHAR, Y., Monte Carlo Simulations of Periodic Pulsed Reactor with Moving Geometry Parts, Annals of Nuclear Energy **85** (2015).

[2.163] BIELEWICZ, M., et al., Measurements of fast neutron spectrum in QUINTA assembly irradiated with 2, 4 and 8 GeV deuterons, Proceedings of Science, XXII International Baldin Seminar on High Energy Physics Problems, JINR, Dubna (2014).

[2.164] SZUTA, M., et al., Lead shielding impact on fast neutron spectrum (>10 MeV) in QUINTA uranium, Proceedings of Science, XXII International Baldin Seminar on High Energy Physics Problems, JINR, Dubna (2014).

[2.165] BIELEWICZ, M., et al., Fast neutron spectrum measurement in the QUINTA assembly of E+T RAW collaboration, XXIV International Seminar on Interaction of Neutrons with Nuclei, Dubna (2016).

[2.166] ADAM, J., et al., Study of secondary neutron interactions with ^{232}Th, ^{129}I, and ^{127}I nuclei with the uranium assembly "QUINTA" at 2, 4, and 8 GeV deuteron beams of the JINR Nuclotron accelerator, FERMILAB-PUB-16-215-APC, Fermi National Accelerator Laboratory, Batavia, IL (2016).

[2.167] FRANA, J., Program DEIMOS32 for Gamma-Ray Spectra Evaluation, J. Radioanal. Nucl. Chem., **257** (2003) 583–587.

[2.168] CHADWICK, M., et al., ENDF/B-VII.1 Nuclear Data for Science and Technology: Cross-Sections, Covariances, Fission Product Yields and Decay Data, Nuclear Data Sheets, **112** (2011) 2887–2996.

[2.169] GOLUTVIN, I., et al., Radiation hardness of silicon detectors for collider experiments, Joint Institute for Nuclear Research (JINR) Preprint No. E14-95-97, Dubna, Russian Federation (1995).

[2.170] GINNEKEN, A., Non Ionizing Energy Deposition in Silicon for Radiation Damage Studies, FN-522, Fermi National Accelerator Laboratory, Batavia, IL (1989).

[2.171] BIELEWICZ, M., et al., Measurements of High Energy Neutron Spectrum (> 10 MeV) by Using Yttrium Foils in a U/Pb Assembly, Nuclear Data Sheets, **119** (2014) 296–298.

[2.172] KONING, A., et al., TALYS: Comprehensive Nuclear Reaction Modeling, in: International Conference on Nuclear Data for Science and Technology, Santa Fe, NM, USA, AIP Conference Proceedings **769** (2004) 1154–1159.

[2.173] KILIM, S., et al., Np-237 incineration study in various beams in ADS setup QUINTA, Nukleonika, **63** (2018) 17–22.

[2.174] SINHA, A., et al, BRAHMMA: A compact experimental accelerator driven subcritical facility using D-T/D-D neutron source, Annals of Nuclear Energy **75** (2015) 590–594.

[2.175] SINHA, A., et al, Experimental subcritical facility driven by D-D/D-T neutron generator at BARC, India, Nuclear Instruments and Methods in Physics Research B **350** (2015) 66–70.

[2.176] SHUKLA, M., et al., Neutron spatial flux profile measurement in compact subcritical system using miniature neutron detectors, Nuclear Instruments and Methods in Physics Research A **772** (2015) 118–123.

[2.177] BAJPAI, S., et al., Evaluation of Spatial Correction Factors for BRAHMMA Subcritical Assembly, Nuclear Science and Engineering **181** (2015) 361–367.

[2.178] ROY, T. et al., Pulsed Neutron Source Measurements in BRAHMMA Accelerator Driven Subcritical System, Nuclear Science & Engineering **184** (2016) 584–590.

[2.179] KUMAR, R., et al., Measurement of Reactivity in a Source Driven Deep Sub-critical System using Neutron Noise Methods, Annals of Nuclear Energy **103** (2017) 315–325.

Chapter 3

CROSS-SECTION MEASUREMENT AND INSTRUMENTATION DEVELOPMENT

3.1. DIAMOND DETECTOR

The operation of a solid-state detector in an environment with a high neutron flux requires significant research and development effort. In particular, for the use within the core of a fission reactor, the detector has to be very compact, resistant to large doses of fast neutrons, prompt (for direct-reading detectors) and almost insensitive to the strong γ background. Among solid state sensors, the best choice for such applications is the diamond detector. With respect to standard silicon detectors, the diamond is more radiation resistant [3.1], it has a lower sensitivity to γ rays ($Z_C/Z_{Si} \sim 0.43$) and it can be operated at high temperatures [3.2], [3.3]. With respect to ^{3}He and CH_4 proportional counters, usually employed at reactor facilities, the solid-state detector can cover a broader energy range while maintaining its compact size and practicable rates. Within the programme of activities towards a better experimental understanding of ADS, the Istituto Nazionale di Fisica Nucleare (INFN) in Genova (Italy) has developed an innovative compact neutron spectrometer based on diamonds. The device can be used to improve the characterization of ADS by means of ordering the measured neutron flux in few energy groups [3.4].

The spectrometer is made of two artificial diamonds enclosing a ^{6}Li neutron converter. By detecting both nuclei emerging from the reaction n + ^{6}Li $\rightarrow \alpha$ + T (where T is a tritium nucleus), the neutron energy is measured for each neutron conversion event through the energy conservation relation. Fast amplifiers are used to preserve the characteristics of the diamond pulses that are typically few nanoseconds in duration.

A first measurement campaign was performed at the Frascati Neutron Generator at 2.5 MeV neutron energy via the deuterium–deuterium (D–D) fusion reaction, showing good agreement in neutron energy reconstruction and efficiency estimate, but with relatively large uncertainties of the order of 20%. The measurement also showed that a relatively small contamination of D–T 14 MeV neutrons in the accelerator can give rise to a significant background under the measured neutron spectra.

The spectrometer was also tested at the thermal column of the TRIGA (Training, Research, Isotopes, General Atomics) reactor at the University of Pavia (Italy) to measure its response to thermal neutrons, thereby calibrating the absolute amount of ^{6}Li in the deposited film. By changing the reactor power by two orders of magnitude, the detector linearity in terms of counting rate was good. It was determined that the energy resolution was dominated by the electronics noise of the amplifier.

An early prototype of the spectrometer was also tested at the fast reactor TAPIRO (located at the ENEA Casaccia Research Centre near Rome, Italy) to measure the neutron spectrum in two different irradiation points: one at 5 cm from the core ('fast' point) and at the edge of the reflector, at 40 cm from the core ('slow' point) [3.5]. Due to rate limitations of the early prototype, that were overcome in more recent versions of the spectrometer, the reactor power was limited to 10–50 W compared to the nominal 5 kW. The neutron spectra were extracted in the two irradiation positions up to 5 MeV with 0.4 MeV bins. The spectra extracted from the measurement appeared to be significantly softer compared to MCNP simulations [3.6], [3.7] of the TAPIRO reactor. A possible explanation of the disagreement was related to the very low reactor power, 100 to 500 times lower than the nominal value. The latest spectrometer prototypes should be able to measure the TAPIRO spectrum at full power [3.8].

The final spectrometer prototype was thoroughly tested at quasi-monochromatic neutron source at the Physikalisch-Technische Bundesanstalt in Braunschweig (Germany). These tests have demonstrated both the incident neutron energy reconstruction with 20 keV uncertainty as well as the agreement (with 5% uncertainties) of the measured and expected detection efficiency as a function of the neutron energy in the range from 0.3 to 2 MeV [3.9]. In these last measurements, it was observed that the highly ionizing α particles appear to lose significantly more energy than expected in the non-sensitive volume of the spectrometer. The non-sensitive volume is essentially attributed to the metallic contacts, and it is assumed to be known from the metal deposition procedure. Such higher, non-measured energy loss limits the overall neutron energy resolution by producing a low energy tail in the α particles energy deposition peak. At the moment, the cause of this phenomenon is unknown, one hypothesis being that it may be due to a dead carbon layer formed at the time of the metal deposition. The Institute of Nuclear Physics (INP) in Tashkent (Uzbekistan) plans to investigate this aspect in more depth in the future and to find a practical solution to improve the energy measurement.

Therefore, apart from the above-mentioned issue with the energy deposition tail, the detector is ready to be used at any research facility, in particular at low-power ADS, with the scope to determine their spectrum in coarse energy groups.

3.2. SILICON DETECTOR

This section presents the activities carried out on three main topics related to neutron detection and flux measurements, namely the measurement of neutron total cross-sections, the measurement of the neutron spectrum by the methods of the time of flight and of the recoil nuclei, and the validation of the experimental techniques by code simulations. The measurement of cross-section data is a relevant issue for the design of nuclear systems. In particular, reliable data at high neutron energy and for heavy nuclei are needed for the assessment of the transmutation properties of ADS. Furthermore, neutron detectors are the basic instruments for the monitoring of subcritical systems.

The versions of the proton recoil method with the silicon (Si) semiconductor detectors and time-of-flight method are considered for the measurement of the neutron yield arising from the interaction of a 14 MeV neutron flux with samples of natural uranium and thorium mixtures. The first one provides detection of the protons recoiled at near-forward angles using the Si semiconductor detectors, whereas the second one provides for the 'tagged' neutrons technique with plastic scintillators used as fast neutron detectors.

3.2.1. Measurement of the neutron total cross-section

The measurement used is based on the transmission method and it has been applied to experimental data obtained using the neutron generator NG-150 at the INP. With this typical transmission technique, the neutron count is measured without (f_0) and with (f) the target. The microscopic cross-section is then determined by Eq. (3.1):

$$\sigma^{tot} = \frac{\log\left(f_0/f\right)}{\left(N_{targ}d_{targ}\right)} \tag{3.1}$$

where N_{targ} is the nuclear density of the material and d_{targ} is the thickness of the target. It is obviously assumed that the target is optically thin, so that the probability of the neutron interaction with target nuclei more than once is negligible. The neutron flux is measured by silicon detectors [3.10]. The schematic of the experimental apparatus is shown in Fig. 3.1. The reaction ^{28}Si(n, α)^{25}Mg inside the detector material is used for the detection of fast neutrons. The extreme right peak area marked in Fig. 3.1 is proportional to the neutron flux crossing the detector. The overall neutron peak is integrated by a separate monitor counter. As an example, Fig. 3.2 shows the detector counts with and without a lead target.

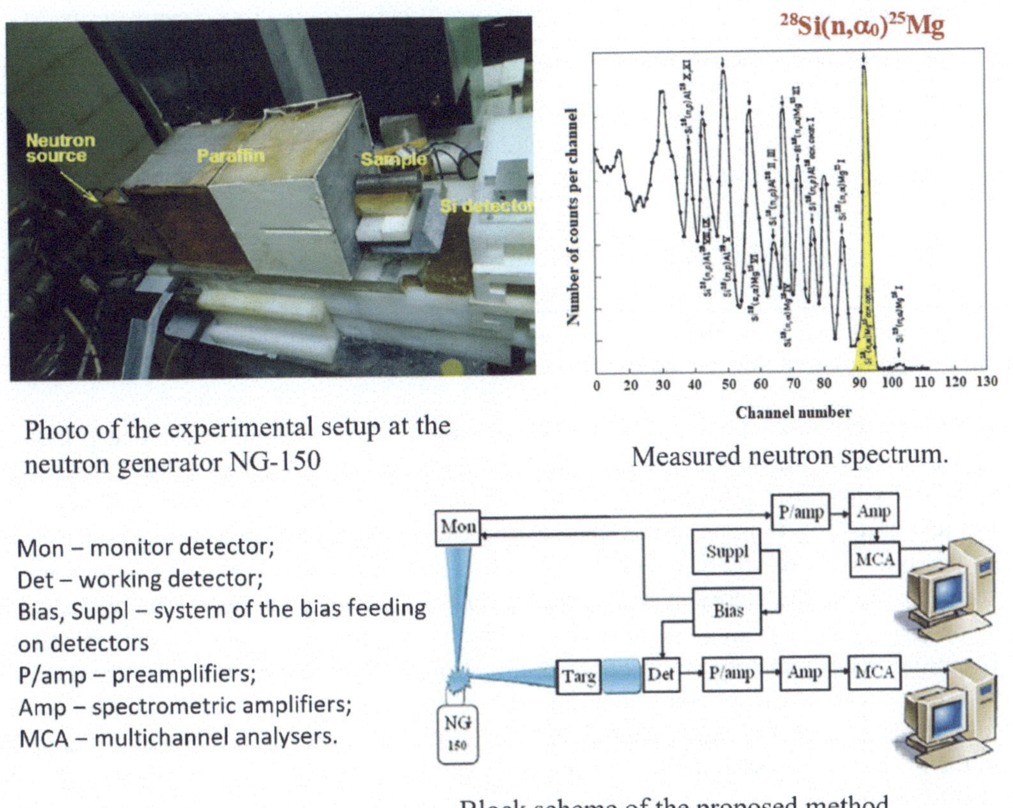

Photo of the experimental setup at the neutron generator NG-150

Measured neutron spectrum.

Mon – monitor detector;
Det – working detector;
Bias, Suppl – system of the bias feeding on detectors
P/amp – preamplifiers;
Amp – spectrometric amplifiers;
MCA – multichannel analysers.

Block scheme of the proposed method

FIG. 3.1. Illustration of the method for measurement of the total neutron cross-section.

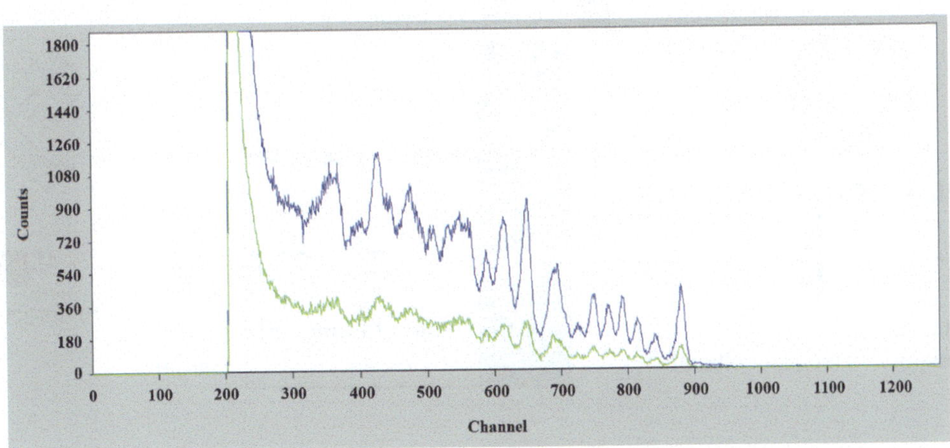

FIG. 3.2. Counts in the detector with (green) and without (blue) a lead target for a 14.2 MeV source.

Table 3.1 presents a comparison of the measured cross-section for various materials with data available in the literature.

TABLE 3.1. COMPARISON OF MEASURED CROSS-SECTIONS WITH AVAILABLE LITERATURE DATA

Nuclide	Thickness (mm)	Density (g/cm^3)	σ_{total} (INP data) (barn)	σ_{total} (literature data) (barn)	Ref.
^{12}C	148	2.25	1.43 ± 0.10	1.30 ± 0.06	[3.11]
^{19}F	148	1.7	2.08 ± 0.15	1.74 ± 0.04	[3.12]
^{32}S	112	1.91	1.77 ± 0.13	1.95 ± 0.03	[3.12]
^{128}Te	37	5.86	4.42 ± 0.26	4.76 ± 0.03	[3.13]
^{115}In	20	7.15	4.50 ± 0.17	4.54 ± 0.02	[3.13]
^{208}Pb	50	11.34	5.38 ± 0.11	5.45 ± 0.02	[3.12]

3.2.2. Neutron energy spectra and yields measurements

The arrangement of the proposed time of flight (TOF) setup is shown in Fig. 3.3. The principle of the registration and measurement of the energy spectrum of the secondary neutrons is as follows. A flux of monochromatic neutrons from the generator NG-150, collimated by a hole in the combined neutron/gamma ray shield (boron and paraffin), falls on the testing sample. An incoming neutron interacting with a sample nucleus can produce one or more secondary neutrons. In this case, the secondary neutron, which escaped at the angle θ with respect to the trajectory of the incoming (primary) neutron, is registered by the scintillation detector. Simultaneously with the primary neutron (formed by the reaction $d + T \rightarrow n + \alpha$ in the neutron target), the α particle is emitted approximately in the opposite direction. Its registration by a strip semiconductor Si detector results in tagging the corresponding neutron. The instants of the occurrence of detection signals in scintillation and α detectors are 'start' and 'stop' signals for TOF. Since the times of flight of the α particle to the detector and the primary neutron to the sample are constant, the remaining time of flight of the secondary neutron to the scintillation detector allows one to determine its speed, i.e., the energy. The coordinates of the beam spot at the neutron target and the strip where the α particle is detected define a trajectory of the primary neutron. Therefore, the location of the scintillation detector gives the scattering angle of the secondary neutron emission.

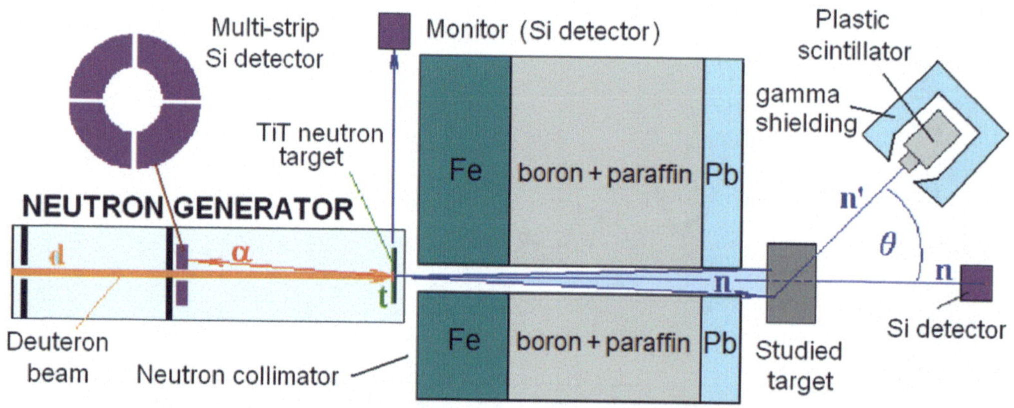

FIG. 3.3. Layout of the TOF arrangement in the hall of NG-150.

The detector of secondary neutrons is an organic scintillator (50 mm × 50 mm) with a photoelectron multiplier of the FEU-30 type. The detector, surrounded by a 60 mm thick lead shield, is mounted on a special platform moving around the sample at a distance of ~1.5 m.

The schematic diagram of the electronics for the developed version of the TOF method is shown in Fig. 3.4. At present, the main electronics modules (in the NIM and CAMAC standard) are assembled and the spectrometric channels and the n-gamma separation mode are being debugged under conditions of large background loads. Preliminary estimates have shown that the time resolution in the implemented TOF variant is ~1.5 ns.

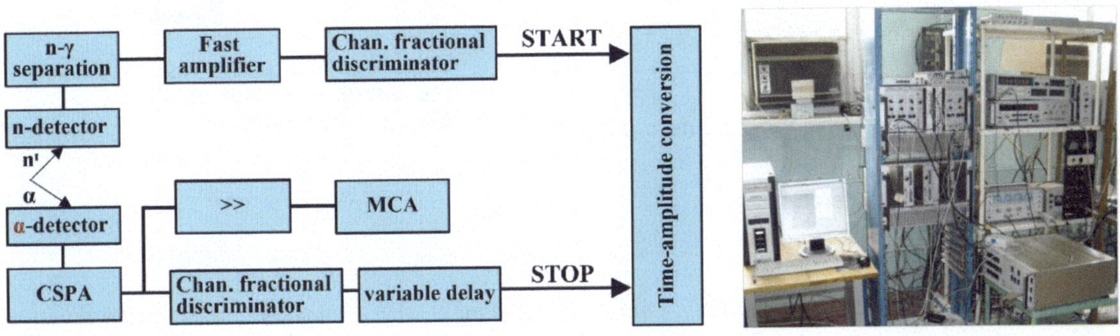

FIG. 3.4. Block diagram of the TOF spectrometer (left) and consoles with electronics modules (right).

The possibility to use semiconductor detector for recoil detection and neutron energy measurement has been analysed. The method has the following advantages:

- High energy resolution;
- Low sensitivity to the γ background;
- Compactness;

and disadvantages:

- Rather small efficiency;
- Limited energy range of measurements;
- Presence of the background reactions in the detector material and structure.

In Fig. 3.5, the scheme of the setup arrangement is shown for neutron energy measuring. It includes one or two totally depleted semiconductor Si detectors (ΔE-detectors) and the detector of full energy losses (E-detector). Ahead of the first detector, the hydrogen-containing (e.g., polyethylene or deutero-polyethylene) thin film is placed. The whole system is surrounded by a layer of carbon (graphite) to prevent entry into the telescope of the charged particles formed by interaction of neutrons with external constructive elements. The system is wholly placed in vacuum to eliminate the energy losses of recoils in air. The charged particle can be registered by the telescope if it produces the signals in all detectors simultaneously. Therefore, a case of the first ΔE-detector is also carefully shielded by ring diaphragms made of graphite to prevent passing of arising charged particles through all ΔE- and E-detectors. Graphite is used because the reaction thresholds of the (n, p) and (n, d) reactions on carbon are quite high.

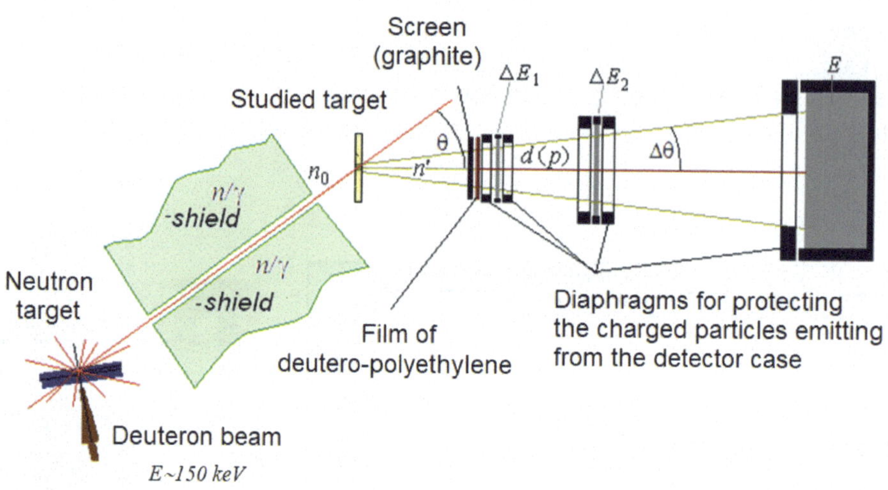

FIG. 3.5. Setup arrangement for fast neutron energy measurement with a telescope of Si semiconductor detectors.

A sum of the particle energy losses in all detectors allows one to determine the energy of the recoil particle, i.e., the energy of the neutron, and the analysis of the 2-D matrix ΔE_1 vs E (or ΔE_2 vs E) allows determining the detected particle type.

If the energy of the incoming monoenergetic neutrons is 14 MeV, then the energies of recoils scattered under an angle close to 0 will be $E_p \approx E_n = 14$ MeV and $E_d \approx 0.89 \cdot E_n = 12.46$ MeV for the recoiled proton and deuteron, accordingly. All the possible background reactions that interfere with the recoil nuclei are the following:

- ^{12}C(n, p)^{12}B; $E_p = 0.7$ MeV ($Q = -12.6$ MeV);
- ^{12}C(n, d)^{11}B; --- ($Q = -13.7$ MeV); E_n below the threshold;
- ^{12}C(n, α)^{12}B; $E_\alpha = 7.9$ MeV ($Q = -5.7$ MeV);
- ^{28}Si(n, α)^{25}Mg; $E_\alpha = 13.5$ MeV ($Q = -2.7$ MeV);
- ^{28}Si(n, p)^{28}Al; $E_p = 10.1$ MeV ($Q = -3.85$ MeV);
- ^{28}Si(n, d)^{27}Al; $E_d = 4.6$ MeV ($Q = -9.35$ MeV).

At present, the INP has developed and tuned the block scheme of a spectrometer for three detector telescopes and the possibility to measure the neutron energy spectrum by such method was demonstrated.

In Fig. 3.6, the 2-D spectrum of the charged particles (ΔE vs E) detected by the two-detectors telescope ($W_{\Delta E}$ = 50 μm, W_E = 700 μm) is shown, together with the corresponding energy spectra. The converter foil of deutero-polyethylene with 20 μm thickness was used at irradiation by the neutron flux with $E_n = 14.1$ MeV. The peak of the recoiled deuterons is well separated.

In conclusion, the method presented can be used for measurements of the spectra of secondary fast neutrons and for studies of yields and angular distributions when a D+T neutron source is used.

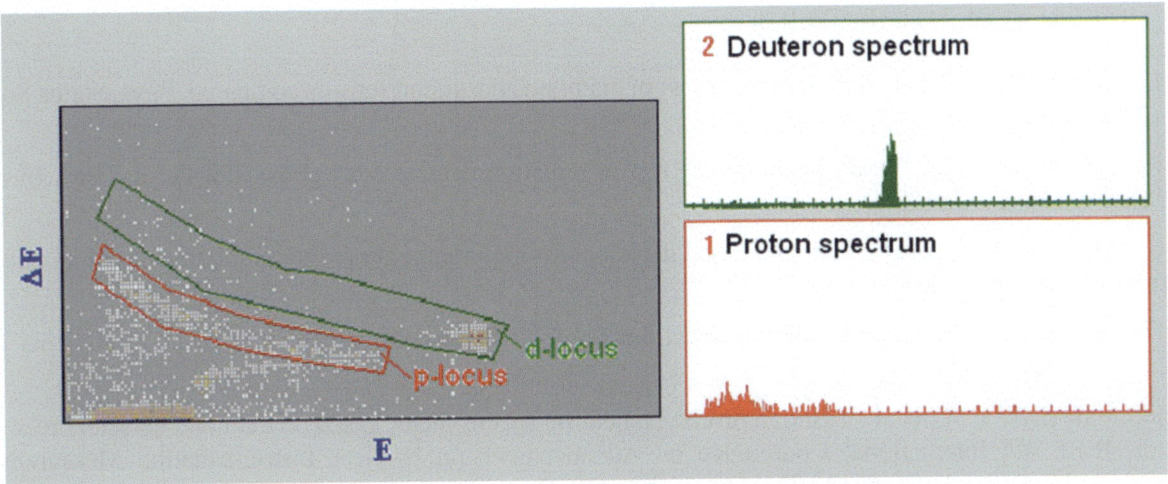

FIG. 3.6. PC-EDE program screen which shows the 2-D spectrum ΔE vs E and the corresponding energy spectrum.

3.3. REFERENCES TO CHAPTER 3

[3.1]   DE BOER, W., et al., Radiation hardness of diamond and silicon sensors compared, Phys. Status Solidi **204** (2007) 3009.

[3.2]   ANGELONE, M., et al., Performance test of radiation detectors developed for ITER-TBM, Fusion Eng. Des. B **136** (2018) 1386.

[3.3]   METCALF, A., et al., Diamond based detectors for high temperature, high radiation environments, J. Instrum. **12** (2017) C01066.

[3.4]   OSIPENKO, M., et al., Neutron spectrometer for fast nuclear reactors, Nuclear Instruments and Methods - A **799** (2015) 207.

[3.5]   OSIPENKO, M., et al., Test of a prototype neutron spectrometer based on diamond detectors in a fast reactor, IEEE 4th International Conference on Advancements in Nuclear Instrumentation Measurement Methods and their Applications (ANIMMA), Lisbon (2015).

[3.6]   BRIESMEISTER, J. (Ed.), MCNP – A General Monte Carlo N-Particle Transport Code, Version 4C, LA-13709-M, Los Alamos National Laboratory, Los Alamos, NM (2000).

[3.7]   APOSTOLAKIS, J., et al., Geometry and physics of the Geant4 toolkit for high and medium energy applications, Radiation Physics and Chemistry **78** (2009) 859–873.

[3.8]   OSIPENKO, M., et al., Upgrade of the compact neutron spectrometer for high flux environments, Nuclear Instruments and Methods - A **883** (2018) 14.

[3.9]   OSIPENKO, M., et al., Calibration of a Li-6 diamond-sandwich spectrometer with quasi-monoenergetic neutrons, Nuclear Instruments and Methods - A **931** (2019) 135.

[3.10]   ARTEMOV, S., et al., Method of the total neutron cross-section measurement by semiconductor Si detector, Proceedings of the International Conference "NUCLEUS-2018 - Fundamental Problems of Nuclear Physics, Atomic Power Engineering and Nuclear Technologies", Voronezh, Russian Federation (2018) 245.

[3.11]   RAPP, M., et al., Beryllium and Graphite Neutron Total Cross-Section Measurements from 0.4 to 20 MeV, Nucl. Sci. Eng. **172** [3] (2012).

[3.12]   ABFALTERER, W., et al., Measurement of neutron total cross-sections up to 560 MeV, Phys. Rev. C **63** (2001) 044608.

[3.13]   Y.V. DUKAREVICH, et al., Total neutron cross-sections of isotopes and the isobaric spin term of the nuclear potential, Nucl. Phys. A **92** [2] (1967).

Chapter 4

ADS APPLICATION FOR POWER PRODUCTION AND TRANSMUTATION OF ACTINIDES AND FISSION PRODUCTS

4.1. CHINESE ACCELERATOR DRIVEN ADVANCED NUCLEAR ENERGY SYSTEM (ADANES)

4.1.1. Introduction

A new nuclear power system concept called ADANES (accelerator driven advanced nuclear energy system) was originally proposed by the Chinese Academy of Sciences. It combines fuel breeding, electric energy generation and transuranic nucleus transmutation in one subcritical system, which utilizes depleted and LEU fuel (with enrichments $\leq$20% in ^{235}U). The subcritical assembly core is driven by a high-power spallation target made by tungsten alloy particles as target material. The subcritical reactor concept is partly designed based on the gas cooled fast reactor with the same structural material (silicon carbide, SiC). The outlet coolant temperature reaches 1200 K for a net plant efficiency of ~40%.

Figure 4.1. depicts the general approach of ADANES. It consists of a burner system and a fuel recycle system. The burner system is based on the existing principles of the ADS.

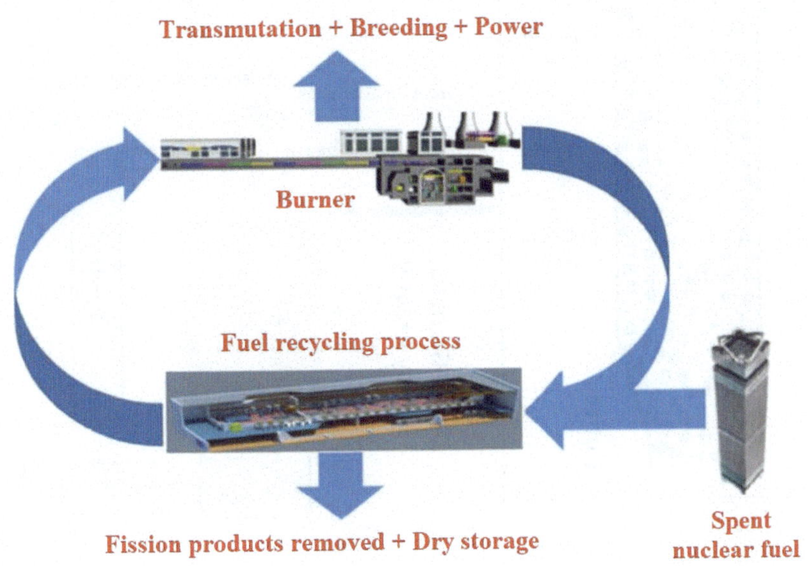

FIG. 4.1. Schematic of the accelerator driven advanced nuclear energy system (ADANES) [4.1].

The ADANES concept is the current main outcome of a global technological development project that includes most of the technological alternatives [4.2] for the accelerator, the high-power neutron target, and the core (coolant and fuel material), as well as the general fuel cycle including reprocessing alternatives. Specific computational tools provide the modelling scheme for the whole system analysis.

The external neutron source in the burner system is a new granular spallation target to increase the potential target power and harden the neutron spectrum. Ceramic materials are used in the reactor core to enhance the system safety and economic performance for a long-period operation. The nuclear raw materials in this system can use depleted uranium, natural uranium, or thorium.

The burner system can provide breeding and burn-up for a long-period operation, and then the spent nuclear fuel from water cooled reactors can be processed in a simple high-temperature dry reprocessing. In this

203

reprocessing, part of the nuclides containing neutron poison in the spent nuclear fuels can be separated and disposed while the remaining nuclides can be reused to produce the carbide fuels. Using this method, only a fraction of the fission products is eliminated, without fine partitioning. The discharged poisonous neutron nuclides, after stockpiling for hundreds of years, can also become part of the available rare-earth resources. This simple high-temperature dry reprocessing in ADANES is economical and significantly reduces the risk of nuclear proliferation, while reducing the amount of nuclear waste to be disposed of.

4.1.2. Gravity driven dense granular target concept

4.1.2.1. Concept design and experimental qualification

A gravity driven dense granular target has been proposed as the target station in the framework of the accelerator driven system under the China ADS project. In the dense granular target, flowing tungsten grains are the target material, which can bring the thermal power out of the target interaction region, impulses by an external loop that lifts the grains again into the target. For the first stage, the China-ADS project proposed to build a test facility at the proton energy of 250 MeV with a beam current of 10 mA [4.3]. A schematic outline of the target is shown in Fig. 4.2.

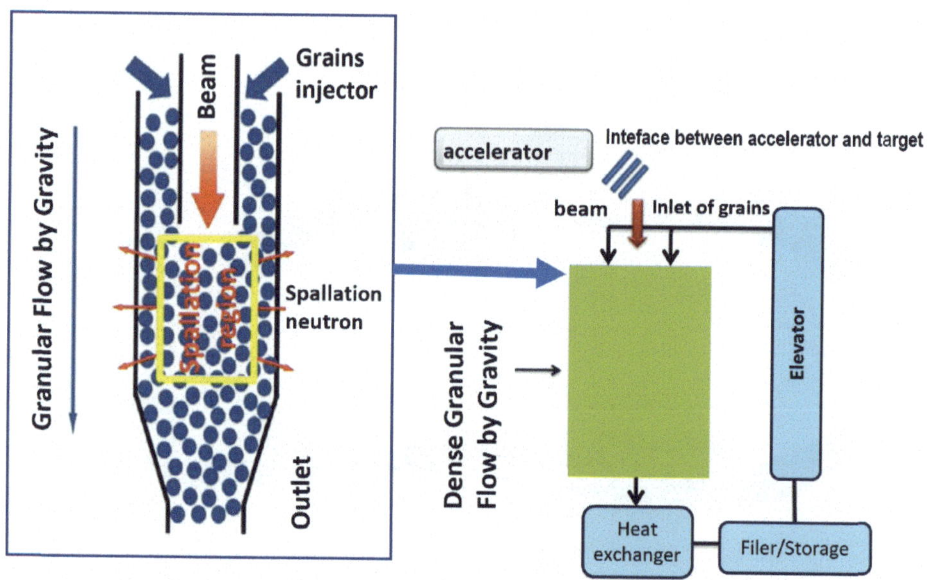

FIG. 4.2. Schematic outline of a gravity driven dense granular flow target [4.3].

A prototype of the target has been set up in the Chinese Academy of Sciences as shown in Fig. 4.3.

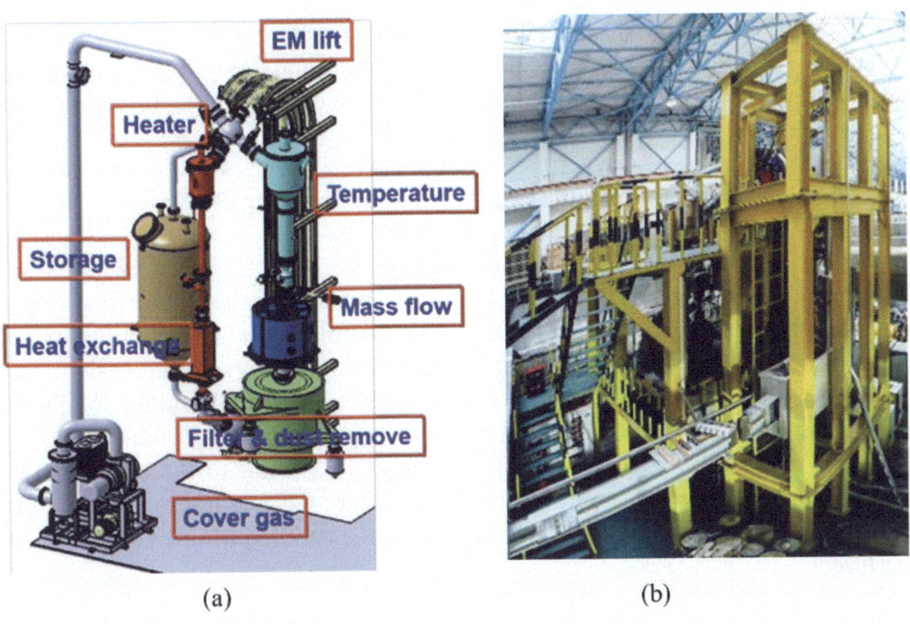

FIG. 4.3. 3-D model (a) and prototype (b) of the gravity driven dense granular flow target.

4.1.2.2. Simulation and computational tools

A proton transport code GMT has been developed, consisting of an intra-nuclear cascade model, an evaporation model, and a transport code for neutrons and other particles. It calculates spallation neutron yield, spatial and energy distribution of the spallation neutrons and leaked neutrons spectrum, particle flux, heat deposition distribution of the standard spallation target. The graphics processing unit (GPU) has emerged as a competitive platform for computing massive parallel problems. The use of stream processors in a general-purpose GPU is an innovative scenario for handling extremely large computations and making high-performance computation affordable. The GMT codes using the Monte Carlo method need huge calculations to estimate N-particles transport in a large-scale irregular ball stacking region. The calculation flow of a new Monte Carlo program for the design of the target station named GMT1.0 is shown in Fig. 4.4.

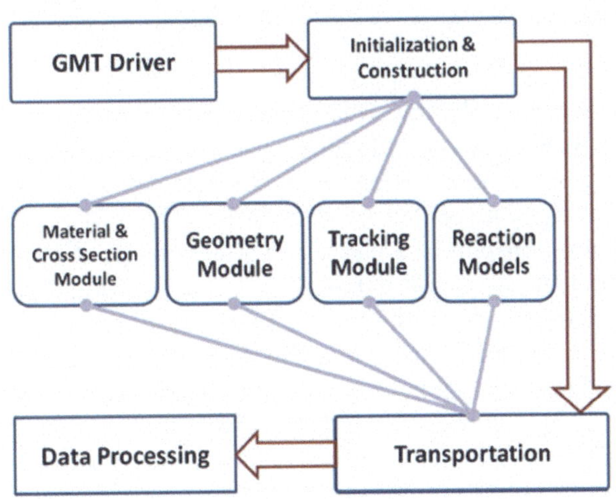

FIG. 4.4. Structure and main components of the GMT1.0 program.

GMT is designed for a massively parallelized simulation of the initiative granular-flow target concept. Based on the combination of the Liège Intranuclear Cascade (INCL) model and the ABLA evaporation/fission model, GMT1.0 integrates a particle transport code and a nuclear reaction code to simulate a spallation target.

To validate the GMT1.0 code, a sequence of calculations of neutronics characteristics and heat-deposit distributions was performed to compare with the experimental data and the predictions of GEANT4 [4.4]. The accuracy of GMT1.0 was observed to be satisfactory. A systematic study of the neutron economy of lead and tungsten targets was performed using GMT1.0. For a large lead target, higher energies tend to be more conducive to obtaining a high neutron economy. For more massive tungsten targets, however, the most economically favourable target diameter seems to be around 20 cm and the most optimal energy seems to be 1.5 GeV. These results are a proof of principle and valuable for the design of the system parameters of the ADS. In addition, the comparison between the GMT results with experimental data for solid as well as granular tungsten target has been performed lately. These quite good agreements are attributed to the INCL4 model, which appears to be one of the most reliable spallation models. Owing to the complexities of the target geometry, the computational cost of the Monte Carlo simulation of particle tracks is highly expensive. Thus, improvement of computational efficiency will be essential for the detailed Monte Carlo simulation studies of the dense granular target.

The use of stream processors in a general-purpose GPU is an innovative scenario for handling extremely large computations and making high-performance computation affordable. The GMT codes using the Monte Carlo method need huge calculations because of lots of samples in the nuclear spallation. In order to speed up the calculations in the spallation target simulation, a GPU-based parallel fast Monte Carlo algorithm will be developed for the INCL code and the evaporation/fission code (ABLA). By this way, the simulation will be significantly accelerated on the GPU. Table 4.1 shows the speedup performance of the GPU version of the ABLA model for four typical excited nuclides used in the simulation.

TABLE 4.1. SPEEDUP PERFORMANCE OF THE GPU VERSION OF THE ABLA MODEL. C++ VERSION RUNS ON INTEL(R) XEON(R) E5410 AND CUDA VERSION RUNS ON KEPLER K20

Nuclides	Excited energy (MeV)	$Time_{c++}$ (μs)	$Time_{CUDA}$ (μs)	Speedup
^{180}W	90.8	937.77	3.543	265.2
^{203}Pb	121.3	1291.6	4.544	284.2
^{197}Pt	334.6	2902.3	9.562	303.5
^{178}Ta	385.8	2917.9	9.937	293.6

Considering the fact that the Monte Carlo simulation of the particle tracks in the granular matter needs to process a massive surface calculation, the computational cost will be expensive. Thus, the optimization of the geometrical construction of a granular target is necessary for fast navigation of particle tracks. Besides, the special design of the particle tracking algorithm is essential for a high Monte Carlo transport efficiency. To perform an efficient Monte Carlo transport of a particle in a granular geometry, fast navigation is necessary. In the program, extremely efficient navigation is based on a 3-D meshing algorithm. Every granule will be attached to a 3-D grid in geometry construction process. Every voxel remembers all granules intersecting it. Thus, fast navigation for tracking particles can be made by local granule searching and boundary calculation in the current voxel. An optimal grid bin number is essential to obtain the highest navigation efficiency. When the voxel is large, the local navigation efficiency will be relatively low because there will be so many granules attached to the current voxel. Conversely, the local navigation in the current voxel tends to be fruitless and then further navigation in the next voxel would be initiated when the voxel is small, or the granules are sparse. Too much fruitless navigation also leads to a low overall efficiency.

With the implementation of delta-tracking algorithm in the GMT simulation of the granular target with massive granules, it can be expected that the tracking process could be accelerated significantly, mainly owing to the decrease in the computational cost of geometrical navigation. The speedup increases sharply along with the decrease of granule diameter. It is shown that the tracking process can be accelerated up to ~40 times for 1 mm granule diameter. In fact, the impressive performance of the delta-tracking algorithm will be more remarkable in the detailed simulation studies of the ADS granular flow spallation target. The simulated results show the speedup performance of the GPU-based ABLA model. It can be seen that a hundred-fold speedup can be achieved (Fig. 4.5). This successful experience in the development of the GPU based nuclear de-excitation model is innovative and demonstrates not only the potential of further accelerating the GMT simulation, but also provides an example of utilizing GPU to perform fast simulation of high energy nuclear reactions. The heterogeneous CPU–GPU algorithm with MPI and CUDA is still under development for future implementation of GPU computation of the spallation model into the GMT program.

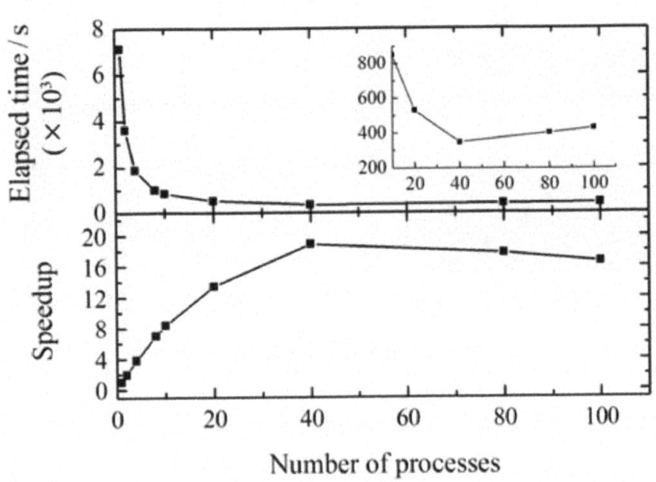

FIG. 4.5. D = 3 mm granular scale calculations, acceleration process number and the number of processes.

In order to improve the computational efficiency of the GMT program, development and application of a message passing interface (MPI) in GMT were studied, to realize random distribution of the large-scale random number in the subprocesses. Rapid reading and writing files were employed instead of the MPI data communication function, which greatly improves the computational efficiency. Different scale calculations were performed to study the relationship of process instance number, speedup to find the maximum acceleration process number and the number of processes when parallel efficiency is highest, which provides a scientific basis for researchers to optimize the computational program between computational resources and computation efficiency. The successful application of MPI in GMT utilizes the computing resources fully and efficiently, improves the computational efficiency, solves the long-time cost and unstable problem of the Monte Carlo method in large-scale event simulations, and plays an important role in the large-scale scanning calculation of the spallation target.

Altogether, the Monte Carlo simulation studies of the granular flow spallation target of the China ADS project are found to be extremely computationally expensive due to the geometrical complexity. The distribution randomness, the small granular dimension, and the multi-layered structure of the granule account for the particularity of a granular geometry. The special designs of the geometrical construction/navigation method and the particle tracking module in the dedicated GMT program are shown to be powerful for the high-efficiency and detailed Monte Carlo simulation of the China ADS granular flow spallation target. Moreover, the speedup performance of the GPU-based reaction model provides the probability of further accelerating the GMT simulation in future.

4.1.2.3. Results of dense granular target neutronic analysis

The neutron and proton leakages, fluxes, energy deposition, residual radioactivity and gamma dose rates has been done using the FLUKA Monte Carlo code for the granular and the thick monolith target. Results shows that the neutron flux and energy deposition in tungsten spheres target are more homogeneous along the axial direction than for a monolith target (Fig. 4.6). It has been estimated that the granular target has more lateral neutron yield and a relatively small number of neutrons in the backward direction. In addition, the total radioactivity is found to be comparatively lower in the granular target, although for some nuclei, the value of their activities are similar for both targets. The above features make the granular target more suitable as an ADS target.

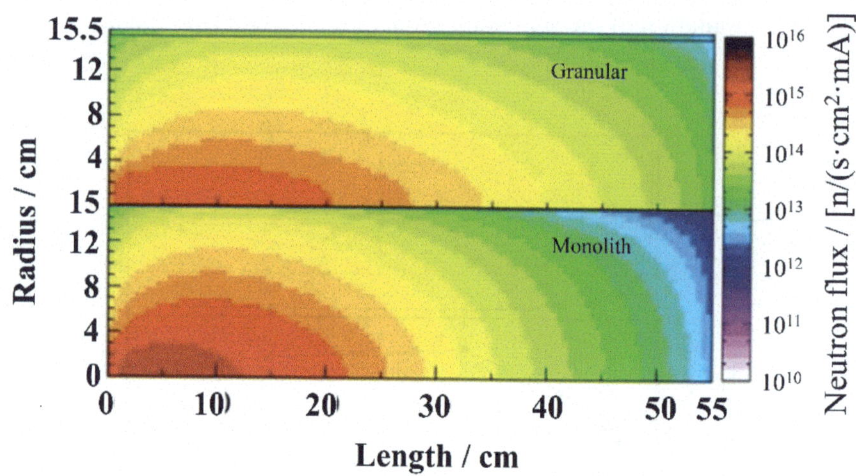

FIG. 4.6. Contour plots of neutron flux in the granular and monolith targets irradiated by 1 GeV protons.

The neutronic performance of uranium carbide (UC), uranium nitride (UN) and uranium dioxide (UO_2) monolith targets and granular targets were evaluated by using the GMT code. Effects of the target thickness and diameter on leaked neutron yield of monolith targets irradiated by 1 GeV protons were studied; the neutron yield and neutron spectra for the dense granular uranium compounds and UC-kernel TRISO type targets were estimated. Results show that, with an optimal dimension of 25×50 cm, uranium compounds target can produce 40% more neutron yield than the tungsten target. Results also show that the designed carbon and SiC layers coated UC-kernel TRISO type pellet target with about 70% kernel volume fraction can produce about 30% more leaked neutron yield than the granular tungsten target in their optimal diameters. Table 4.2 presents a description of the TRISO pellets.

TABLE 4.2. CHARACTERISTICS OF THE TRISO TYPE PELLET MATERIAL (WITH A DIAMETER OF 2 mm)

Pellet type	C diameter (mm)	Cladding layer material and thickness	Core volume fraction (%)	Average granular density
Case 1	1.6	SiC (0.2)	51.2	8.52
Case 2	1.6	C (0.1) & SiC	51.2	8.05
Case 3	1.7	SiC (0.15)	61.4	9.59
Case 4	1.7	C (0.075) & SiC	61.4	9.21
Case 5	1.8	SiC (0.1)	72.9	0.80
Case 6	1.8	C (0.05) & SiC	72.9	10.53

In Fig. 4.7, the neutron yield comparison between UC and tungsten granular target as a function of the target diameter is shown. With growing UC volume fraction, the neutron yield is increasing. Neutron yields for all TRISO pellet targets reach a maximum when their diameters are about 40 cm. In addition, for the pellet targets with the same UC volume fraction, adding a carbon layer and reducing the thickness of SiC layer can enhance the neutron yield. This is mainly due to the fact that the density and the neutron capture cross-section of carbon are lower than those of SiC.

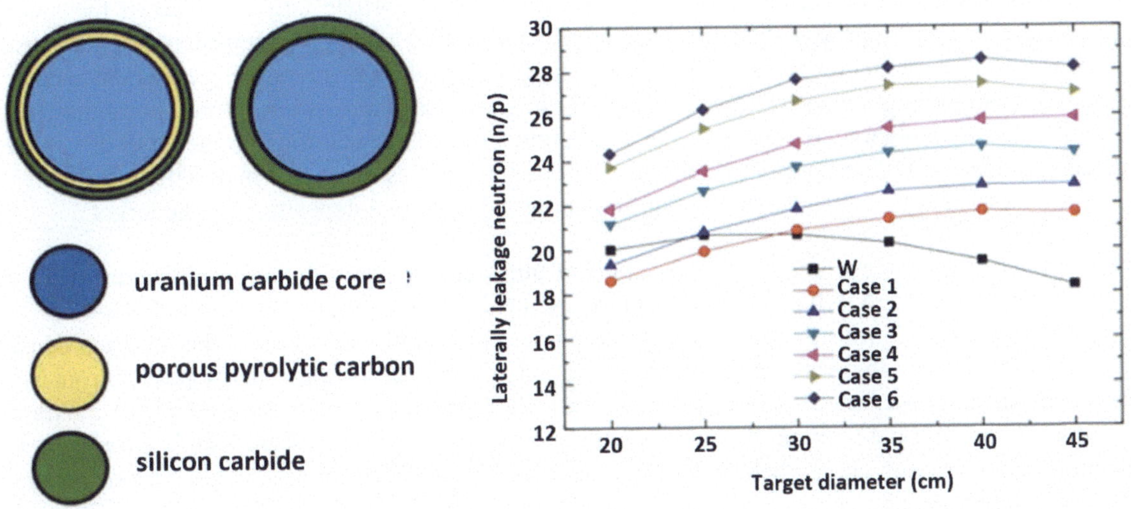

FIG. 4.7. Neutron yields for different TRISO type granular targets as a function of the target diameter.

The radiation damage and the activation of the dense granular flow spallation target has been estimating by using the Monte Carlo program GMT and FLUKA at the proton energy of 250 MeV. The results show that the flowing tungsten granular target can afford a stable external neutron source, and the granule diameter have no obvious effect on the neutronic performance when the granule diameter is lower than 1 cm. The discussion implies that the beam profile diameter influences slightly the neutronics of the spallation target, while target diameter and volume fraction have obvious effect on the neutronic performance.

A prediction was made for the radiation damage of tungsten grains and its structure material with stainless steel 316L of bare target and coupling target with subcritical blanket, with a proton beam power of 2.5 MW and exposure in one operation year of 5000 h. Results show that the maximum displacement per atom (DPA) for target vessel and beam tube, respectively, is about 1.0 and 1.6 DPA/year in the bare target, and increases to 2.6 and 2.8 DPA/year in the ADS system (Fig. 4.8). The average DPA for tungsten grains is relatively low, and the helium and hydrogen production is of minor concern. The activity and afterheat of granular target were also calculated, and it is found that the decline rate of radioactivity and afterheat with cooling time is decreasing with increasing irradiation time.

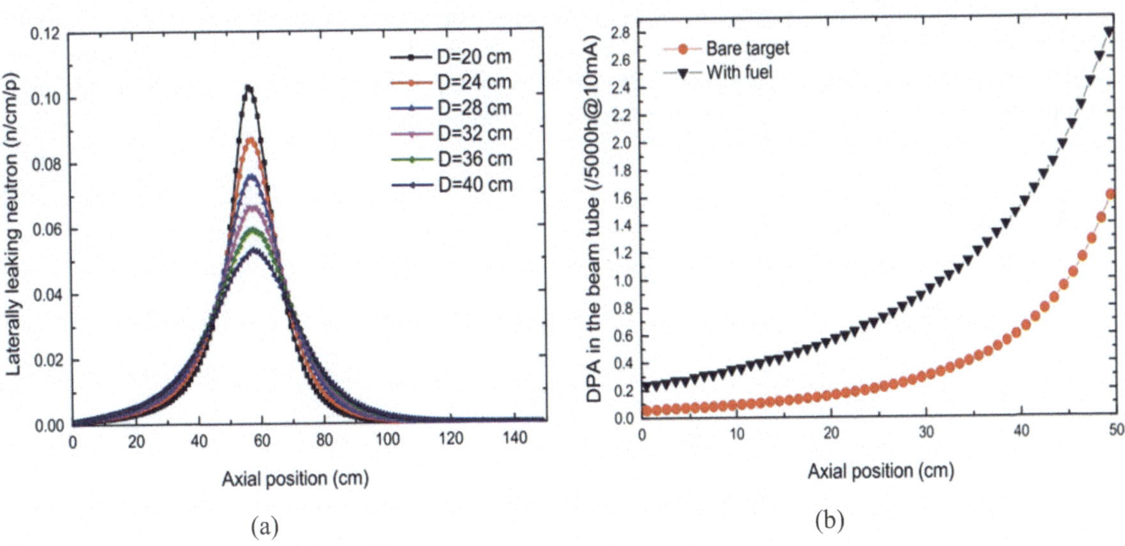

FIG. 4.8. Target vs axial position: neutron generation (a), and DPA distribution in the beam tube (b).

Additional analysis has been done for a low-density porous tungsten particle target instead of the natural tungsten. The neutronic performance of the bare target have been studied with porous tungsten granule density using Monte Carlo code. A model for ADS was then set up to evaluate the neutron sources performance caused by density variation of porous tungsten granule. The results show that there is an optimal combination of tungsten granular density and target diameter to make the utmost of leakage neutrons. In the ADS driven by a tungsten dense granular target, an appropriate density porous tungsten as the target material can effectively increase the source efficiency, reduce the beam requirement and the thermal deposition on the target.

The neutron yields and its average energy as a function of tungsten density at the target diameter of 50 cm were studied and shown in Fig. 4.9. It can be seen that the escaped neutron yield peaks around R/N=0.65. The mean energy of the leaked neutrons increases with the reduction of granular density. As the tungsten density R/N decreases from 1 to 0.65, there are more neutrons escaped with higher average energy per neutron. Most of the leaked neutrons are distributed around (0.001 – 1500) MeV, and they peak at about 0.5 MeV. With the decreasing tungsten density, the fraction of neutrons with energy above 0.5 MeV significantly increases, and the proportion of neutrons with energy below 0.5 MeV decreases. The spectrum of leaked neutrons becomes harder with low granular density.

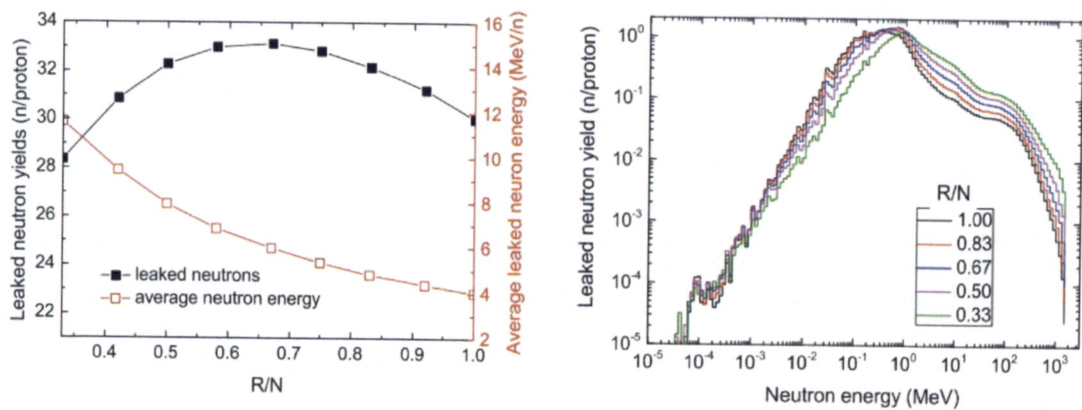

FIG. 4.9. Variation of neutron yield and their average energy with tungsten density (R/N = 1 thick tungsten) and neutron energy spectrum.

Neutron yields and mean neutron energy of dense granular targets with various tungsten densities has been calculated by GMT code. The proton beam energy is 1.5 GeV. As shown in Fig. 4.10, the maximum neutron yields at different target diameters almost keep a constant of 33 n/p, and the tungsten density corresponding to the maximum neutron yield decreases with the increase of target diameter.

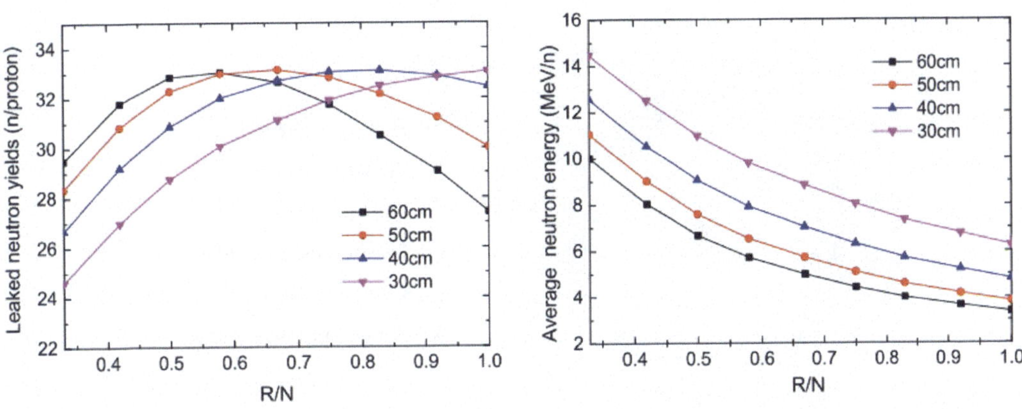

FIG. 4.10. Neutron yield and its average energy for different target diameters.

4.1.3. Conceptual design of a granular flow cooled ceramic fast reactor

The concept of an accelerator driven ceramic fast reactor (ADCFR) is proposed in the ADANES concept. The main components of the reactor are presented in Fig. 4.11. This reactor is designed to operate continuously throughout a 40-year core life, without fuel loading or shuffling. The ADCFR consists of a high-power superconducting linear accelerator, a spallation-target, and a ceramic-coolant fast reactor. The neutrons to sustain the nuclear fission in the reactor are produced by a particle accelerator via spallation in a gravity driven dense granular target. Waste transmutation, breeding, and power production are built into the ADCFR. Ceramic materials are proposed in the core of the reactor to improve system safety and economic performance during long-time breeding and burnup.

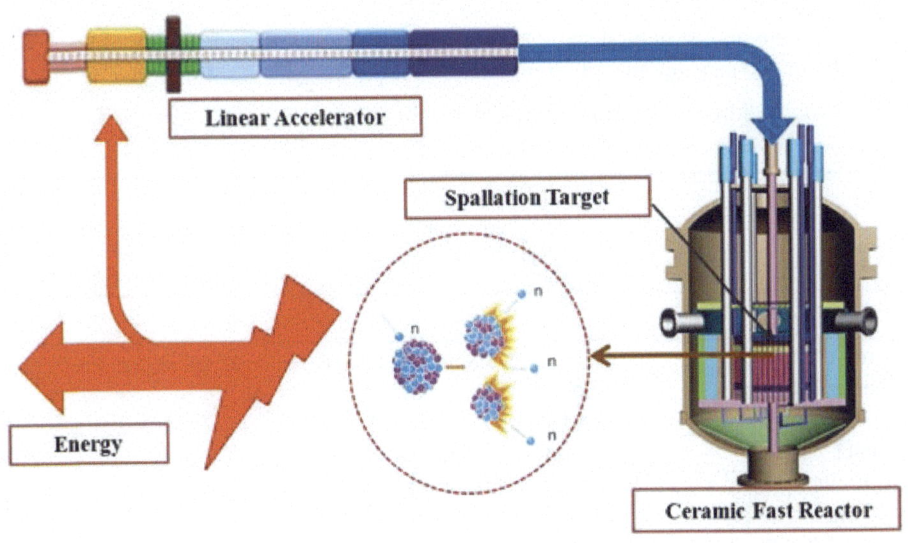

FIG. 4.11. Main components of an accelerator driven ceramic fast reactor.

Figure 4.12 shows a cross-section of the core of the ADCFR, with the central position of the spallation target, and the core assemblies. A material analysis was performed to evaluate the suitable material combination for the core. Ceramic based materials are proposed as they usually possess high temperature resistance, good neutron characteristics, high corrosion resistance, radiation resistance, high thermal conductivity, high strength, and good stability.

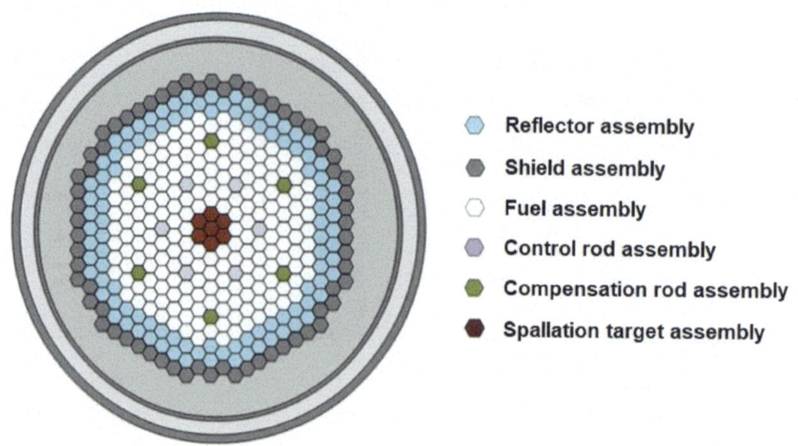

FIG. 4.12. Schematic of radial cross-sectional view of the reactor core of the ADCFR.

The material selection for the ADCFR core is summarized in Table 4.3. The main design parameters of the ADCFR are depicted in Table 4.4.

TABLE 4.3. CHARACTERISTICS AND MATERIAL SELECTION FOR THE COMPONENTS OF THE ADCFR CORE

Component	Main features	Optional materials
Nuclear fuel	High melting point (>2700 K), high density (>10 g cm^{-3}), high burnup, high thermal conductivity (> 10 W m^{-1} K^{-1})	UO_2, UC, UN
Coolant	High melting point (>2000 K), high heat capacity and thermal conductivity, chemical stability, low pressure	Gas–solid two-phase granular flow, Al_2O_3, He
Structural material	Antiradiation, structural stability, low neutron absorption and thermal expansion, high melting point (>3000 K)	Al_2O_3, SiC, BeO, Zr_3Si_2
Absorption and control material	Strong neutron absorption, chemical stability	B_4C

TABLE 4.4. MAIN DESIGN PARAMETERS OF THE ADCFR

Parameter	Value
Core dimensions	Diameter 2.8 m, height 3 m
Thermal power	0.5–2 GW
Power density	~40 W/cm^3
Fuel	Uranium carbide
Enrichment of ^{235}U	~10%
Spallation target	Tungsten granular target
Accumulation rate	~50%
Beam	Protons, 1.5 GeV at 20 mA
Coolant	Ceramic granule, gas-granule two-phase flow
System pressure	Atmospheric or negative pressure
Operation lifetime	>40 years
Inlet temperature	300°C
Outlet temperature	>700°C

The estimated neutron energy by neutronic calculation is mainly between 1 keV and 10 MeV. The relative neutron density peak is about 0.1 MeV (Fig. 4.13). The neutron spectrum is fast, having the ability to breed and transmute, and can be operated for long periods without fuel shuffling or supplementation.

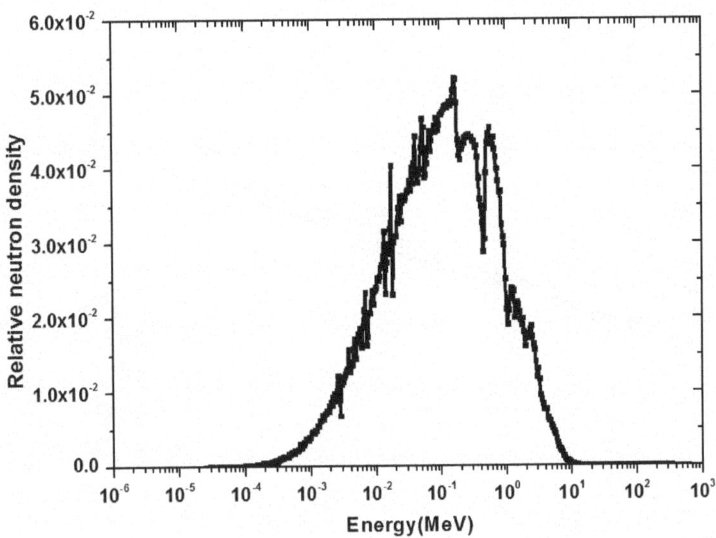

FIG. 4.13. Neutron spectrum distribution of ADCFR.

Figure 4.14 shows the effective neutron multiplication factor k_{eff} as a function of the operation time along a typical life cycle (~40 years) without refuelling or fuel shuffling. At early operation stages, ADCFR can operate under a subcritical mode driven by the external neutron source and k_{eff} initially set to 0.98 (the initial proton beam density is <10 mA). Nearly 5 years later, k_{eff} reaches around ~1.0, i.e., it can begin to operate under a critical condition because of excess reactivity. After 15 years, k_{eff} reaches its peak, before it gradually decreases for more than 20 years.

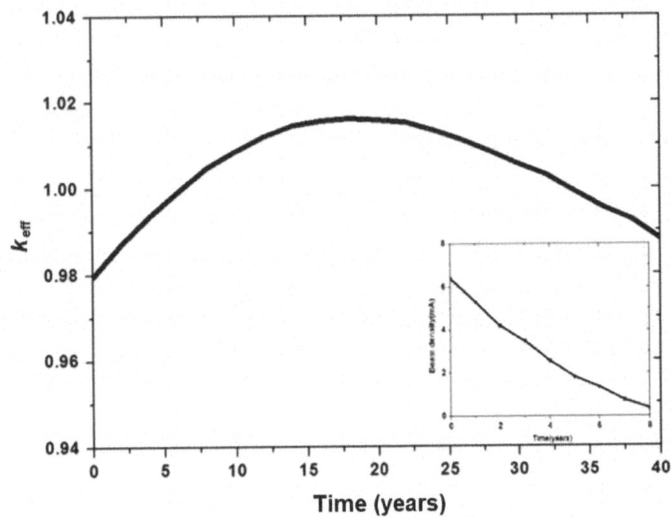

FIG. 4.14. Evolution of keff during a 40-year life cycle of an ADCFR, and of the beam intensity during a first subcritical period.

Different coolant materials – Al_2O_3, ZrO_2, SiC, MgO, and Zr_3Si_2 – were selected to compare neutron performance. These coolants are granular materials, where the filling ratio in the coolant piping is ~0.6.

Figure 4.15 shows the time-dependent distribution of the ideal k_{eff} for different coolant materials. The initial k_{eff} of 0.95 is fixed, operating the accelerator-driven subcritical core with a proton beam energy of 1.2 GeV and a core total power of 1 GW. The required beam intensity is less than 20 mA. It can be seen that the k_{eff} curve is gradually increased over time. These five kinds of ceramic materials have breeding properties.

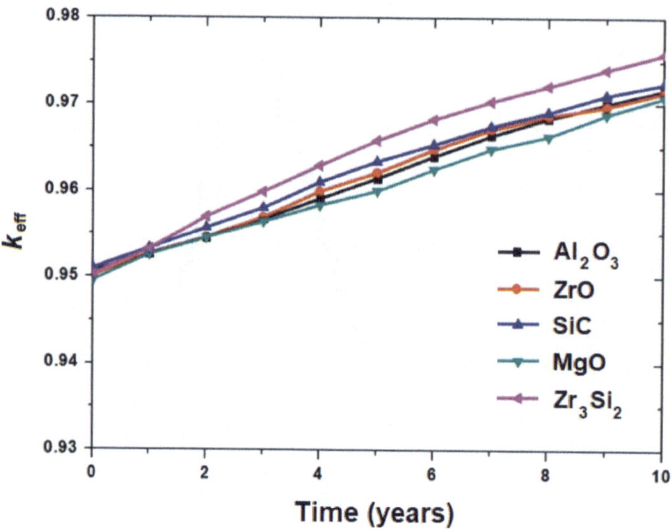

FIG. 4.15. Time-dependent distribution of the ideal keff for five different coolant materials.

4.1.4. Summary of the ADANES features

The main reactor features of the reactor system that has been presented can be summarized with the following characteristics:

- Two-phase gas-solid flow coolant: composed of gravity-driven granular materials and helium, combining the advantages of a gas and solid phase: good thermal conductivity and inertia, high efficiency heat exchange, stability and safety;
- High temperature operation due to the utilization of ceramic material, withstanding temperatures up to 1600°C, with excellent radiation resistance, operational safety and reliability. Very good response in the case of a loss of coolant accident;
- Subcritical and critical operation modes;
- Economic viability: The high safety performance of ceramic materials can simplify the safety design, and long operational cycles reduces operation and maintenance costs. The power efficiency is very high due to the high operation temperature, extending potential co-applications as cogeneration, hydrogen or isotope production. Initial operation in subcritical mode, followed by critical mode enables the utilization of multiple cores per beam accelerator.

4.1.5. Additional supporting activities

4.1.5.1. Accelerator development

The Institute of Modern Physics (IMP) of the Chinese Academy of Sciences has built the first international demonstration prototype of the front of Superconducting Proton Linac (Fig. 4.16). In January 2019, the continuous wave proton beam achieved 100 hours operation test with a power of more than 30 kW. The beam energy was 15.8–16.3 MeV and the beam intensity was 2.0–2.1 mA. The test result is the highest beam power and the longest running time record of the continuous beam high current linac in the world, which is in the international leading level.

214

FIG. 4.16. Superconducting Proton Linac at the IMP.

4.1.5.2. Fuel reprocessing

Reprocessing the spent nuclear fuel is one of the key issues for a sustainable nuclear energy concept. Fission products would be removed by a dry-processing process. The following strategy is being developed for the separation of minor actinides and lanthanides. Water-satured ionic liquid betainium bis(trifluoromethylsulfonyl)imide, (Hbet)(Tf$_2$N), can dissolve lanthanide oxide from simulated spent nuclear fuel with a dissolution ratio of 100% at 40°C (Fig. 4.17). However, the dissolution of uranium is almost negligible (<1%) under the same conditions. This process of separation displays an outstanding performance to separate some key fission product oxides, such as Ln$_2$O$_3$, and allows the recovery of actinides (AcO$_2$) as a group in solid form. A small volume of liquid high level radioactive waste was generated during this process and there was no need for dissolving spent nuclear fuel since only some fission products were dissolved selectively and separated.

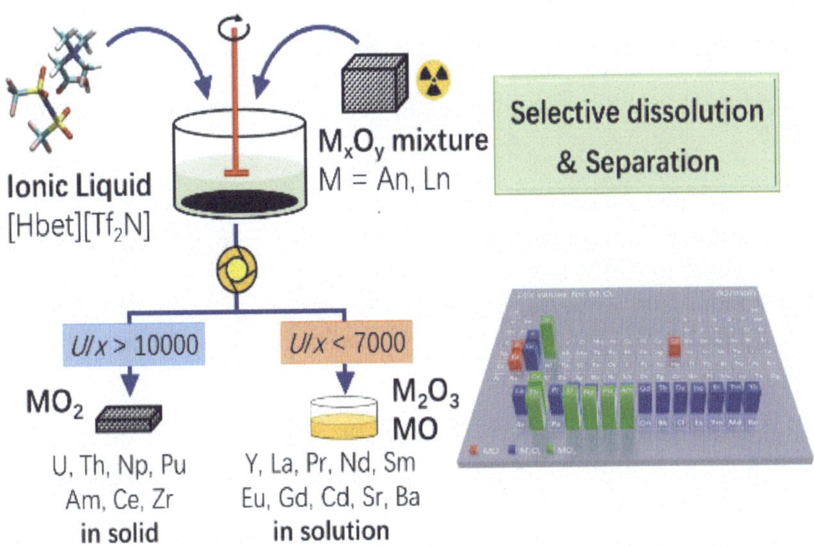

FIG. 4.17. Reprocessing of spent nuclear fuel based on the selective dissolution and separation using (Hbet)(Tf2N).

This work represents the first case for efficient fission products removal by selective dissolution, avoiding the complete dissolution of spent nuclear fuel, the producing of the large high-level radioactive waste and reducing environmental hazards. Moreover, this approach would enhance proliferation resistance significantly because sole plutonium would not be separated. The present method provides a simple, environmental friendly and high efficient fission products separation strategy from spent nuclear fuel and introduces a new separation approach to the reprocessing of spent nuclear fuel.

Uranium carbide (UC) ceramic fuel has been considered as a potential nuclear fuel for the next generation of fast neutron reactors, especially for the ADS, as ADANES. Based on the accelerator driven recycling of used nuclear fuel, UC can recrystallize with plutonium and minor actinides.

Recently, homogeneous ceramic nuclear microspheres and UC powder have been successfully prepared at IMP (Fig. 4.18), using an improved microwave-assisted rapid internal gelation process and Pechini-type in situ polymerizable complex method respectively.

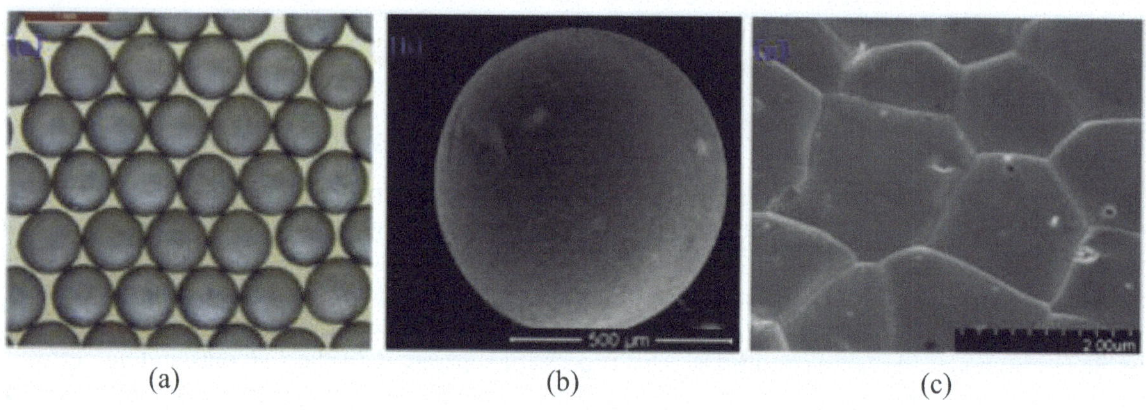

(a)   (b)   (c)

FIG. 4.18. Ceramic UC spheres fabricated by the improved microwave-assisted rapid internal gelation process. (a) Optical image of the ceramic UC microspheres; (b) Scanning electron microscopy image of the ceramic UC microspheres; (c) Micro-morphology of the ceramic UC microspheres.

4.1.5.3. Material irradiation facility

Research on materials related to cladding and the reactor core, achieving a series of notable advancements, has been conducted by IMP.

With respect to module structure materials, a multistage structure mold was designed, and SiC/ZrC composite nanopowder was synthesized by the improved hot pressing sacrificial template method with hydrothermal precursor. There was no sintering aid at low temperature and high-efficiency densification. The densification of ceramic material was more than 98%, and the oxidation resistance was up to 1200°C. At the same time, a large-scale hot pressing mold was developed, and high-quality SiC based core components were gradually enlarged.

With respect to cladding structure materials, a controllable synthesis of high-purity boron free sintering assistant was realized, and low neutron toxic SiC ceramics with a density ≥98.2% was obtained, which has excellent mechanical, thermal and radiation resistance. The forming process of low neutron toxic SiC ceramic tube blank (600 mm long) was realized. A thermodynamic evaluation of the composite ceramics and the evaluation of the radiation resistance of the components in the composite ceramics were carried out. The experimental device for precursor synthesis was designed and built, the research and development platform of SiC fibre was built, and the test method of SiC fibre's main performance was established. The technology of high brittle fibre circular knitting was explored, the optimization of weaving angle and the hybrid weaving of carbon fibre and optical fibre were realized, and the carbon fibre cladding tube was trial manufactured.

4.2. JAPANESE ADS FOR TRANSMUTATION OF MINOR ACTINIDES

4.2.1. Introduction

A long-term programme, called OMEGA, has started since 1988 in Japan for research and development of new technologies for partitioning and transmutation of minor actinides and fission products. The main aims of the programme are: (1) exploring the possibility to transmute long-lived nuclides to short-lived or stable ones and to utilize some nuclides as resources; and (2) widening options of the future waste management. Under the OMEGA programme, the Japan Atomic Energy Agency (JAEA) (the former Japan Atomic Energy Research Institute (JAERI)) proposed the concept of double-strata fuel cycle, in which partitioning and transmutation are carried out in a dedicated small-scale fuel cycle attached to the commercial fuel cycle. For the dedicated transmutation system, JAEA has been proceeding with the research and development on the ADS.

4.2.2. Benchmark specification

4.2.2.1. General concept

As a primary option, JAEA is proposing a lead–bismuth eutectic (LBE) cooled ADS [4.5], [4.6], [4.7]. There are some advantages in system with LBE coolant in comparison with sodium-cooled one, although thermal property is inferior to that of sodium. The chemical activity of LBE is lower than that of sodium. The neutron slowing down power of LBE is smaller than that of sodium, and hence neutron spectrum becomes hard in the LBE cooled core. The hard neutron spectrum is preferable for the transmutation of minor actinides by fission reactions. In the ADS, LBE can be used simultaneously as the coolant and a spallation target.

4.2.2.2. Core geometry

The core layout of the dedicated ADS is shown in Fig. 4.19. The benchmark calculation is carried out for a cylindrical R–Z model with the dimensions given in Fig. 4.19.

FIG. 4.19. Core configuration of the dedicated ADS.

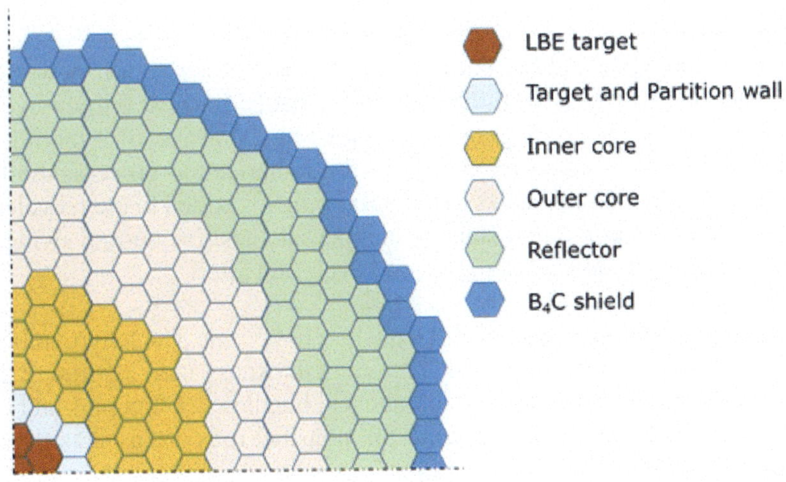

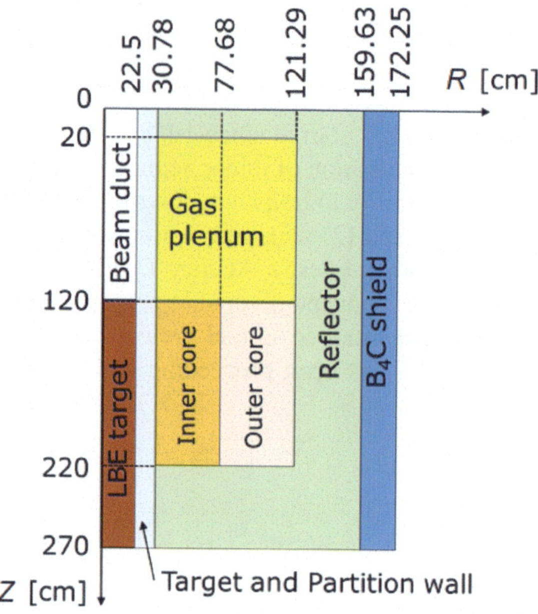

FIG. 4.20. R–Z calculation model of JAEA proposed LBE cooled ADS. The mesh size is about 5 cm, both for R and Z directions.

Buffer assemblies are loaded between the spallation target and the fuel region to reduce the radial power peaking factor. The buffer assembly has the same hexagonal cross-section as the fuel, but it has empty duct structure. The buffer assemblies prevent cross flow of the coolant from the fuel assembly to the target region. The core is radially divided into two zones with different initial plutonium loading for smoothing the radial power distribution. Surrounding the core, LBE reflector, stainless steel shielding, and boron carbide (B4C) shielding are loaded.

The proton Linac with a proton energy of 1.5 GeV is used for the accelerator. A proton beam duct is inserted along the core central axis. A Gaussian beam profile with $1\sigma = 11.16$ cm is assumed in this benchmark calculation.

The core thermal power is 800 MW, and the burnup period is 600 effective full power days (EFPD). After each burnup cycle, all fuels are removed from the core and reloaded for next burnup cycle after cooling and reprocessing. The time period for the cooling and the reprocessing consists of 9-month unloading, 1.5-year cooling, 1-year refabrication, and 9-month reloading. In the refabrication process, the fission products are removed and only minor actinides of equal mass to the burnup fuel is added to the recycled fuel. Plutonium is used as mixture of minor actinides at the first stage to compensate positive burnup swing of minor actinides, though plutonium is not additionally loaded to the second cycle and after.

The fuel assembly is hexagonal duct-less type, and the pitch is 134.5 mm. The total number of fuel assemblies is 276. The cladding tube is made of T91 steel, with 0.5 mm in thickness and 7.65 mm in diameter. The active fuel length is 100 cm. Each fuel assembly has no duct tube and bundles 121 fuel pins and 6 non-fuel rods. The triangular pitch of 11.48 mm is fixed by grid spacers. Six non-fuel rods are used as the tie rods to keep the mechanical integrity of the assembly.

4.2.2.3. Fuel composition

For the core fuel, a mixture of mono-nitride of minor actinides and plutonium is used, with a pellet density of 95% of the theoretical density and a smear density of 85% of the theoretical density. As the inert matrix, zirconium nitride (ZrN) is used with the fuel. Nitrogen with ^{15}N (assumed as 100% enrichment) is used for both aforementioned nitrides. The isotopic compositions of minor actinides and plutonium were determined considering the following reprocessing scheme. The spent PWR fuel of 45 GWd/t burnup is reprocessed after 7 years cooling, and minor actinides and plutonium are recovered. Before fabrication of the ADS fuel, an additional 3-year period after recovery of minor actinides and plutonium is assumed. The volume fractions in the core region are 0.2172, 0.1043, 0.6274 and 0.0510 for fuel, structure material, coolant and helium bond, respectively. This coolant volume ratio corresponds to the ratio of fuel pin pitch to fuel pin diameter of 1.5. Homogenized atomic number densities for inner and outer core regions are listed in Table 4.5.

TABLE 4.5. INITIAL HOMOGENEOUS ATOMIC NUMBER DENSITIES OF INNER AND OUTER CORE REGIONS ($\times 10^{24}$/cm^3)

Nuclide	Inner core	Outer core	Nuclide	Inner core	Outer core
U-234	2.852E-07	3.557E-07	V-51	1.903E-05	1.903E-05
U-236	7.069E-08	8.816E-08	Cr-50	3.655E-05	3.655E-05
Np-237	9.053E-04	1.129E-03	Cr-52	7.046E-04	7.046E-04
Pu-238	1.668E-05	2.081E-05	Cr-53	7.990E-05	7.990E-05
Pu-239	3.802E-04	4.742E-04	Cr-54	1.988E-05	1.988E-05
Pu-240	1.740E-04	2.171E-04	Mn-55	4.422E-05	4.422E-05
Pu-241	7.510E-05	9.366E-05	Fe-54	4.538E-04	4.538E-04
Pu-242	4.798E-05	5.984E-05	Fe-56	7.054E-03	7.054E-03
Am-241	6.027E-04	7.517E-04	Fe-57	1.834E-04	1.834E-04
Am-242m	1.117E-06	1.393E-06	Ni-58	5.652E-06	5.652E-06
Am-243	2.500E-04	3.118E-04	Ni-60	2.165E-06	2.165E-06
Cm-243	5.609E-07	6.995E-07	Ni-61	9.569E-08	9.569E-08
Cm-244	7.584E-05	9.459E-05	Ni-62	2.981E-07	2.981E-07
Cm-245	7.352E-06	9.169E-06	Ni-64	6.840E-08	6.840E-08
Cm-246	7.571E-07	9.442E-07	Nb-93	4.184E-06	4.184E-06
N-15	6.412E-03	6.243E-03	Mo-92	7.516E-06	7.516E-06
Zr-90	1.993E-03	1.584E-03	Mo-94	4.685E-06	4.685E-06
Zr-91	4.347E-04	3.454E-04	Mo-95	8.063E-06	8.063E-06
Zr-92	6.644E-04	5.279E-04	Mo-96	8.448E-06	8.448E-06
Zr-94	6.733E-04	5.350E-04	Mo-97	4.837E-06	4.837E-06
Zr-96	1.085E-04	8.618E-05	Mo-98	1.222E-05	1.222E-05
He-4	4.297E-06	4.297E-06	Mo-100	4.877E-06	4.877E-06
C-nat	4.045E-05	4.045E-05	Pb-204	1.167E-04	1.167E-04
N-14	1.735E-05	1.735E-05	Pb-206	2.008E-03	2.008E-03
Si-28	8.683E-05	8.683E-05	Pb-207	1.842E-03	1.842E-03
P-31	3.137E-06	3.137E-06	Pb-208	4.367E-03	4.367E-03
S-32	1.520E-06	1.520E-06	Bi-209	1.030E-02	1.030E-02
V-50	4.864E-08	4.864E-08			

The initial plutonium loading is 27.4 wt% for both the inner and outer core regions. The amounts of the inert matrix for the inner and outer core regions are 38.9 wt% and 28.9 wt%, respectively. The operating temperature of the fuel is 983 K. The buffer region is assumed to be composed of T91 steel and LBE, and their volume ratios are 30% and 70%, respectively. The target temperature is 653 K. The temperature of gas plenum is 703 K. The temperatures of other regions are 603 K. Homogenized number densities for each region are listed in Table 4.6.

TABLE 4.6. HOMOGENEOUS ATOMIC NUMBER DENSITIES ($\times 10^{24}/cm^3$)

Nuclide	LBE target	Gas plenum	Target and wall	SUS reflector	B_4C shield
He-4		2.257E-05			
C		4.045E-05	3.686E-04	7.753E-05	1.483E-02
B-10					1.180E-02
B-11					4.751E-02
N-14		1.735E-05	1.581E-04	3.325E-05	3.325E-05
Si-28		8.683E-05	7.913E-04	1.664E-04	1.664E-04
P-31		3.137E-06	2.859E-05	6.013E-06	6.013E-06
S-32		1.520E-06	1.385E-05	2.913E-06	2.913E-06
V-50		4.864E-08	4.432E-07	9.322E-08	9.322E-08
V-51		1.903E-05	1.734E-04	3.647E-05	3.647E-05
Cr-50		3.655E-05	3.331E-04	7.005E-05	7.005E-05
Cr-52		7.046E-04	6.422E-03	1.351E-03	1.351E-03
Cr-53		7.990E-05	7.281E-04	1.531E-04	1.531E-04
Cr-54		1.988E-05	1.812E-04	3.810E-05	3.810E-05
Mn-55		4.422E-05	4.030E-04	8.475E-05	8.475E-05
Fe-54		4.538E-04	4.136E-03	8.698E-04	8.698E-04
Fe-56		7.054E-03	6.429E-02	1.352E-02	1.352E-02
Fe-57		1.834E-04	1.671E-03	3.515E-04	3.515E-04
Ni-58		5.652E-06	5.151E-05	1.083E-05	1.083E-05
Ni-60		2.165E-06	1.973E-05	4.149E-06	4.149E-06
Ni-61		9.569E-08	8.720E-07	1.834E-07	1.834E-07
Ni-62		2.981E-07	2.717E-06	5.714E-07	5.714E-07
Ni-64		6.840E-08	6.234E-07	1.311E-07	1.311E-07
Nb-93		4.184E-06	3.813E-05	8.019E-06	8.019E-06
Mo-92		7.516E-06	6.850E-05	1.441E-05	1.441E-05
Mo-94		4.685E-06	4.269E-05	8.979E-06	8.979E-06
Mo-95		8.063E-06	7.348E-05	1.545E-05	1.545E-05
Mo-96		8.448E-06	7.699E-05	1.619E-05	1.619E-05
Mo-97		4.837E-06	4.408E-05	9.270E-06	9.270E-06
Mo-98		1.222E-05	1.114E-04	2.342E-05	2.342E-05
Mo-100		4.877E-06	4.445E-05	9.348E-06	9.348E-06
Pb-204	1.860E-04	1.167E-04	9.120E-06	1.488E-04	3.724E-05
Pb-206	3.201E-03	2.008E-03	1.570E-04	2.561E-03	6.410E-04
Pb-207	2.935E-03	1.842E-03	1.440E-04	2.348E-03	5.878E-04
Pb-208	6.960E-03	4.367E-03	3.413E-04	5.568E-03	1.394E-03
Bi-209	1.642E-02	1.030E-02	8.054E-04	1.314E-02	3.289E-03

(a) Burnup

At the first phase of the benchmark calculation, burnup calculation with an external neutron source for the first cycle is required. The cycle length is 600 EFPDs, and the core thermal power is 800 MW.

(b) Sensitivity and uncertainty analyses

As the second phase of the benchmark calculation, sensitivity and uncertainty analyses are required. The sensitivity coefficients of the k_{eff} value at BOC and the burnup reactivity are required as a function of neutron energy. Major reactions (capture, fission, elastic scattering, inelastic scattering, (n, 2n), nu, mu-bar and chi) for TRU nuclides, LBE nuclides and ^{15}N need to be investigated. Uncertainties deduced from covariance data prepared in nuclear data library for these reactions and nuclides are also requested.

4.2.3. Requested benchmark results

The list of parameters to be calculated in the benchmark is given here below:

- Effective neutron multiplication factor k_{eff} at BOC and EOC;
- Radial fission reaction distributions ($Z = 170$ cm) and axial fission reaction distributions ($R = 30.78$ cm and 77.68 cm) at BOC and EOC;
- Proton beam current for the core thermal power of 800 MW at BOC and EOC;
- Fuel isotopic composition at EOC;
- Sensitivity coefficients (capture, fission, elastic scattering, inelastic scattering, (n, 2n), nu, mu-bar and chi) of keff and burnup reactivity for TRU nuclides, LBE nuclides and 15N as a function of neutron energy;
- Uncertainty deduced from covariance data for each parameter, reaction and nuclide referred to in point 5.

In Section 4.2.4, the JAEA results are provided. The comparison between the results of JAEA and KIT is presented in Section 4.2.5. In the KIT-JAEA comparative studies, only points 1, 2 and 4 are considered, i.e., the effective neutron multiplication factor k_{eff}, the fission reaction distributions at BOC and EOC, and the fuel isotopic composition at EOC. The proton beam current (point 3) was mainly determined by k_{eff}. The comparisons related to sensitivity and uncertainty evaluations (points 5 and 6) are options for future studies.

4.2.4. Calculation options and results

The JAEA calculations were performed using the codes and methods described in Table 4.7. In the high-energy range (above 20 MeV), the Monte-Carlo code PHITS was used for particle (proton and neutron) transport calculations. An analytical model, INCL, provided the particle cross-sections. PHITS computed the source of neutrons at low energies, below 20 MeV, after collision of particles of higher energy with the target. These low energy neutrons gave the neutron source term for the analyses performed with the ADS3D code system. This system employed the PARTISN code for neutron transport simulations, a burnup code for computing variations in fuel isotopic compositions during reactor operation.

TABLE 4.7. CALCULATION OPTIONS FOR THE JAEA BENCHMARK

High energy part (proton and neutron)	Calculation code	PHITS [4.8]
	Method	Monte-Carlo
	Upper energy	1.5 GeV
	Lower energy	20 MeV
	Treatment of neutrons for transport calculation	Neutrons below 20 MeV were used in the transport calculation.
	Physical model	INCL [4.8]
	Usage of nuclear data library	No
	Nuclear data library	n.a.
	Upper energy	n.a.
	Lower energy	n.a.
Neutron transport part (usually below 20 MeV)	Calculation code	ADS3D (PARTISN) [4.9]
	Method	Deterministic
	Nuclear data library	JENDF-4.0 [4.10]
	Cell calculation code (if used)	SLAROM-UF [4.11]
	Energy group	73
	Upper energy	20 MeV
	Lower energy	1E-5 eV
Burnup calculation	Calculation code	ADS3D (BURNUP)
	Treatment of fission products	Each nuclide
	Burnup chain	ChainJ40 [4.12]
	Time step	100 days
	Treatment of thermal power	Just input 800 MW
Sensitivity and uncertainty analyses	Calculation code	MARBLE (SAGEP/CITATION)
	Nuclear data library for covariance data	JENDL-4.0
	Energy group	70
	Upper energy	20 MeV
	Lower energy	10 μeV

Note: ENDF-B/VII.1 and JEFF-3.2 were also used (except for S/U analyses).

Tables 4.8 to 4.11 present the results of calculations performed by JAEA.

TABLE 4.8. CALCULATED EFFECTIVE NEUTRON MULTIPLICATION FACTOR, BURNUP REACTIVITY AND PROTON BEAM CURRENT AT BOC AND EOC

	BOC	EOC
k_{eff}	0.97996	0.97157
Burnup reactivity (pcm)	838	838
Proton beam current (mA) for the core thermal power of 800 MW	9.33	12.27

TABLE 4.9. RADIAL FISSION REACTION DISTRIBUTION (Z = 170 cm)

Radial position (cm)	BOC ($cm^{-1}s^{-1}$)	EOC ($cm^{-1}s^{-1}$)
30.78	7.749E+12	9.555E+12
35.8	7.540E+12	8.920E+12
40.8	7.369E+12	8.169E+12
45.8	7.362E+12	7.987E+12
50.8	7.374E+12	7.846E+12
55.8	7.361E+12	7.553E+12
60.8	7.342E+12	7.451E+12
65.8	7.300E+12	7.287E+12
70.8	7.235E+12	7.110E+12
77.68 (as outer core)	8.871E+12	8.437E+12
80.7	8.837E+12	8.364E+12
85.7	8.672E+12	8.174E+12
90.7	8.347E+12	7.794E+12
95.7	7.905E+12	7.330E+12
100.7	7.360E+12	6.794E+12
105.7	6.602E+12	6.177E+12
110.7	6.015E+12	5.704E+12
115.7	5.260E+12	4.997E+12
121.29	4.424E+12	4.322E+12

TABLE 4.10. AXIAL FISSION REACTION DISTRIBUTION

Axial position (cm)	BOC ($cm^{-1}s^{-1}$)		EOC ($cm^{-1}s^{-1}$)	
	$R = 30.78$ cm	$R = 77.68$ cm	$R = 30.78$ cm	$R = 77.68$ cm
120	5.781E+12	4.990E+12	7.644E+12	5.538E+12
125	5.747E+12	5.194E+12	7.428E+12	5.580E+12
130	5.986E+12	5.782E+12	7.591E+12	5.994E+12
135	6.547E+12	6.435E+12	8.188E+12	6.498E+12
140	7.040E+12	7.045E+12	8.803E+12	7.059E+12
145	7.434E+12	7.587E+12	9.224E+12	7.445E+12
150	7.727E+12	8.030E+12	9.573E+12	7.834E+12
155	7.876E+12	8.391E+12	9.742E+12	8.082E+12
160	7.931E+12	8.651E+12	9.788E+12	8.300E+12
165	7.883E+12	8.811E+12	9.748E+12	8.398E+12
170	7.749E+12	8.871E+12	9.555E+12	8.437E+12
175	7.549E+12	8.828E+12	9.344E+12	8.392E+12
180	7.282E+12	8.687E+12	8.984E+12	8.252E+12
185	6.971E+12	8.451E+12	8.638E+12	8.070E+12
190	6.613E+12	8.121E+12	8.167E+12	7.762E+12
195	6.224E+12	7.705E+12	7.723E+12	7.448E+12
200	5.810E+12	7.210E+12	7.188E+12	6.988E+12
205	5.374E+12	6.641E+12	6.695E+12	6.553E+12
210	4.946E+12	6.011E+12	6.148E+12	6.215E+12
215	4.534E+12	5.346E+12	5.768E+12	5.728E+12
220	4.387E+12	4.704E+12	5.537E+12	5.104E+12

TABLE 4.11. FUEL ISOTOPE COMPOSITION AT EOC ($\times 10^{24}/cm^3$)

Isotope	Inner core		Outer core	
	Number density	Δ(EOC – BOC)	Number density	Δ(EOC – BOC)
^{234}U	1.742E-06	1.458E-06	1.671E-06	1.316E-06
^{236}U	1.061E-07	3.571E-08	1.305E-07	4.270E-08
^{237}Np	6.575E-04	-2.491E-04	9.194E-04	-2.114E-04
^{238}Pu	2.240E-04	2.074E-04	1.970E-04	1.763E-04
^{239}Pu	2.583E-04	-1.204E-04	3.627E-04	-1.096E-04
^{240}Pu	1.718E-04	-1.558E-06	2.153E-04	-9.460E-07
^{241}Pu	5.185E-05	-2.296E-05	7.026E-05	-2.303E-05
^{242}Pu	6.583E-05	1.804E-05	7.445E-05	1.485E-05
^{241}Am	4.283E-04	-1.753E-04	6.042E-04	-1.486E-04
^{242m}Am	1.635E-05	1.523E-05	1.536E-05	1.397E-05
^{243}Am	1.915E-04	-5.886E-05	2.622E-04	-4.998E-05
^{242}Cm	3.257E-05	3.257E-05	2.773E-05	2.773E-05
^{243}Cm	2.732E-06	2.171E-06	1.804E-06	1.103E-06
^{244}Cm	1.046E-04	2.864E-05	1.158E-04	2.103E-05
^{245}Cm	1.115E-05	3.783E-06	1.166E-05	2.476E-06
^{246}Cm	1.355E-06	5.973E-07	1.354E-06	4.086E-07

The uncertainty deduced from covariance data for k_{eff} at BOC amounts to 1.04%.

4.2.5. Discussion and conclusions

The following calculation results of JAEA and KIT are compared: the effective neutron multiplication factor k_{eff} (Fig. 4.21), the fission reaction distributions at BOC and EOC (Figs 4.22–4.24), and the fuel isotopic composition at EOC (Tables 4.12 and 4.13). In these calculations, homogeneous material compositions in each reactor subassembly (i.e. fuel homogeneously mixed with structure and coolant) in each reactor region is considered.

33-group effective macroscopic cross-sections were generated from an ultra-fine-group ECCO data library (JEFF3.1, 1968 energy groups) of ERANOS. The 2-D (RZ) and the 3-D (HEX-Z) ERANOS neutron transport models have been used.

For the 2-D model, the external radius of the target region was set to 18.7 cm in the 2-D (RZ) model, corresponding to the 13.45 cm pitch. Meaningful differences between KIT and JAEA (ADS3D code, 73 groups) for results at BOC can be noticed as well as for those on the reactivity swing. The radial and axial fission rate distributions are also compared.

In general, the KIT and JAEA results are in reasonable agreement. Compared to the JAEA values, the KIT values show a smaller Np consumption (Table 4.12). The Pu and Am consumptions are slightly larger (in absolute values) as compared to those of JAEA when ENDF7.1 and JEFF3.2 libraries are employed (Table 4.12). These deviations are related to computations for ^{238}Pu and ^{242m}Am isotopes. One may see a good agreement between KIT and JAEA for the consumption rate of other Pu and Am isotopes.

The KIT and JAEA results on isotopic inventories are in good agreement (Table 4.13), except for ^{242}Cm, for which no value has been provided by JAEA.

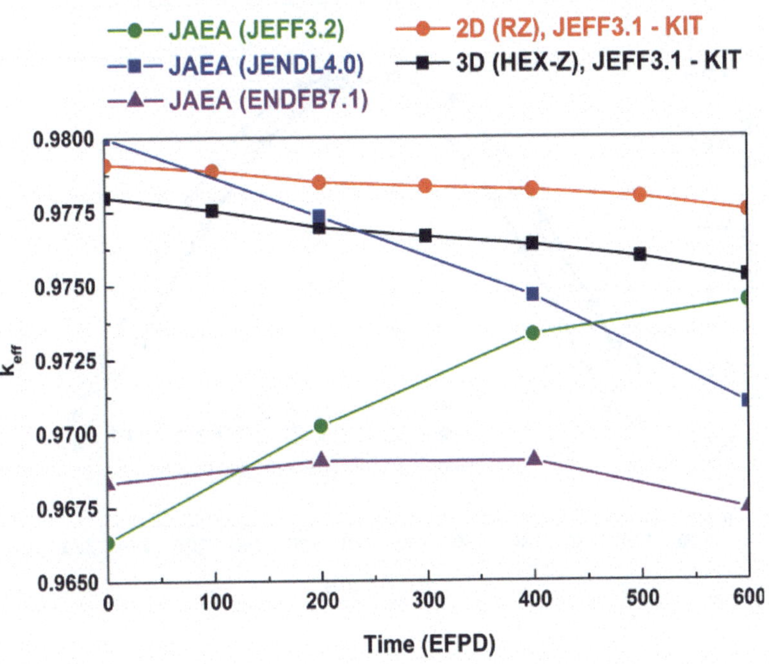

FIG. 4.21. Variation of the effective neutron multiplication factor keff as a function of the irradiation time, computed at JAEA and KIT with different calculation options.

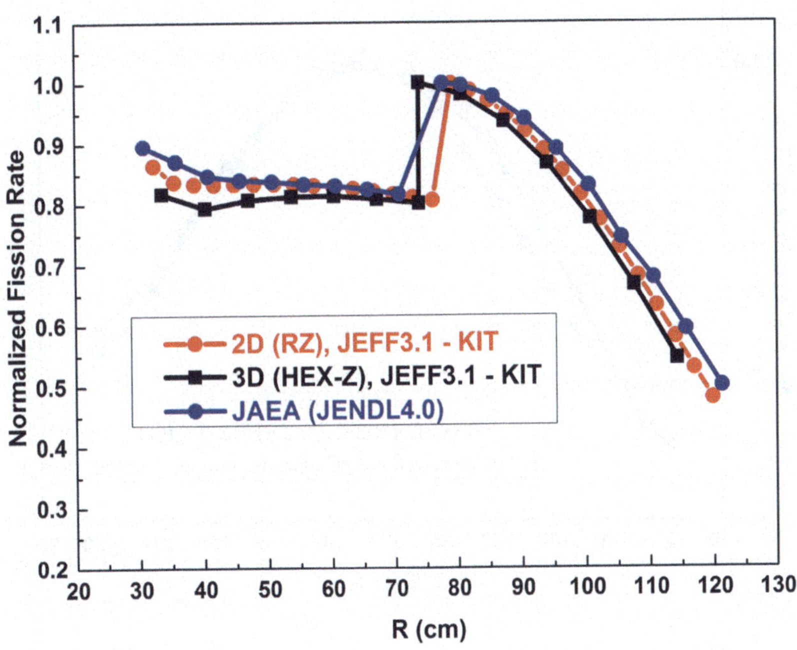

FIG. 4.22. Radial fission rate distributions at Z = 170 (fissile region mid-plane) computed at JAEA and KIT with different calculation options.

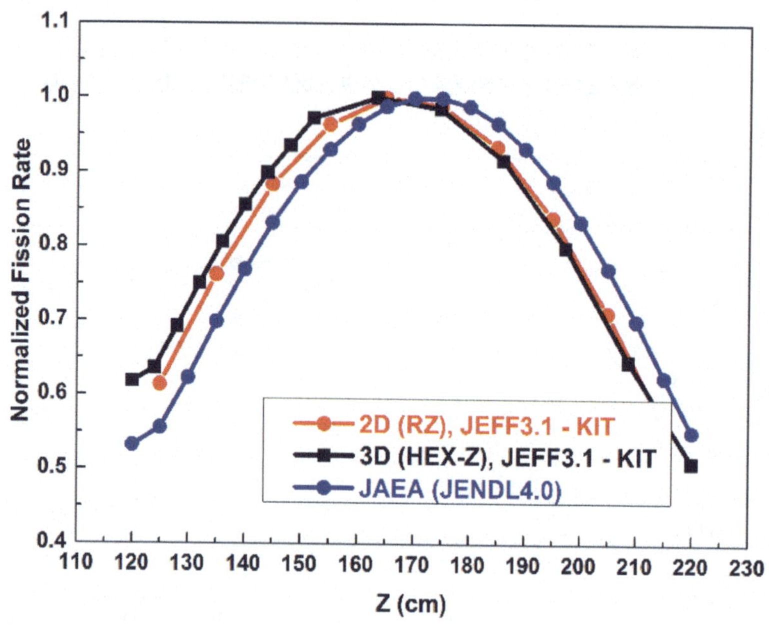

FIG. 4.23. Axial fission rate distributions at R = 30.78 (the innermost SA ring in the core, near the target) computed at JAEA and KIT with different calculation options.

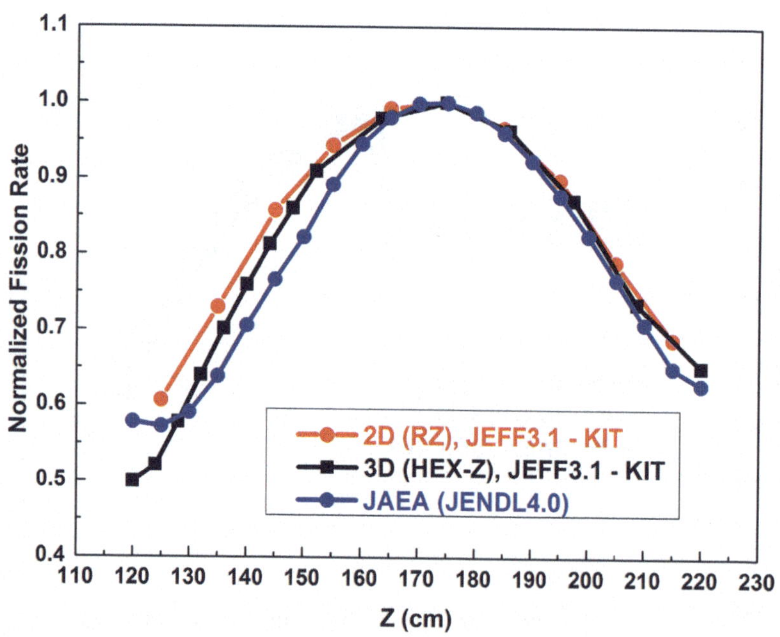

FIG. 4.24. Axial fission rate distributions at R = 77.68 (at the boundary between inner and outer cores) computed at JAEA and KIT with different calculation options.

TABLE 4.12. TRANSURANIC INVENTORY VARIATIONS BETWEEN BOC AND EOC COMPUTED AT KIT AND JAEA

Element	KIT (2D, JEFF3.1)	KIT (3D, JEFF3.1)	JAEA (JENDL4.0)	JAEA (ENDF7.1)	JAEA (JEFF3.2)
Pu	9.4	9.5	9.6	10.7	10.4
Am	-29.7	-29.8	-29.6	-30.8	-31.1
Np	-31.4	-31.4	-33.2	-33.8	-33.2
Cm	9.2	9.2	4.4	4.8	4.6
Total	-42.6	-42.5	-48.9	-49.1	-49.3

TABLE 4.13. TRANSURANIC INVENTORY VARIATIONS BETWEEN BOC AND EOC COMPUTED AT KIT AND JAEA FOR TRANSURANIC ISOTOPES [a]

Isotope	KIT (2D, JEFF3.1)	KIT (3D, JEFF3.1)	JAEA (JENDL4.0)	JAEA (ENDF7.1)	JAEA (JEFF3.2)
^{238}Pu	26.9	26.9	27.8	28.3	28.4
^{239}Pu	-16.4	-16.4	-16.9	-16.5	-16.9
^{240}Pu	-0.1	0.0	-0.2	-0.2	-0.3
^{241}Pu	-3.5	-3.5	-3.5	-3.5	-3.5
^{242}Pu	2.5	2.6	2.4	2.5	2.6
^{241}Am	-23.1	-23.2	-23.7	-24.4	-25.1
^{242m}Am	1.4	1.4	2.2	2.2	2.3
^{243}Am	-8.0	-8.0	-8.0	-8.6	-8.3
^{242}Cm	4.7	4.7	–	–	–
^{243}Cm	0.1	0.1	0.2	0.2	0.2
^{244}Cm	3.5	3.5	3.6	4.1	3.9
^{245}Cm	0.8	0.8	0.4	0.5	0.4
^{246}Cm	0.1	0.1	0.1	0.1	0.1

[a] The results for ^{237}Np are the same as for Np and not given

4.3. REFERENCES TO CHAPTER 4

[4.1] YAN, X. et al., Concept of an Accelerator-Driven Advanced Nuclear Energy System, Energies 10 [7] (2017) 944–956.

[4.2] YANG, L., Development of C-ADS, International Symposium on New Horizons of Partitioning and Transmutation Technologies with Accelerator System, Tokyo, Japan (2018).

[4.3] YANG, L., ZHAN, W., New concept for ADS spallation target: Gravity-driven dense granular flow target, Science China Technological Sciences **58** [10] (2015) 1705–1711.

[4.4] AGOSTINELLI, S., et al., GEANT4 – a simulation toolkit, Nuclear Instruments and Methods in Physics Research Section A: Accelerators, Spectrometers, Detectors and Associated Equipment **506** [3] (2003) 250–303.

[4.5] TSUJIMOTO, K., et.al., Neutronics Design for Lead-Bismuth Cooled Accelerator-Driven System for Transmutation of Minor Actinide, J. Nucl. Sci. Technol. **1** [1] (2004) 21–36.

[4.6] NISHIHARA, K., et al., Investigation of nuclear data accuracy for the accelerator-driven system with minor actinide fuel, in: Proceedings of 11th OECD/NEA Information Exchange Meeting on Actinide and Fission Product Partitioning and Transmutation, San Francisco, CA, USA (2010).

[4.7] SUGAWARA, T., et al., Impact of impurity in transmutation cycle on neutronics design of revised accelerator-driven system, Annals of Nuclear Energy **111** (2018) 449–459.

[4.8] SATO, T., et al., Particle and Heavy Ion Transport Code System PHITS, Version 2.52, J. Nucl. Sci. Technol. **50** [9] (2013) 913–923.

[4.9] SUGAWARA, T., et al., Development of three-dimensional reactor analysis code system for accelerator-driven system, ADS3D and its application with subcriticality adjustment mechanism, J. Nucl. Sci. Technol. **53** [12] (2016) 2018–2027.

[4.10] SHIBATA, K., et al., JENDL-4.0: A New Library for Nuclear Science and Engineering, J. Nucl. Sci. Technol. **48** [1] (2011) 1–30.

[4.11] HAZAMA, T., et al., Development of a fine and ultra-fine group cell calculation code SLAROM-UF for fast reactor analyses, J. Nucl. Sci. Technol. **43** [8] (2006).

[4.12] OKUMURA, K., et al., Analyses of assay data of LWR spent nuclear fuels with a continuous-energy Monte Carlo code MVP and JENDL-4.0 for inventory estimation of ^{79}Se, ^{99}Tc, ^{126}Sn and ^{135}Cs, Progress in Nucl. Sci. Technol. **2** (2011) 369–374.

Chapter 5

PHYSICS DEVELOPMENT

This Chapter presents the methods that were developed during the CRP. These methods have been used to carry out some of the evaluations that are presented in the previous Sections.

Section 5.1 illustrates the methodology set up for the analysis of fuel burnup in spallation source driven multiplying systems. To carry out the burnup analysis of a subcritical system, the source term to be introduced in the neutronic calculations needs to be correctly defined to obtain a consistent level for the power generated. The procedure for the source definition and the power normalization is here described, together with the procedure adopted for the simulation of the ANL conceptual design. Although developed for a specific application, such procedures are quite general and may be applied to any subcritical system.

Section 5.2 is devoted to the MAρTA method for the prediction of the system reactivity from neutron detector signals. This method is based on the mathematical properties of point kinetic differential equations. Also, the following Section 5.3 presents a new method for the reactivity estimation, based on the behaviour of a spectral index that can be measured with proper neutron detectors.

The standard methods for the reactivity determination based on the point kinetic model are affected by spatial and spectral effects that need to be compensated by proper corrections. Section 5.4 presents two techniques that were studied during the project to mitigate spatial and spectral effects.

Section 5.5 addresses some problematics connected to the physically consistent definition of the kinetic parameters for a subcritical system. The problems of noise analysis methods and the techniques developed to filter statistical fluctuations are discussed in Section 5.6. Finally, Section 5.7 illustrates a novel comprehensive scheme for the design of an accelerator driven system using fuels containing fissile and fertile components.

5.1. METHODOLOGY FOR ADS FUEL BURNUP ANALYSIS

5.1.1. Introduction

In ADS systems, the neutron multiplications of the subcritical core and the neutron flux fields within the core continuously varies as nuclear fissions, nuclear absorptions, and nuclear decay processes change the fuel compositions. Fuel burnup simulations solve both the coupled time-dependent neutron transport equations for the neutron flux field and the time dependent Bateman equations (transmutation/depletion equations) for the nuclide number densities. The nuclide number densities within the core change slowly within days or years. The neutron flux field responses quickly to the core configurations within seconds or reaching equilibrium within few minutes. The nuclide number densities and the neutron flux field are loosely coupled. For example, in many of the Monte Carlo fuel burnup codes [5.1]–[5.7], the transport equations are solved at a series of static time points. The overall calculation scheme for the fuel burnup simulation is a predictor–corrector scheme, in which the fuel number densities are first predicted by solving the transmutation equations using reaction rates calculated from the previous time step. The transport equations are then solved again using the predicted fuel number densities to provide better estimates of the reaction rates. Finally, the Bateman equations are solved for correcting the fuel number densities using the updated reaction rates. In some cases, the predictor–corrector step may require fine burnup steps to accurately capture the nonlinear flux variations locally induced by neutron absorbers such as gadolinium. Higher-order depletion schemes, such as the quadratic depletion method, have also been recently developed for enabling the simulation with longer burnup steps [5.8].

The critical fission reactors are self-sustained and are operated without external neutron sources. Generally, Monte Carlo fuel burnup codes are suited for analysing the fuel depletion in critical fission reactors, in which the critical eigenvalue equations are solved for neutron flux field and its magnitude is determined from the user specified fission power. However, not every fuel burnup code can be used for ADS systems. The ADS system needs extra sources particles to support its fission chain reaction. Therefore, the fixed-source neutron transport equations need to be solved for the neutron flux field. Its absolute value is proportional to both the external source strength and the neutron multiplications in the subcritical fission assembly. In ADS systems, if the total fission power is kept constant, the reactivity loss due to the fuel burnup has to be compensated by additional means. For liquid fuel systems such as molten salt fuel or other slurry fuels in the subcritical blanket, external fuel feeding, or on-line fuel reprocessing can be used for controlling the power [5.9]–[5.12]. For solid fuel systems, the reactor power can also be maintained by reshuffling the fuel, by increasing the external source strength, or by adjusting the control rod positions [5.13]. Depending on the type of the ADS system and the different operation modes, different approaches may be considered to simulate the fuel burnup in an ADS system.

There are many ADS conceptual designs which use the spallation neutron target to drive its subcritical fission blanket either for generating power, or for serving as spent fuel transmuters [5.10], [5.12] and [5.14]–[5.20]. The fuel burnup analyses for these conceptual designs are often performed in a two-step approach. In the first step, a high energy physics code such as the high energy physics modules in MCNP6 is used to simulate the spallation process and to generate the neutron source. In the second step, the neutron source is used with a general fuel burnup code to calculate the fuel depletion in the subcritical fission blanket. The reason of adopting the two-step approach is due to the lack of high energy physics module in many of the fuel burnup codes developed for critical fission reactors. MCNP6 has the high-energy physics module. However, within the code, the critical eigenvalue equations are always solved for the neutron flux field in its fuel burnup simulations. Therefore, the MCNP6 alone is not able to simulate the fuel burnup in the ADS systems.

There are many different approaches of modelling the fuel burnup in the ADS system. The objective of Section 5.1 is to address two key points that were learned within this CRP. The discussion is particularly related to the fuel burnup simulations for the ADS systems, which are driven by spallation neutron sources. Without paying attention to these key points, the analysis shows that it is possible to underestimate the total fuel burnup within the subcritical assembly by 10% or more [5.18]. The fuel burnup analysis performed in the ANL ADS conceptual design is used as an example for illustrating two key points. Although some of the details are related to the use of the two Monte Carlo fuel burnup codes SERPENT and MCB5 in this example, the discussion is applicable to similar work using other fuel burnup codes, which do not contain the high energy physics module. A complete two-step procedure of simulating the fuel burnup in the ANL ADS conceptual is discussed at the end of this section.

5.1.2. Source term for fuel burnup analysis in spallation neutron source driven subcritical system

The first key point to be discussed is the spallation neutron source used in solving the fixed-source neutron transport equations for the ADS systems. In the fuel burnup analysis performed for the ANL ADS conceptual design, the neutron source was generated by modelling the 1 GeV protons striking the liquid lead target. The high energy physics module in MCNP6 was used to simulate the spallation process. The subcritical fuel assembly was not modelled in this simulation and vacuum boundaries were imposed on the outer boundary of the target zone. In this simulation, the spallation neutrons' positions, energies, and flying directions were all saved in an external volume source file once they were born. These spallation neutrons were further transported in the target zone. Any neutron tracks that leaked out of the target boundary were also recorded into the surface neutron source file using the WSR card in MCNP6. E of the two source files can be used for the fuel burnup simulations. At each burnup step, the starting neutron source particle was randomly sampled from the neutron particles saved in one of these two source files. It is assumed that the biases in the later Monte Carlo fuel burnup steps due to the usage of the same source file are small, since the source neutron from the source file will start with a different random number.

In the ANL conceptual design, the MCNP6 high energy physics simulations were performed with about half million histories. On average there were about 15.6 neutrons per 1 GeV proton particle being tallied at their born site. These source neutrons were recorded in the volume source file. Its energy spectrum is plotted in Fig. 5.1. Among them, about 16% of the neutrons have energy higher than 20 MeV, and about 2% of the neutrons have energy higher than 200 MeV. The spallation source neutrons were transported within the target region after they were born and created more neutrons through the (n, xn) interactions. On average, for every 1 GeV proton, about 29.5 neutrons leaked out the target zone. These neutrons were saved in the surface source file. The energy spectrum of these source neutrons is also plotted in Fig. 5.1. It shows that the fraction of high energy neutrons in the surface source file has been significantly reduced. For example, there is only about 1.6% of the neutrons with energies higher than 20 MeV, and about 0.1% of the neutrons with energies higher than 200 MeV. In the current ENDF/B cross-section data files for neutrons, the upper energy boundary is 20 MeV for most nuclides. Some nuclides such as most of the lead isotopes have extended cross-section data with upper energy boundary of 150 MeV or 200 MeV. Despite MCNP6 cannot model the fuel burnup in the ANL ADS conceptual design, it can simulate the coupled system at steady state. In this step, the proton–neutron–photon coupled simulation modelled the spallation process as well as source neutrons leaking out of the target zone and their multiplication in the subcritical core. The neutron target zone and the subcritical assembly were both included in the model. MCNP6 used its high-energy physics models to generate neutron cross-sections if the neutron energy is higher than the upper limit of the ENDF/B cross-section data files. The total nuclear heating generated in the ANL subcritical core was calculated to be 2.67 GW using the +F6 tally assuming that the proton beam power is 25 MW.

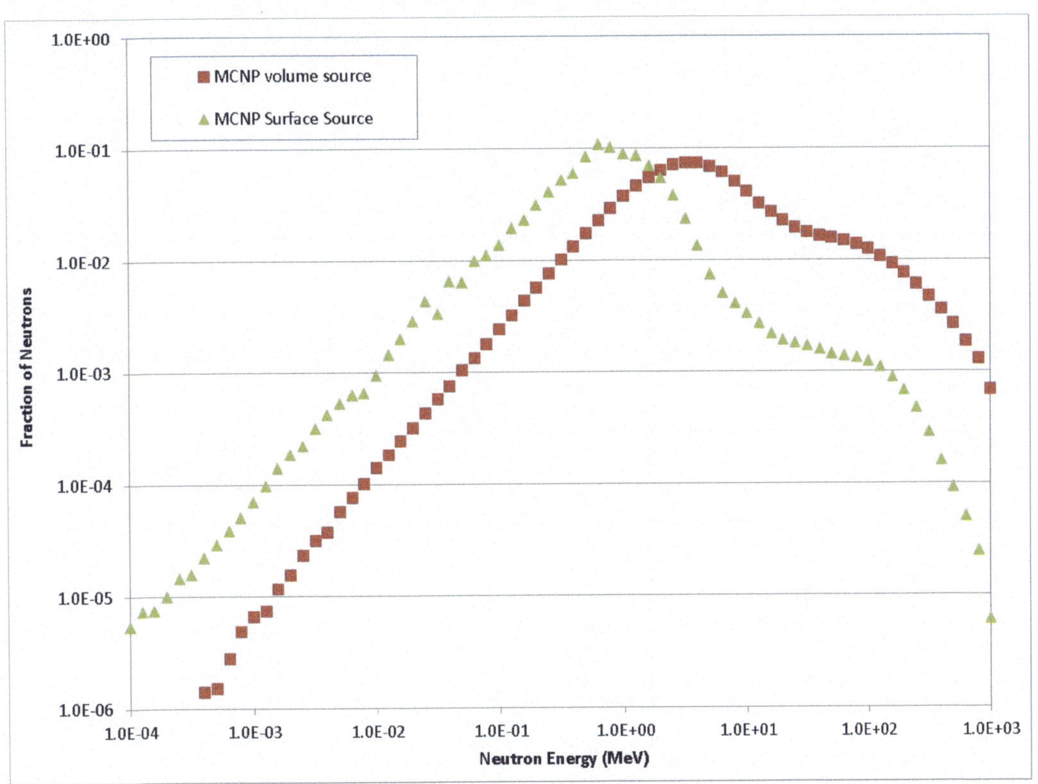

FIG. 5.1. MCNP6 calculated neutron fractions of neutron sources for ANL ADS conceptual design.

MCB5 and SERPENT are the two Monte Carlo codes tested for simulating the fuel burnup in the ANL ADS subcritical core. Neither code has physics models for high-energy neutrons. MCB5 uses the cross-sections at the upper energy boundary if the neutron energy is above that [5.21]. SERPENT ignores the high-energy neutrons by setting the cross-sections to zero [5.22]. For the ANL ADS conceptual design, both SERPENT and MCB5 have used the source files to calculate the total fission power for the initial core at the steady state. SERPENT simulation using the volume source file generates 2.13 GW, and SERPENT simulation using the surface source generates 2.41 GW. These calculations underestimate the total power by about 20% and 10%, respectively. Similarly, the total fission power calculated from MCB5 is 2.14 GW using the volume source file. It is like the SERPENT result and is also significantly less than the total power calculated from MCNP6. In both MCB5 and SERPENT there are different options that will impact the calculated total fission power. These options and their impact will be addressed in the next section for the second key point. The total fission powers obtained from the two codes were adjusted so that they can be compared with the MCNP6 +F6 tally result. The comparison of the calculated total fission power from the three different codes indicates that for ADS systems driven by spallation neutron source, the contributions of the high-energy neutrons to the total fission power are significant. These high energy neutrons can neither be ignored nor be treated as low energy neutrons without proper normalizations in the fixed-source neutron transport calculation.

For further investigation, total neutrons fluxes in the ANL ADS subcritical assembly were also recalculated assuming that the subcritical system was driven by a point neutron source with an energy of 10, 20, 100, 150, 200, or 300 MeV, respectively. These external point sources were monoenergetic and isotropic and were placed at the centre of the target zone. Four independent simulations were performed using MCB5, SERPENT, MCNP6 with the high energy physics module enabled, and MCNP6 with the high energy physics module disabled for every neutron source. Table 5.1 shows that, for the cases with 10 MeV or 20 MeV neutron source energy, all four calculations agree with each other very well. When the neutron source energy increases, smaller neutron fluxes were obtained from the MCB5 simulation, the SERPENT simulation, and the MCNP6 simulation with high energy physics model disabled. In addition, the MCB5 code and the MCNP6 code without the physics model consistently provide similar results for all the cases. The total fission power calculated by SERPENT or MCB5 is smaller than the MCNP6 proton–neutron–photon coupled simulation with the high energy physics module enabled. This is due to the lack of neutron cross-section data for the high energy source in the fixed-source transport calculations.

TABLE 5.1. COMPARISON OF THE TOTAL NEUTRON FLUXES OF THE ANL BUNDLE CONCEPTUAL CONFIGURATION CALCULATED WITH MCNP6, SERPENT, AND MCB5 CODES USING A POINT NEUTRON SOURCE WITH DIFFERENT ENERGIES

Neutron source energy (MeV)	MCNP6 Flux (SD)			MCB5			SERPENT		
	I Physics model enabled	II Physics model disabled	(I–II)/I (%)	Flux (SD)	(I–MCB5)/I (%)	(II–MCB5)/II (%)	Flux (SD)	(I–SERP)/I (%)	(II–SERP)/II (%)
10	2.9214E4 (0.4%)	2.9282E4 (0.4%)	−0.23	2.9732E4 (0.4%)	−1.77	−1.54	2.9513E4 (0.5%)	−1.02	−0.79
20	4.8417E4 (0.3%)	4.8494E4 (0.4%)	−0.16	4.8391E4 (0.4%)	0.05	0.21	4.8215E4 (0.7%)	0.42	0.57
100	1.5509E5 (0.4%)	1.5436E5 (0.6%)	0.47	1.5193E5 (0.4%)	2.04	1.58	1.4941E5 (0.7%)	3.67	3.21
150	2.1222E5 (0.5%)	2.0989E5 (0.5%)	1.1	2.0891E5 (0.4%)	1.56	0.47	2.0642E5 (0.9%)	2.73	1.65
200	2.8651E5 (0.4%)	2.3437E5 (0.5%)	18.2	2.3388E5 (0.5%)	18.37	0.21	–	–	–
300	4.1643E5 (0.4%)	2.3480E5 (0.5%)	43.6	2.3499E5 (0.5%)	43.57	−0.08	–	–	–

In an ADS system, the effective neutron multiplication factor k_{eff} of the subcritical fission assembly needs to be sufficiently away from the critical condition to provide enough safety margin to operate the system. On the other hand, k_{eff} also needs to be high enough to provide good neutron economy. In the ANL conceptual design, k_{eff} is set to 0.98 during all the operation time. Therefore, the number of neutrons created by each neutron source particle can roughly be estimated to be around 50. In the subcritical core, most of the neutrons are fission neutrons. The external source neutrons only make up a small fraction around 2% of the total neutron population. The fraction of the extremely high energy source neutrons is even smaller. Figure 5.2 plotted several neutron energy spectra in the ANL subcritical core obtained from the MCNP6 coupled simulation with high energy physics module enabled, or from the SERPENT simulation coupling with the volume source file. Figure 5.2 shows that the two neutron spectra are almost identical, and the high energy neutrons have negligible contributions to the neutron energy spectrum within the subcritical core. SERPENT also calculated neutron spectrum using the surface source file and it is almost identical to the spectrum using the volume source file. Therefore, it is not included in Fig. 5.2.

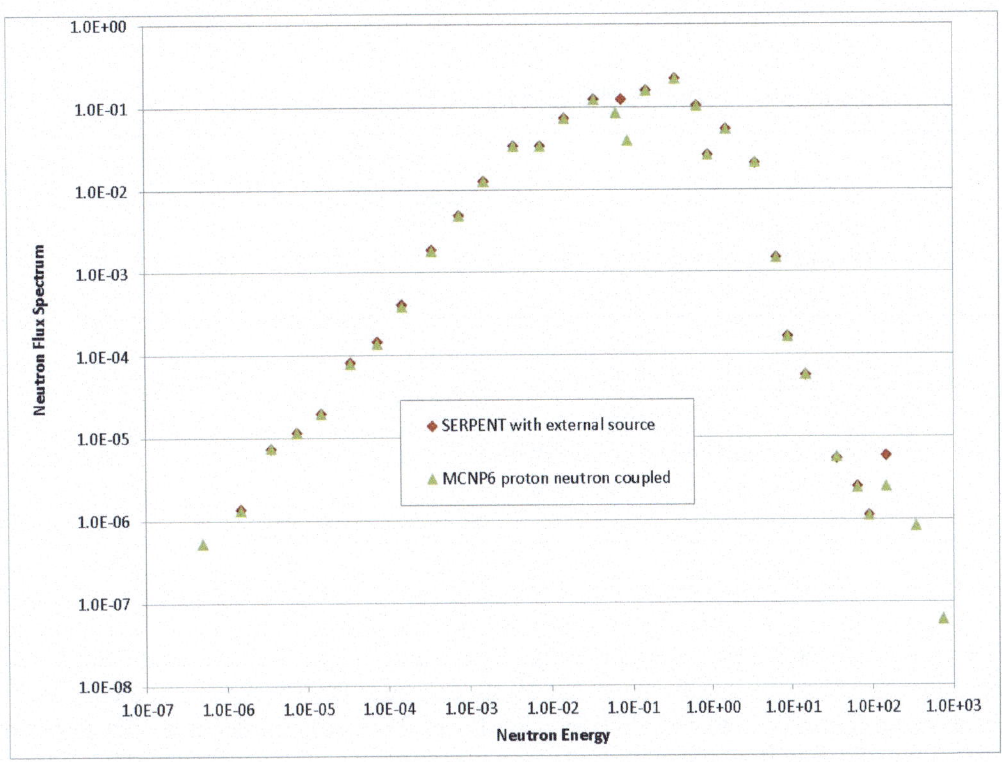

FIG. 5.2. Comparison of the calculated energy spectra in the ANL ADS subcritical fission assembly by the MCNP6 proton neutron coupled simulation, and SERPENT fixed source simulation using the volume source.

In the fuel burnup analysis, the transmutation chain of each nuclide is determined by the reaction rates from various reactions. Particularly the probability for the nuclide moves to different branches is determined by the ratios of the different reaction rates or the ratio of the reaction rate to the decay constants. These reaction rates are determined by the neutron spectrum and the magnitude of the neutron flux field or the reactor core fission power within the core. Therefore, the MCB5 and SERPENT codes can still be used to perform the fuel burnup calculations since the fixed-source transport calculations correctly captured the neutron energy spectrum within the subcritical core. However, the magnitude of the neutron fluxes or the total fission power which is used to determine the absolute value of the reaction rates was not calculated correctly. The results from the fixed-source transport calculation cannot be normalized to the external source strength. Other means are required to provide the proper fission power. For the ANL ADS conceptual design, the total fission power is determined from the MCNP6 proton–neutron–photon coupled simulation with the high energy physics modules enabled.

5.1.3. Power normalization for fuel burnup analysis in spallation source driven subcritical systems

The MCNP6 proton–neutron–photon coupled simulation used the +F6 tally to calculate the total fission power in the subcritical core. It includes the nuclear heating from all the particles, i.e., energies released from fission, neutron scattering, neutron capture as well as prompt gamma scattering [5.1]. The kinetic energy in the proton beam is expected to be mainly deposited in the target region and is not included in this tally. In the ENDF-102 format, the fission energy is defined into several components. Table 5.2 lists the value from the EDNF/B-VII.1 library for each component of the 235U fissioned with thermal neutrons [5.23]. The kinetic energy carried away by the anti-neutrinos are not recoverable and is excluded in the +F6 tally. In MCNP6, there are different options to model the delayed neutrons and delayed gammas. By the default option in MCNP6, the energies carried by the delayed gammas and delayed betas are also not included. The gamma heating from the neutron and prompt gamma scattering interactions are included in the +F6 tally while transporting the prompt gammas in the coupled simulations. Compared with the large energy component released from fission, most of the gamma heating from these scattering events is small in the energy range of keV. Therefore, in the ANL ADS conceptual design, the calculated 2.67 GW power was treated as the total fission energy promptly released in the core.

TABLE 5.2. ENDF/B-VII.1 FISSION ENERGY COMPONENTS OF THE ^{235}U FISSION REACTION [5.23]

Energy component		Fission energy (MeV)	MCNPX/ MCNP6	MCB5	SERPENT
Prompt released energy	Kinetic energy of the fission products	169.13	✓	✓	✓
	Kinetic energy of prompt neutrons	4.84	✓	✓	✓
	Kinetic energy of prompt gammas	6.60	✓	–	✓
Decay heat	Kinetic energy of delayed neutrons	0.0074	✓	✓	✓
	Kinetic energy of delayed gammas	6.33	–	✓	✓
	Kinetic energy of delayed betas	6.50	–	✓	✓
Gamma capture	Gamma ray energy from neutron capture	–	✓	–	–
Unrecoverable	Energy carried away by anti-neutrinos	8.75	–	–	✓
Total energy (MeV)		202.2	calculated	186.8	202.3

There are also a few options which can affect the way MCB5 to calculate the total nuclear heating in the fixed-source neutron transport calculation. By default, the fission power is calculated using KERMA factors. It assumes the prompt gammas are locally deposited all its energies. It also includes the decay heat from the fission products using the data from the ORIGEN library [5.24]. Since no gamma transport simulation is performed, the gamma heating from the scattering event is not included in the calculated fission power. Table 5.2 marked the different components included in the MCB5 default calculation which assumes that the decay heat calculated by the MCB5 using data from the ORIGEN library will hold similar values to the component in the ENDF/B-VII. In MCB5, the total amount of fission energy (Q-value) per fission can be adjusted from the user input. The fission power tally can also be limited to tally burnable zones only. These options may be useful for some fuel burnup calculations, but they complicate the calculation of the total fission power. In the ANL ADS fuel burnup analysis, the default option in MCB5 is used to obtain the fission power. Therefore, the total fission power calculated by MCB5 can be adjusted by the factor of 180.6/186.8 to compare with the fission power calculated by MCNP6 +F6 tally. Where the 186.8 MeV is the total fission energy components included in Table 5.2 for the MCB5 fixed-source calculation and the 180.6 MeV is the total fission energy components for the MCNP6 +F6 tally. In the MCB5 fuel burnup analysis for the ANL ADS conceptual design, the fission power of 2.67 GW was specified to match the tallied fission power from the MCNP6 simulation using the high energy physics module. The fission power components shown in Table 5.2 depend on the fissile isotope as well the incident neutron energy of the fission reaction. In the ANL study, it is assumed that for other isotopes or neutrons at other energies, the relative differences between the MCB5 code and the MCNP6 +F6 tally are similar to ^{235}U.

SERPENT is also flexible in calculating the fission power for fuel burnup simulations. By default, the fission Q-value is used to count for the nuclear heating from the fission interaction. The components included in the Q-value are listed in Table 5.2. It has included energies taken away by neutrinos. It also assumes that all gamma energies have been locally deposited. Since no gamma transport is performed, the gamma heating from scattering events is not included. Similarly, to compare with the MCNP6 +F6 tally result, a factor of 180.6/202.3 can be used to adjust the total fission power calculated from the fixed-source calculation in SERPENT. The fuel burnup analysis performed for the ANL ADS conceptual design used the default option and default Q-values in the cross-section libraries. The total fission power was specified to be about 2.99 GW to match the fission rates calculated by MCNP6 +F6 tally result using the high energy physics module.

Overall, different fuel burnup codes may have different methodologies implemented to model the nuclear heating generated in the reactor core. Therefore, it is the user's responsibility to make sure that the fission power within the core has been modelled correctly. The numbers listed in Table 5.2 show that the different options can overestimate or underestimate the total burnup by more than 10%.

5.1.4. Fuel burnup procedure for simulating the ANL ADS conceptual design

The procedure of using MCB5 or SERPENT to simulate the fuel burnup in the ANL ADS system is illustrated in Fig. 5.3, which was designed to operate in the continuous or batch fuel feeding mode. The source terms were generated separately using the MCNP6 code with the high energy physics module. A shell job script was developed to implement the procedure and oversee the fuel burnup calculations at each burnup.

At each burnup step, the eigenvalue k_0 was first calculated at the beginning of the burnup step. The fuel was depleted during the burnup step by MCB5 or SERPENT utilizing the source terms from MCNP6. Then the critical eigenvalue k_1 was calculated using the fuel composition at the end of the fuel burnup step. A python script was called from the shell script to calculate the feed factor f which requires to compensate the reactivity losses within this burnup step. The fuel compositions are then updated. The eigenvalue of the subcritical fission assembly with the updated fuel composition was recalculated to give a better estimation of f, and the fuel composition was updated again for the next burnup step.

For the continuous fuelling mode, as shown in Fig. 5.3(b), the Monte Carlo fuel burnup simulations have the same procedure to obtain f for each burnup step. To simulate the continuous fuelling, a second fuel burnup calculation was performed with an updated SERPENT which has the continuous feed vector included in its burnup simulations [5.11], [5.25]. The fuel composition at the end of the burnup step was directly used as the initial fuel composition for the next burnup step.

Overall, Fig. 5.3 provides a general procedure of performing the fuel burnup analysis for the Accelerator-Driven subcritical system. Depending on the different types of reactor systems and different types of neutron source, the simulation may not follow the same steps as shown in the figure. However, comparing the fuel burnup analysis done for critical fission reactors, the fuel burnup simulation performed for ADS systems is more complicated. The way of connecting the source term with the subcritical core in the fuel burnup analysis is important and needs to be modelled correctly. In addition, there are generally more steps required in the fuel burnup analysis for the ADS system than for a traditional critical. Not only the fixed-source neutron transport equation needs to be solved within the predictor–corrector scheme, but the critical eigenvalue equation also needs to be solved, and sometimes even solved multiple times to facilitate the fuel reprocessing modelling, or to provide clear guidance on the safety margin of the ADS system at different fuel burnup stages.

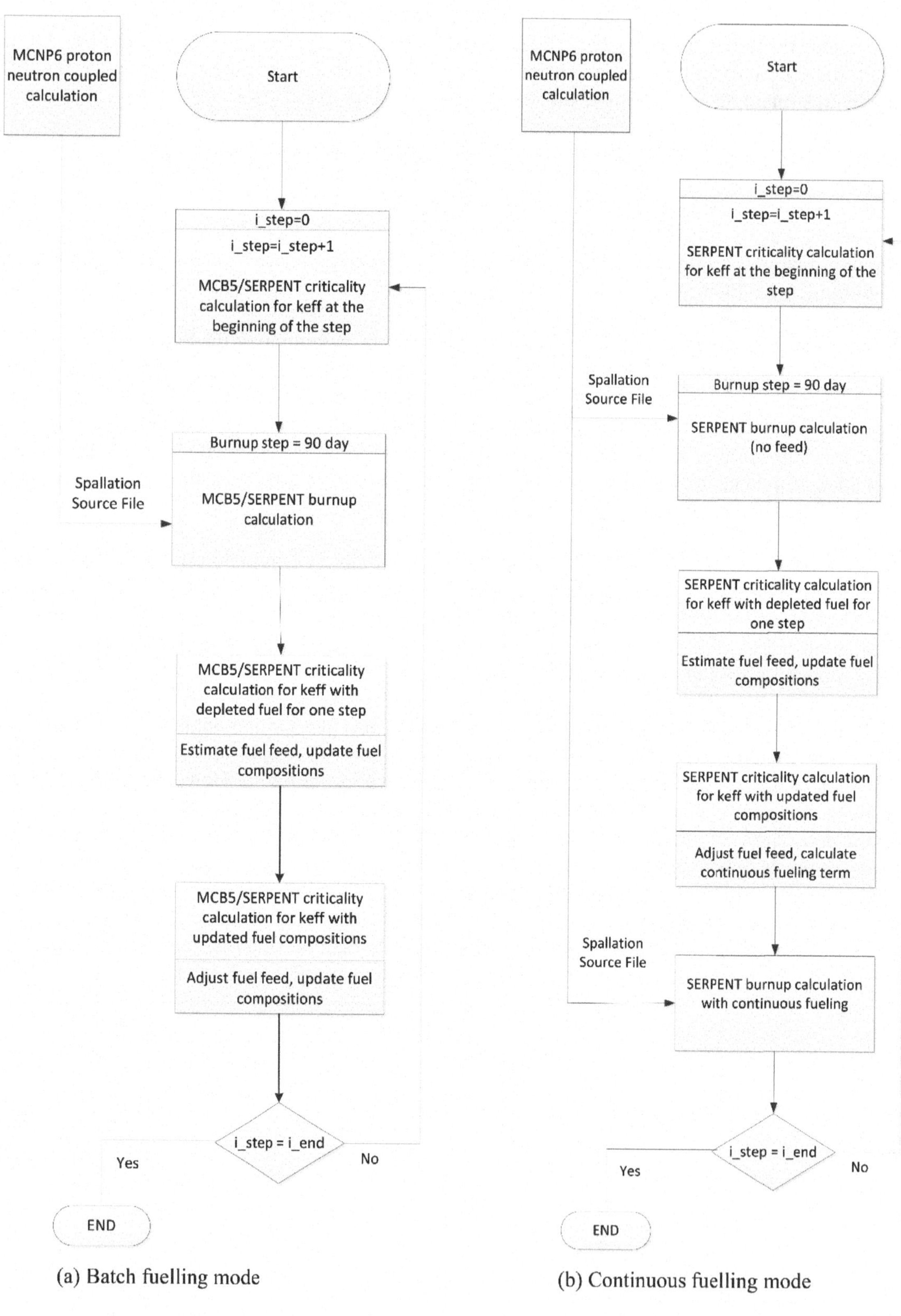

FIG. 5.3. Monte Carlo fuel burnup flow chart for the ANL ADS conceptual design.

5.2.   THE MAρTA METHOD

The Monitoring Algorithm for Reactivity Transient Analysis (MAρTA) method was developed to reconstruct the value of the reactivity of a system during the operation of a nuclear multiplying system using the signal from a neutron detector.

The determination of the reactivity (and, more generally, of any system integral parameters) constitute an inverse problem in reactor physics, as the unknown parameter is recovered from some information obtained by neutronic measurements carried out on the system. Some methods are designed to be used only for a specific type of experiment and, thus, cannot be used for the continuous monitoring of the system during operation. These methods require a change in the external source or an active operation on the control apparatus. The MAρTA method is passive, and it can be applied to both off-line experiments and on-line real-time continuous monitoring during any transient evolutions associated to both normal and accident situations, whatever the initiating cause may be.

All inverse methods require a reference model, that is usually constituted by mathematical equations for the state of the system, which involve the parameter of interest. The inverse algorithm allows to retrieve the parameter from the knowledge of the state of the system that is supposed to be obtained from direct measurements. The MAρTA method is based on the point kinetic model and, hence, the measurement from a local detector is supposed to well represent the global behaviour of the whole system. This basic assumption establishes the limitations of the method. When spatial and spectral effects are relevant, some corrections will be needed, as is discussed in a following section.

The point kinetic model is constituted by a system of first-order ordinary equations, involving the neutron flux amplitude (or the system power) and the effective delayed neutron concentrations [5.26]. The roots of the method are posed on general mathematical properties of first-order differential systems of equations. In fact, it is possible to connect the instantaneous values of the state variables (in the specific problem of neutron kinetics they are the power P and the neutron precursor concentrations C_i , $i = 1, 2, ..., R$, R being the number of delayed neutron families adopted) and the inverse of the fundamental time constant ω of the system in the transient asymptotic exponential evolution that would be reached if no further change in the system parameters would occur [5.27]. In neutron kinetics, it is well known that the instantaneous value of ω is directly related to the instantaneous value of the reactivity ρ through the in-hour equation [5.26]. The information on the power is assumed to be available from flux measurements inside the system by neutron detector counts.

Equations (5.1) and (5.2) report the explicit relationships between the state variables and the time constant for source-free reactors [5.28]:

$$\frac{dP(t)}{dt} - \omega P(t) + \sum_{i=1}^{R} \frac{\lambda_i}{\omega + \lambda_i} \left[\frac{dC_i}{dt} - \omega C_i \right] = 0 \tag{5.1}$$

and for source-driven systems [5.29]:

$$\frac{dP(t)}{dt} - \omega P(t) + \sum_{i=1}^{R} \frac{\lambda_i}{\omega + \lambda_i} \left[\frac{dC_i}{dt} - \omega C_i \right] - S_0 = 0 \tag{5.2}$$

In the latter case, the source needs to be normalized on the basis of the knowledge of the initial configuration, as follows:

$$P_0 = \frac{S_0 \Lambda}{\rho_0} \tag{5.3}$$

Equations (5.1) and (5.2) establish a relationship among the state variables and the time constant ω. The power P is assumed to be measurable using neutron detectors, while the effective delayed neutron concentration can be reconstructed by formally integrating the point equations for the delayed neutrons:

$$C_i(t) = C_i(0)e^{-\lambda_i t} + \frac{\beta_i}{\Lambda}\int_0^t d\tau P(\tau)e^{-\lambda_i(t-\tau)} \tag{5.4}$$

At last, the reactivity ρ is determined using the in-hour equation (Eq. (5.5)):

$$\rho = \omega\Lambda + \sum_{i=1}^{R}\frac{\omega\beta_i}{\omega + \lambda_i} \tag{5.5}$$

In the above equation, the quantities β_i and Λ are the effective delayed neutron fractions and the mean prompt effective generation time, respectively. These integral parameters have to be known to implement the method. It is therefore assumed that these quantities are available from independent measurements or from numerical evaluations and it is crucial to have accurate and consistent values to have good prediction of the system reactivity.

The computation of ω from either Eq. (5.1) or (5.2) requires evaluating the derivative of the power signal and of the effective delayed neutron concentrations (Eq. (5.4), involving a time integral). Therefore, appropriate discrete differentiation and quadrature formulae have to be used. It is to be expected that the experimental noise and uncertainties can strongly affect the accuracy of these evaluations, especially with regards to derivatives. Some filtering techniques [5.30] to reduce the noise are used, in conjunction with an algorithm specifically developed for the differentiation of noisy data [5.31][5.32].

In conclusion, the reactivity estimation procedure can be summarized in the following steps:

1. Measurement of the neutron flux by localized detectors at discrete values of the time: it constitutes the sample of the signal $P(t)$;

2. Filtering/smoothing of the signal by noise reduction techniques;

3. Estimation of the asymptotic time constant corresponding to the instantaneous value of the reactivity of the system by MAρTA;

4. Evaluation of the reactivity by the in-hour equation.

For testing purposes, the technique has been applied to different transient situations, simulated by codes adopting various neutronic models, such as point kinetics and multigroup diffusion. These computational exercises have allowed to test the accuracy of the prediction and to assess the effect of localized spatial and spectral phenomena [5.28], [5.29]. A second step of the testing has been carried out by using experimental data available in the literature [5.33]. In this case, real data for experiments performed on power reactors have been used. The results have proven the applicability and robustness of the algorithm also in the presence of non-linear feed-back phenomena.

MAρTA has been then applied to real experiments carried out on subcritical source-driven configurations [5.34] and, more recently, to KUCA experiments (see Sections 2.1 and 2.2 for details). The results also in these cases have proven the adequacy of the method to predict the system reactivity in transient situations.

At last, sensitivity analyses have been performed to evaluate the effects of the uncertainty on the knowledge of the parameters that are needed for the application of the method, e.g., the effective delayed neutron fraction β, the effective mean prompt generation time Λ, and the initial subcriticality level ρ_0, in the case of source-driven systems. The results have proven that it is very important to have accurate values of these parameters. In particular β plays an important role as it determines the scale of the system reactivity.

5.3. SPECTRAL INDEX FOR REACTIVITY MEASUREMENT

This section describes a semiempirical method that could be useful to monitor the subcriticality level in a fast nuclear system.

The effective neutron multiplication factor k_{eff} is traditionally considered as an essential parameter to determine and to monitor the (sub)criticality level for a neutron multiplying system: this because – from the pioneering times and ideas of Fermi and Wigner [5.35] – the description of the system in terms of point kinetics for the evolution of the neutron population and of the delayed neutron precursors has become so familiar to overcome the manifest difficulties both in measuring k_{eff} as well as in precisely interpreting its values for a subcritical system in a model independent way.

For the case of subcritical fast systems sustained by an external neutron source (i.e., an ADS), it has been suggested [5.39] that the spectral ratio R in Eq. (5.6):

$$R = \frac{\int_{E_{th}}^{\infty} \Phi(E)\,dE}{\int_{0}^{\infty} \Phi(E)\,dE} \tag{5.6}$$

could have a simple relation with k_{eff} if measured in a position where the contribution from neutrons coming directly from the source is negligible. The rationale for this suggestion derives mainly from the fact that in such positions at least the numerator in Eq. (5.6) is a direct measure of how many fissions are taking place, provided the threshold E_{th} selects energies covering the main region of the fission neutron spectrum. Since no specific model to interpret Eq. (5.6) is introduced here, the possible relation $R(k_{eff})$ can be determined and studied only on phenomenological grounds; moreover, for the same reasons, its specific form can in principle depend on the system under examination: then R is expected to be a useful observable for the purpose of monitoring, simple to be locally measured, after a calibration of the detector has been performed.

Other methods proposed to determine k_{eff} instead either rely on the ability to deliver to the system specific beam time structures [5.37] or need complicated post-analysis of the data because time-dependent local flux measurements will yield different k_{eff} estimates that require to be spatially corrected, owing to the simultaneous presence during transients of many normal modes.

On the contrary of what could have been expected, MCNPX simulations have shown that $R(k_{eff})$ depends very little on parameters such as geometry (e.g., rod dimensions, fuel to coolant volume ratio), but more on others, like material composition (fuel and coolant). Four cylindrical systems (models) were used to perform this check; all of them are surrounded by a 120 cm thick lead reflector:

1. A lead type fast reactor core designed as a (1.6 cm × 1.6 cm × 90 cm) hexahedral type lattice containing the 0.37 cm radius fuel rods surrounded by a 0.05 cm thick steel cladding and embedded in solid lead. The fuel is a mixture of 238U and 235U; keff is varied by increasing the 235U enrichment up to 30%.

2. A lead type fast reactor core designed as a (2.3 cm × 2.3 cm × 90 cm) hexahedral type lattice containing 1 cm radius fuel rods surrounded by a 0.1 cm thick steel cladding and embedded in lead. The fuel composition is the same as in (1), with relative 235U enrichment varied up to 12%.

3. A fast waste transmuter core designed as a (1.14 cm × 1.14 cm × 90 cm) hexahedral lattice filled with 0.5 cm radius rods surrounded by a 0.05 cm thick steel cladding and embedded in helium gas. The fuel, a Futurix6 type of ceramic actinide mixture [5.38], can be varied again by changing the plutonium isotopes mass percentage. In this case, it was found that no material can be interposed between the fuel elements to obtain a relatively high value of keff.

4. A sodium type fast reactor core with the same geometry as in (1), with liquid sodium replacing lead and using MOX [5.39] as fuel: keff is varied by changing up to 30% the plutonium isotopes mass percentage in the mixture.

In Fig. 5.4, the function $R(k_{eff})$ is shown for systems (1), (2) and (3). System (4), with sodium as a coolant, has a different slope instead.

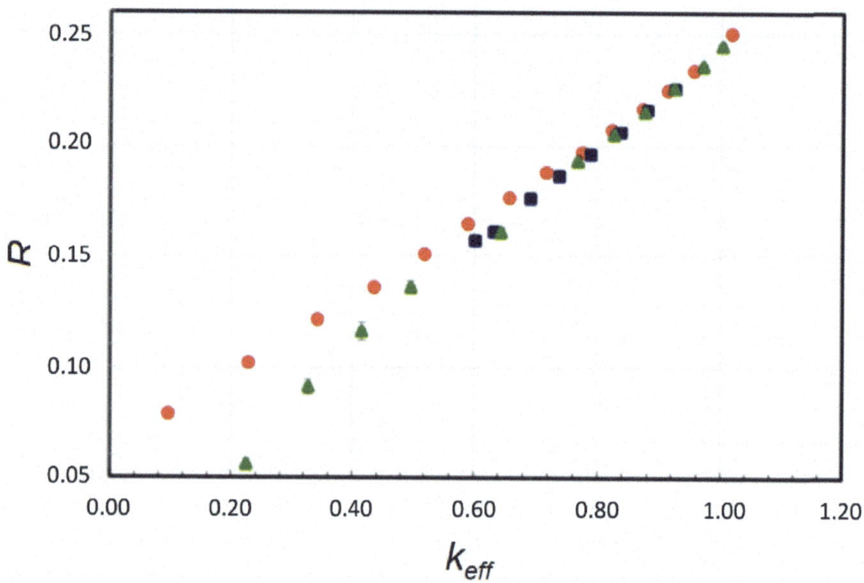

FIG. 5.4. Function R(keff) for a boundary rod for systems (1) (red circles), (2) (green triangles) and (3) (blue squares). Error bars are smaller or equal to the marker size. The last red point for keff = 1.00 was obtained with a k-code run of MCNPX, while all the others were obtained with fixed source runs. It has been assumed that Eth = 0.5 MeV.

The results obtained can be summarized as follows: The spectral ratio R shows a smooth, monotonically increasing dependence (see Fig. 5.4), which can be reproduced in the whole range, inside errors, by a simple quadratic, almost linear, function; the data above k_{eff} = 0.7 can be represented by a linear function. The calculation was repeated for a peripheral rod with a realistic core shape without observing significant modifications. Remarkably, the slopes of the curves are the same within three digits.

The fact that the spectral ratio R does not depend on the geometrical detail of the system, is a good indication of the robustness of the model. In conclusion, for different models of fast subcritical reactors, R approaches an almost linear function of k_{eff}; the functions for the various reactor models are quite similar, with slight differences in slope and approach to the critical limit, indicating a kind of convergence between different types of cores.

The spectral ratio R can thus be used as an observable for estimating k_{eff}, provided the curve has been computed for a realistic reactor model and checked against experimental data (calibration). Clearly, a sudden change of the detected value of R during operations would signal the presence of an anomalous condition.

More details on the robustness of the method can also be found in Ref. [5.40]. It is acknowledged that the presented approach to vary k_{eff} by increasing the amount of fissile material in the fuel mix is not what would be performed in a real experiment. Indeed, in a subcritical system, k_{eff} would be varied for example by extracting or inserting a control rod; While this aspect has not been investigated in the present work, information is available that some results have been presented in this respect by other research groups.

5.4.   REDUCTION OF SPATIAL AND SPECTRAL EFFECTS

All the inverse methods that have been developed and applied for the experimental prediction of a multiplying system reactivity are based on the interpretation of the signal of a single neutron detector. This is also the case for the measurement of integral parameters. Therefore, the information retrieved is localized in space and it depends on the detector characteristics with respect to the neutron energy. The most widely used methods are based on the point kinetic equations as a reference model. Consequently, signals from detectors localized at different positions in the phase space of the system lead to different predictions for the reactivity. To overcome this drawback, methods have been studied to correct for spatial and spectral effects.

The correction procedures are based on various ideas to filter out the contribution of higher order harmonics in the signal or on the use of computational simulations of the whole system to establish correction factors for the local effects. The perfect correction procedure, when applied to the predictions recovered from different detectors, ought to thus lead to the same reactivity value for all of them.

Bell and Glasstone [5.41] performed a theoretical study and derived a factor, also known as Bell-Glasstone spatial correction factor, which helps to overcome the spatial effects in the determination of the reactivity with an inverse approach, such as, for example, the Sjöstrand area ratio method. This factor can be evaluated using Monte Carlo simulations and can be used to correct the measured reactivity. The corrected reactivity value is obtained by multiplying the measured reactivity value by the spatial correction factor.

As an example, Fig. 5.5 shows the evaluated spatial correction factors for axial and radial experimental channels in the BRAHMMA core [5.42], considering both a D–D and a D–T neutron source at the centre of the system. The complete data set is presented in Ref. [5.43]. Correction factors for the entire BRAHMMA core have also been evaluated at 1600 experimental locations. This may facilitate the determination of locations where little or no correction is required while measuring reactivity. Figure 5.6 shows the variation of the spatial correction factor for a single layer A. A1 to A13 denote the fuel locations along the vertical direction.

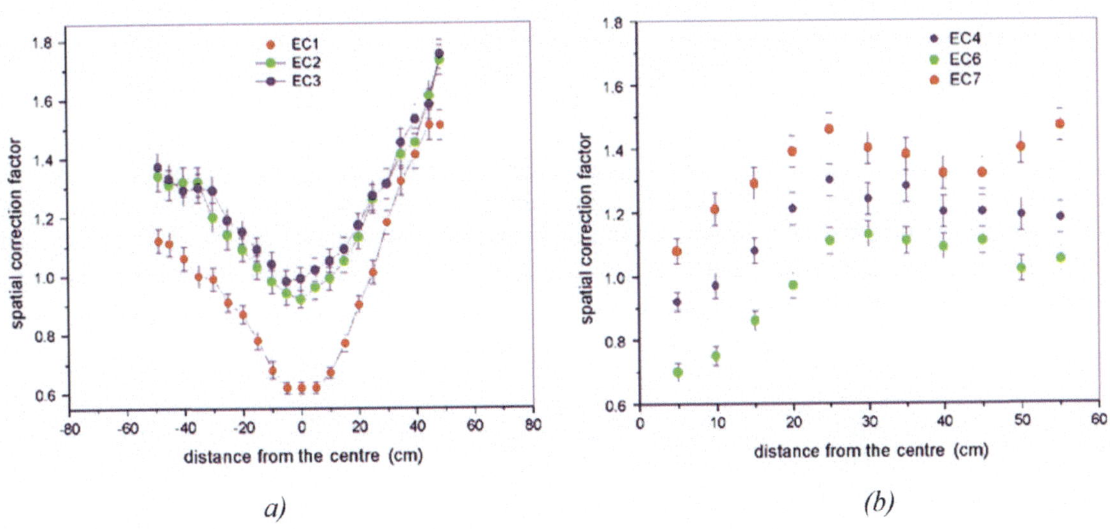

FIG. 5.5. Bell-Glasstone spatial correction factors for (a) axial and (b) radial different experimental channels (EC) in the BRAHMMA core.

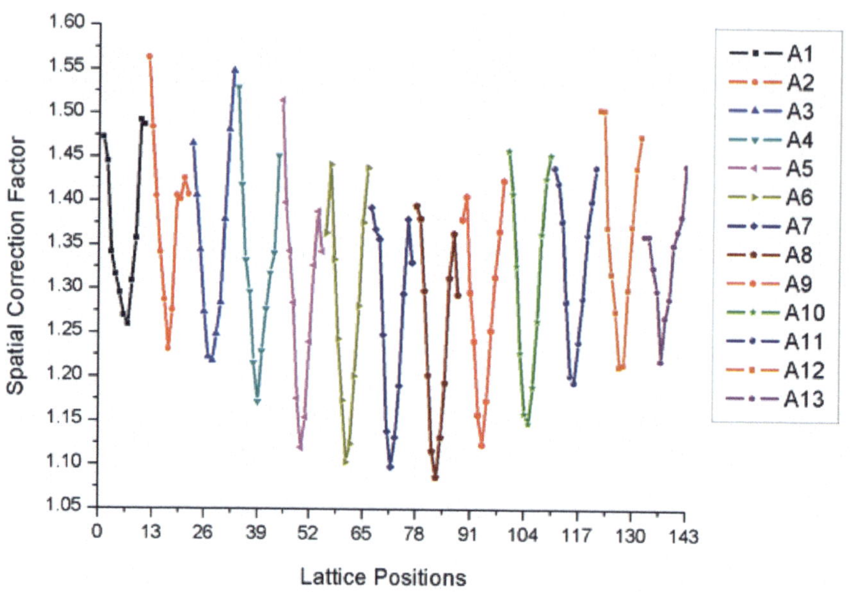

FIG. 5.6. Spatial variation of the correction factor for the first layer of the core.

An alternative philosophy to account for spatial and spectral effects may also be proposed. Suppose the system is equipped with several detectors at different phase-space locations. Their signals provide some information on the global behaviour of the system, which is what point kinetic equations aim to describe, as it is clearly evident when looking at the fundamental separation-projection procedure proposed in Ref. [5.44]. To construct a signal that is more representative of the evolution of the neutron flux amplitude, one may consider a proper combination of the signals of each detector, to be carried out prior to the interpretation. Denoting with $\Phi_i(t)$ the signal for detector i, assuming a linear combination of the signals, one can write:

$$P(t) = \sum_{i=1}^{I} w_i \Phi_i(t) \tag{5.7}$$

where w_i are appropriate weights assigned to each detector and P is the global signal to be used for the interpretation. The choice of the weights plays an important role. Using a physical intuition, one can choose to weigh the signals according to the local neutron importance, as the importance is a measure of the capability of a neutron to contribute to the chain reaction and, hence, to the energy production once injected at a given phase-space point. Consequently, assuming:

$$w_i = \Phi_i^+ \tag{5.8}$$

it is expected that the combination is more representative of the global behaviour of the system than the signal of a single detector [5.45]. The neutron importance Φ_i^+ can be determined computationally by the solution of the adjoint problem for the system being considered [5.46].

The proposal has been tested with a set of computational experiments. The response to a pulsed source was simulated by the Monte Carlo code MCNP and the signals at detectors located at different positions together with various of their combinations were analysed by the MAρTA method [5.29]. Different systems were considered, some of which characterized by strongly decoupled zones in order to strongly enhance spatial effects as a challenge to the interpretation tool. In general, the use of the combination of signals according to Eq. (5.7) using the weights given by Eq. (5.8) improves the quality of the reactivity prediction [5.47], [5.48].

However, there are two aspects that may have a relevant influence on the quality of the results. At first, it is worth observing that the quality of the results depends on the values of the integral parameters β and Λ.

Therefore, these quantities need to be carefully measured or computationally evaluated. It is also clear that if the signal from a very good detector that is representative of the global behaviour of the system is combined with that of a detector that is affected by space effects the prediction will deteriorate. In the choice of the detectors to be used in the combination, the physical insight may serve as a guide. For example, detectors too close to the source or too close to material discontinuities need to certainly be avoided. The best and most representative detectors need to count neutrons that have gained a good knowledge of the multiplication process through the fission reactions. However, their identification may not be straightforward especially for very complicated systems. In conclusion, it is of great importance to develop a solid and robust optimization of the detector distribution inside the multiplying system.

5.5. INTEGRAL PARAMETERS FOR SOURCE DRIVEN SYSTEMS

The study of multiplying systems requires the solution of the equations that constitute the model describing the neutron motion in material media. The reference model is based on the Boltzmann approach, in which the collision contributions to the particle balance are evaluated through statistical principles. However, the neutron Boltzmann transport equation cannot be solved directly for realistic configurations. Alternatively, Monte Carlo methods can be used, although they are computationally very intensive and are affected by statistical uncertainty. For engineering applications, the transport model is approximated, obtaining simplified formulations of the neutron balance, such as the classical diffusion equation. At last, numerical methods with the domain discretization or the series expansions need to be applied to obtain numerical results for a detailed study of the systems of interest. All this process introduces a distortion of the physics that needs to be kept under control with great attention.

The most drastic simplification of the neutronic model amounts to eliminating all the dependences on space, energy, and direction and, therefore, obtaining a purely time-dependent model. This can be done by different mathematical procedures that, of course, are based on the weighted integrations on all the phase variables. The resulting set of equations are known to constitute the point reactor kinetics model. These equations establish a balance of some global physical quantity, such as the total number of neutrons, the fission power, or the total neutron importance, and they may highlight different physical features of the multiplication process. The physical characteristics of the system are all included within a set of few parameters that are known as integral (or kinetic) parameters. In the standard point kinetic model, the effective prompt neutron lifetime, the reactivity, the effective delayed neutron fractions, and the effective source are introduced. Each one of them provides a very relevant physical information on the global characteristics of the system. For instance, the effective neutron lifetime determines the time scale of the neutronic evolution, which is essential to perceive the time response of the system to a perturbation or to a neutron source pulse. The reactivity is the effective measure of the distance of the system from the critical state. Therefore, it is directly connected to the eigenvalue of the steady state neutronic model, i.e., the effective multiplication factor. However, its consistent definition in a transient situation (dynamic reactivity) requires the introduction of the actual neutronic shape and a projection on a weighting function, hence it depends on the evolution of the system. This aspect is considered, for instance, in the quasi-static procedures for transient analyses. The effective delayed neutron fractions have also a prominent role in establishing the contribution of delayed emissions to the multiplication process. Their values depend, of course, on the fissile material, but also on the neutron distribution and energy spectrum.

Although the point models are too simplified for the effective design of reactors, they may be very useful to provide a comprehensive physical information for many engineering assessments and for the interpretation of experiments. Therefore, it is important to have consistent procedures for their evaluation by computational tools. Furthermore, it is also important to develop techniques for their experimental estimations. Usually experiments provide some information on the evolution of the neutron population, hence the determination of the parameters mathematically amounts to the solution of an inverse problem.

The definition of integral parameters depends on the procedure followed to obtain the point model. The integration process (projection) is usually carried out using a proper weighting function. If this function is unitary, the equations will yield a physical model for the evolution of the total number of neutrons in the system. However, a more physically effective model is obtained if the weighting function is chosen as the neutron importance. This procedure is unique and consistent only if the importance is evaluated for a critical reference system, because it is determined as the solution of the unique adjoint transport equation (or any of its approximations). However, for subcritical source-driven systems, the adjoint model requires the introduction of a proper source, whose choice is not unique and somewhat arbitrary, as it may depend on the objective of the analysis. As a consequence, the whole problem of integral parameters needs to be theoretically re-formulated and physically discussed for ADS [5.49]. It is important to recognize that for the modelling of ADS a re-thinking of the whole matter of integral parameters is necessary and no straight generalization of the standard theoretical approach for reference critical systems is possible. As a general statement, one may say that integral parameters need to be problem-tailored, having in mind the purpose for which one intends to use them. Although this issue is particularly important for ADS, the discussion on the significance of integral parameters goes back to the early times of the history of nuclear energy (see e.g., the contribution to the discussion by A.F. Henry in the proceedings of the conference on pulsed neutron research [5.50]). For a more comprehensive overview of the question, one may consult the list of references reported in [5.49], [5.51].

5.6.   FORMULATION OF THE ROSSI-α METHOD

In the measurement methodology for β_{eff}, the Nelson number method based on the Rossi-α method [5.52] provides the advantage of reducing the parameters that are considered difficult to obtain experimentally, including detection efficiency, fission rate at the core centre, and the number of neutrons. The methodology assumes, however, that β_{eff} is measured near the critical state and by locating the external neutron source at the core centre. It is also applicable to the PNS experiments by modifying the formulation of the Rossi-α method. In the PNS experiments, the formulation by the Rossi-α method is already available for processing neutron signals by the methodology used in the analysis with the pulsed neutron source [5.53], [5.54].

In the Rossi-α method, the joint probability $P(t_1,t_2)$ between two neutron signals detected at times t_1 and t_2 is evaluated by categorizing the same fission chain reactions into correlated probability P_C and the different fission chain reactions and the neutron sources into uncorrelated probability P_U, as follows:

$$P(t_1, t_2)dt_1 dt_2 = P_c(t_1, t_2)dt_1 dt_2 + P_u(t_1, t_2)dt_1 dt_2 \tag{5.9}$$

P_C in the existence of the pulsed neutron source can then be expressed, as follows:

$$P_c(t_1, t_2) = g\left[\frac{\lambda_d \lambda_f \langle v_p(v_p-1)\rangle}{2\alpha}\right] e^{-\alpha(t_2-t_1)} dt_1 dt_2 \tag{5.10}$$

where α is the prompt neutron decay const$\langle v_p(v_p-1)\rangle$: second moment of the prompt neutron multiplicity distribution for induced prompt fission neutrons, λ_d the detection efficiency for a neutron, λ_f the detection efficiency of a fission reaction, and g the correction factor taking into account the variation in the probability of detecting correlated counts originating from neutrons of different worth [5.52]. For uncorrelated probability P_U, the formulation of its signal is sensitive to the pulsed shape of the external neutron source [5.53]. In the present study, the shape was regarded as the Gaussian function, and the uncorrelated probability is represented by the sum a constant term $P_{U, const}$ and a trigonometric term $P_{U, trig}$ as shown in Eq. (5.11):

$$
\begin{aligned}
P_U(t_1, t_2)&dt_1 dt_2 \\
&= P_{U,const}(t_1, t_2)dt_1 dt_2 + P_{U,trig}(t_1, t_2)dt_1 dt_2 \\
&= \frac{\lambda_d g * S\Lambda\sqrt{2\pi}\,\sigma}{(-\rho)T_0} dt_1 dt_1 \\
&+ \frac{\lambda_d g * S(-\rho)T_0}{2\Lambda\sqrt{2\pi}\,\sigma} \sum_{n=1}^{\infty} \frac{1}{a^2 + (2n\pi/T_0)^2} e^{-(2n\pi/T_0)^2\sigma^2} \cos\{(2n\pi/T_0)(t_2 - t_1)\} dt_1 dt_2
\end{aligned}
\tag{5.11}
$$

where S is the intensity of source neutrons, T_0 the pulsed period, ρ the reactivity, Λ the generation time and σ the pulsed width and $g*$ the correction factor for spatial and energy distribution of the source neutrons [5.52]. With the result of the Rossi-α method in the PNS experiments, the intensity of P_C and P_U is obtained from the fitting by Eq. (5.9). Since P_C is predicted to decay rapidly, however, the fitting is considered difficult for obtaining both intensities together. Accordingly, the uncorrelated terms were deduced by first fitting in the region, where P_C is sufficiently decayed in Eq. (5.11), with fitting parameters B, D and σ as shown in Eq. (2.5):

$$P_v(t_1, t_2)dt_1 dt_2 = B dt_1 dt_2 + D \sum_{n=1}^{1000} \frac{1}{\alpha^2 + \left(\frac{2n\pi}{T_0}\right)^2} e^{-\left(\frac{2n\pi}{T_0}\right)^2 \sigma^2} \cos\left\{\left(\frac{2n\pi}{T_0}\right)(t_2 - t_1)\right\} d\pi_1 dt_2 \tag{2.5}$$

where B is the fitting parameter for the constant term $P_{U,\ const}$ (Eq. (5.12)):

$$B = \frac{\lambda_d\ g * S\Lambda \sqrt{2\pi}\ \sigma}{(-\rho)T_0} \tag{5.12}$$

The upper value of summation was set as 1000, leaving a margin from the saturation of the fitting results by setting about 300 in the upper value. With the fitting results of B, D and σ, P_C is deduced by subtracting uncorrelated terms from the result of the Rossi-α method in the PNS experiments shown in Eq. (5.9), as follows:

$$P_C(t_1, t_2)dt_1 dt_2$$

$$= P(t_1, t_2)dt_1 dt_2 - \left[B\ dt_1 dt_2 + D \sum_{n=1}^{1000} \frac{1}{\alpha^2 + (2n\pi/T_0)^2} e^{-(2n\pi/T_0)^2\sigma^2} \cos\{(2n\pi/T_0)(t_2 - t_1)\}\ dt_1 dt_2\right] \tag{5.13}$$

Here, the intensity of the correlated probability C is obtained by fitting Eq. (5.13) as follows:

$$P_C(t_1, t_2)dt_1 dt_2 = C e^{-\alpha(t_2 - t_1)} dt_1 dt_2 \tag{5.14}$$

where C is expressed as follows:

$$C = g\left[\frac{\lambda_d \lambda_f \langle v_p (v_p - 1)\rangle}{2\alpha}\right] \tag{5.15}$$

5.7. ESTIMATION OF β_{eff}

For the estimation of β_{eff}, the parameters B and C obtained by the fitting with Eqs (2.5) and (5.15), respectively, are used with the value of α, which is defined with the use of the prompt multiplication factor k_p and the neutron lifetime l (Eq. (5.16)):

$$\alpha = \frac{\beta_{eff} - \rho}{\Lambda} = \frac{1 - k_p}{l} = \frac{1 - (1 - \beta_{eff})k_{eff}}{l} = \frac{\beta_{eff}}{l}\frac{1 - \rho_s}{1 - \rho_s\beta_{eff}} \tag{5.16}$$

where k_{eff} is the effective multiplication factor and $\rho_\$$ the reactivity in dollar units. Further, λ_f and Λ can be rewritten with the use of β_{eff} and k_{eff} (Eqs (5.17) and (5.18)):

$$\lambda_f = \frac{1}{l_f} = \frac{k_p}{l\langle v_p\rangle} = \frac{(1 - \beta_{eff})k_{eff}}{l\langle v_p\rangle} \tag{5.17}$$

$$\Lambda = \frac{l}{k_{eff}} \tag{5.18}$$

where l_f is the mean time between fission iterations, $\langle v_p \rangle$ the average number of prompt neutrons released per fission. With the use of Eqs (5.16), (5.17) and (5.18), the intensity of correlated probability C in Eq. (5.15) can be expressed as follows:

$$C = \frac{g\lambda_d\langle v_p(v_p-1)\rangle k_p}{2\alpha l\langle v_p\rangle} = \frac{g\lambda_d\langle v_p(v_p-1)\rangle k_p}{2\alpha\Lambda\langle v_p\rangle} = \frac{g\lambda_d\langle v_p(v_p-1)\rangle(1-\beta_{eff})k_{eff}}{2(1-\rho_s)\beta_{eff}\langle v_p\rangle k_{eff}} = \frac{g\lambda_d\langle v_p(v_p-1)\rangle(1-\beta_{eff})}{2(1-\rho_s)\beta_{eff}\langle v_p\rangle} \tag{5.19}$$

Also, B shown in Eq. (5.12) can be rewritten with the use of ρ_s and α (Eq. (5.20)):

$$B = \frac{\lambda_d\, g * S\Lambda\sqrt{2\pi}\sigma}{(-\rho)T_0} = \frac{\lambda_d\, g * S(1-\rho_s)\sqrt{2\pi}\sigma}{(-\rho)T_0\alpha} \tag{5.20}$$

Here, the Nelson number N is defined by the combination of α, B in Eq. (5.20) and C in Eq. (5.19), as follows:

$$N = \left[\frac{2\sqrt{2\pi}\,\sigma g * S\langle v_p\rangle}{g\langle v_p(v_p-1)\rangle T_0}\right]\left[\frac{C}{\alpha B}\right] = \frac{(1-\beta_{ef})(-\rho_s)}{\beta_{eff}(1-\rho_s)^2} \tag{5.21}$$

The correction factors g and $g*$ were obtained with by the following calculations [5.52]:

$$g = \frac{\int P(r)dr \int P(r)I^2(r)dr}{(\int P(r)I(r)dr)^2} \tag{5.22}$$

$$g* = \frac{\int S(r)I_q(r)dr \int P(r)dr}{\int P(r)I(r)dr \int S(r)dr} \tag{5.23}$$

$$P(r) = \int \Sigma_f(r,E)\phi(r,E)dE \tag{5.24}$$

$$I(r) = \int \chi(r,E)\phi^+(r,E)dE \tag{5.25}$$

$$I_q(r) = \int \chi_q(r,E)\phi^+(r,E)dE \tag{5.26}$$

where ϕ and ϕ^+ are the forward and adjoint fluxes, respectively, Σ_f the macroscopic fission cross-section, and χ and χ_q the fission spectrum and the spectrum of the external neutron source, respectively. From the relation between N and β_{eff} shown in Eq. (5.21), β_{eff} is obtained experimentally as follows:

$$\beta_{eff} = \frac{-\rho_s}{N(1-\rho_s)^2 - \rho_s} \tag{5.27}$$

The procedure for the deduction of β_{eff} in the PNS experiments is shown in Fig. 5.7.

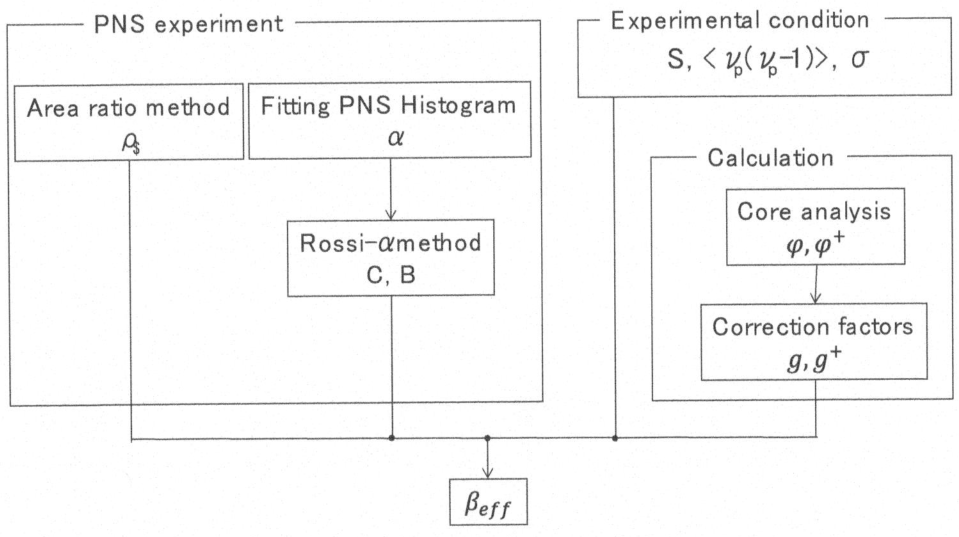

FIG. 5.7. Procedure for deduction of β_{eff} in PNS experiments.

5.8. FILTERING FOR PULSE MEASUREMENT (HUNGARY)

As described in Section 2.1.2.4, Feynman-α and Rossi-α measurements were performed at the KUCA facility with a pulsed neutron source (PNS) to measure kinetic parameters. The Rossi-α method was also applied to determine delayed neutron fraction (β_{eff}) as described in Section 2.1.4.2. The presence of the source pulsing considerably complicates the theory of neutron noise measurements because correlations with the pulsing frequency appear. This results in complicated formulas like Eqs. (2.4), (5.11), (2.5) and (5.13), where parameters as the prompt decay constant α and the intensities of uncorrelated and correlated terms B and C, respectively, are obtained via complex non-linear fitting processes. Although the results are promising, an alternative way of evaluating PNS neutron noise measurements was investigated at BME (Hungary), involving the filtering of the pulsing effect from the raw measurement data.

5.8.1. Description of the filtering method

The filtering technique is based on the concept that the effect of pulsing on noise measurements is due to the artificial correlations introduced by the deterministic pulsing. If the pulsing were random with Poisson-distribution, as in the experiments of Kitamura [5.52] where an isotopic source triggered a neutron generator, the artificial correlations would disappear and the simpler formulas of noise measurements with an isotopic (Poisson) source can be applied. Data processing allows to simulate this effect if the recorded file is cut into parts at the pulses and the pulses, together with the subsequent counts, are redistributed randomly, as it can be seen in Fig. 5.8. Scaling may be needed to avoid empty intervals, which means that the original measurement time is contracted by an appropriate scaling factor and the pulses are redistributed to this time interval.

The random redistribution will clearly eliminate the correlations on longer times then the pulsing period, while the much shorter correlations will be preserved. The question is how this filtering effects the correlations in the range of the pulsing period, what kind of bias may be introduced to the measurements and whether it can be corrected. While detailed theoretical investigations are in progress, simulation results and application for selected KUCA measurements are presented here to assess the applicability of the method.

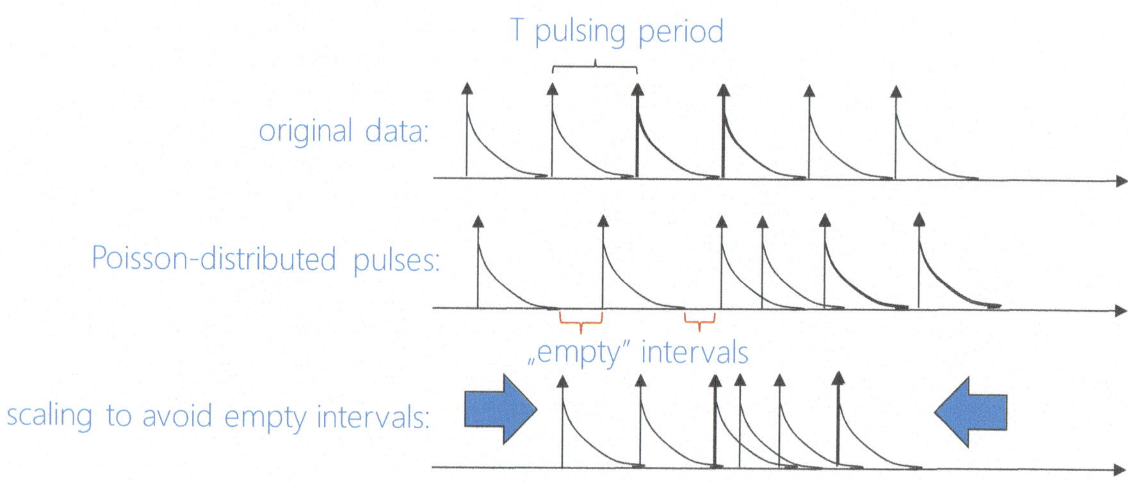

FIG. 5.8. Illustration of the filtering method based on randomization.

5.8.2. Simulations

Simulated noise measurement data files were produced for an infinite homogenous reactor model (including one delayed neutron group) with the assumption of both isotopic (as reference) and pulsed sources, and Feynman-α evaluations were performed on them after applying the filtering method in the case of the pulsed source.

Simulation parameters were set in a way to produce prompt decay constants between 500 and 5000 s^{-1}. Pulsing frequencies were considered from 10 to 100 Hz, while the intensity of individual pulses was the same. By the comparison of the obtained curves to the reference and the fitted α-value to the theoretical one, a scoping study was performed on the applicability of the method in different circumstances. Examples for α-values of 500 s^{-1}, 2500 s^{-1} and 5000 s^{-1} can be seen in Figs 5.9–5.11. As one can see, the filtering method introduces a bias depending on the pulsing frequency, but higher α-values are less affected by it.

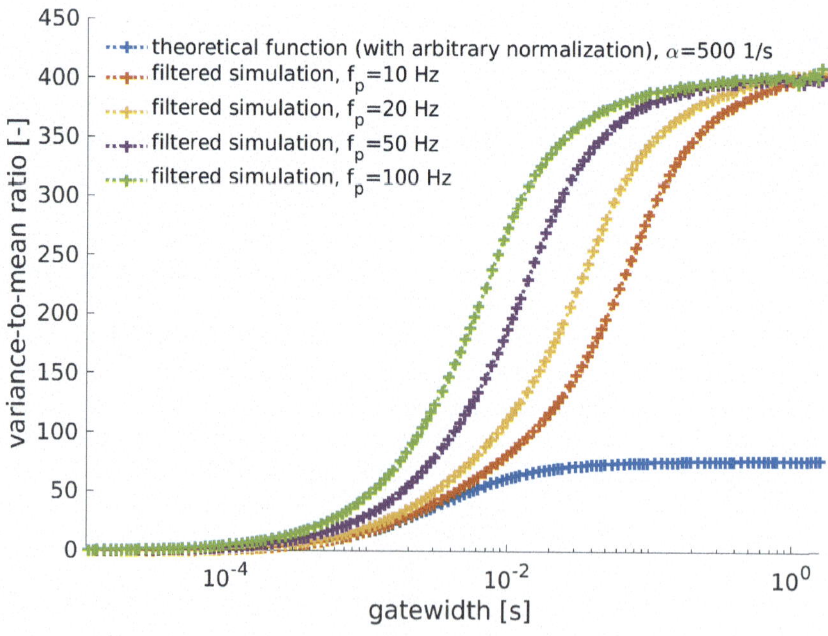

FIG. 5.9. Evaluated variance-to-mean (Feynman) curves of simulated measurements with α=500 s^{-1} after applying the filtering technique (no scaling).

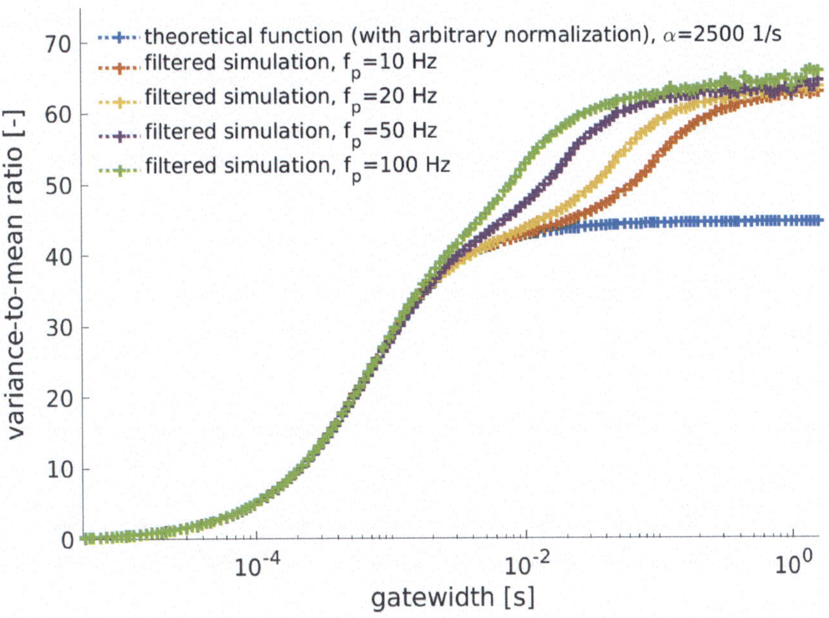

FIG. 5.10. Evaluated variance-to-mean (Feynman) curves of simulated measurements with α=2500 s⁻¹ after applying the filtering technique (no scaling).

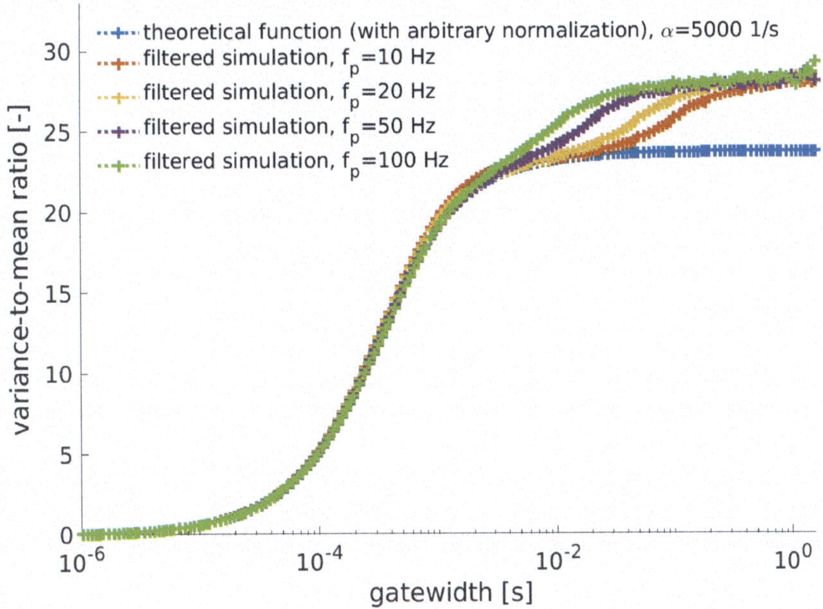

FIG. 5.11. Evaluated variance-to-mean (Feynman) curves of simulated measurements with α=5000 s⁻¹ after applying the filtering technique (no scaling).

This conclusion is also reinforced by Fig.5.12 where the observed deviation of the fitted α-values from the theoretical α is presented for the different configurations. As one can observe, the bias decreases for all frequencies as the α-value increases, and the lower the frequency the lower the bias.

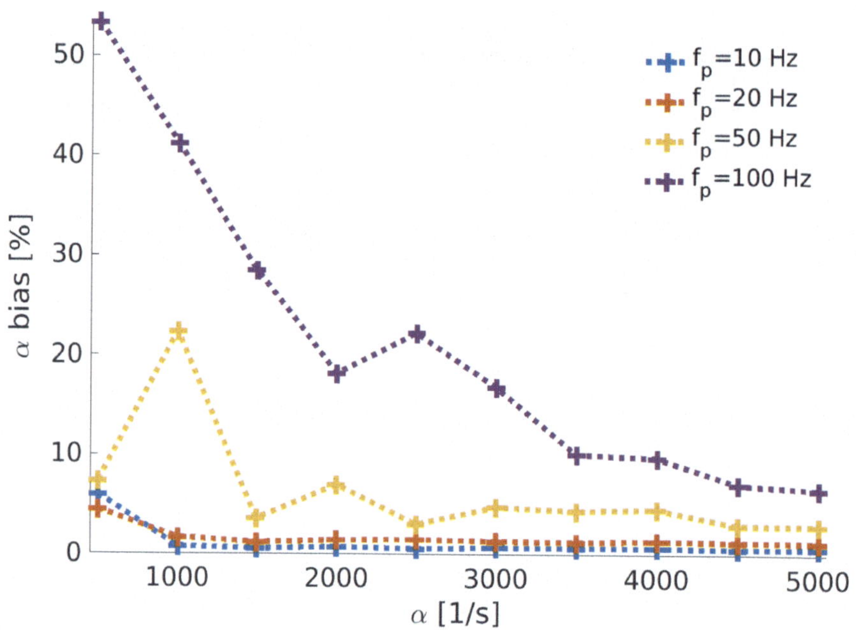

FIG. 5.12 Absolute value of the difference between fitted α-values and the theoretical α in case of filtered pulsed simulations.

5.8.3.   Experimental data

In order to test the methodology on real measurement data, 125 datasets were chosen from the experimental benchmark of the KUCA ADS pulsed source measurement series [5.53], [5.54]. The measurements were performed in different configurations at different subcriticality levels, with k_{eff} values between about 0.9 and 0.99, resulting in different α-values in a range of approximately 400 s^{-1} to 2000 s^{-1}. The applied pulsing frequencies (f_p) varied between 10–100 Hz.

The experimental configurations – including material compositions, assembly structure, zone configurations, as well as detector positions – are described in detail in Sections 2.1 and 2.2.

5.8.4.   Evaluation and results

In the current study no scaling was applied during the filtering process, as the effect of the scaling is not investigated in sufficient detail at this point.

After applying the filtering method on the raw measurement data, Feynman-α evaluations were performed on the filtered datasets. The α-values were estimated from the least-squares fitting of up to three Feynman-α terms with or without an additional linear term, described by the function shown in Eq. (5.28):

$$Y(t_g) = At_g + \sum_{i=1}^{N} B_i \left(1 - \frac{1 - \exp(-\alpha_i t_g)}{\alpha_i t_g} \right) \tag{5.28}$$

where t_g is the gate width, A is the coefficient of the linear term, B_i are the saturation values of the corresponding Feynman-α terms and N is the number of Feynman-α terms. Examples of the fitted curves can be seen in Figs 5.13–5.15, and the results of the least-squares fitting for all datasets are collected in Tables 5.3 and 5.4.

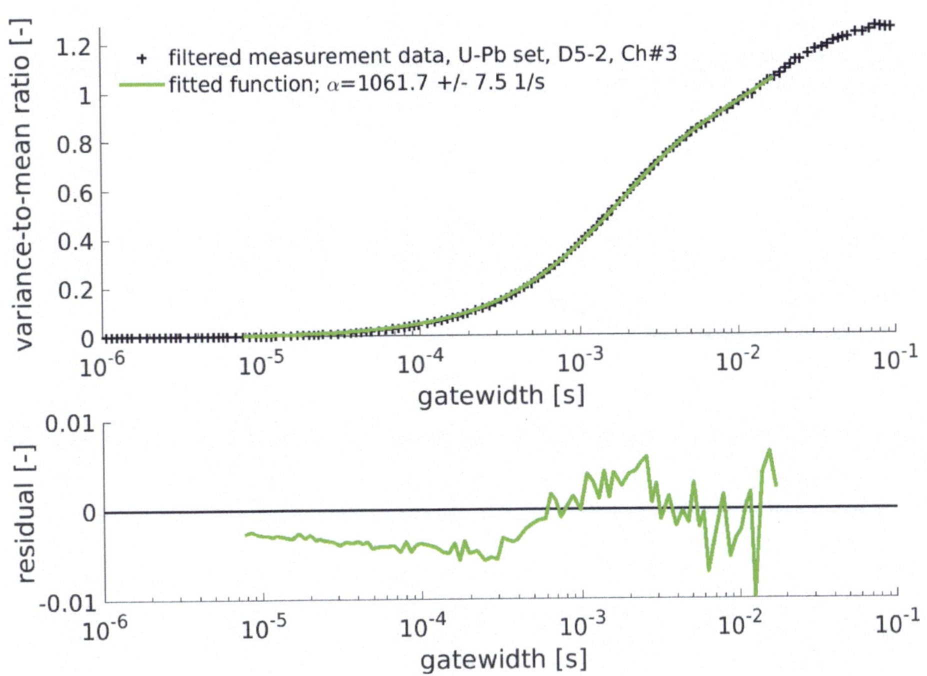

FIG. 5.13. *Evaluated variance-to-mean (Feynman) curve of measurement case D5-2 Ch#3, U–Pb benchmark [5.54] (f_p =50Hz, k_{eff}= 0.9556, α_{pulsed} = 1138.4 ± 8.2 s^{-1}), and the result of the least-squares fit.*

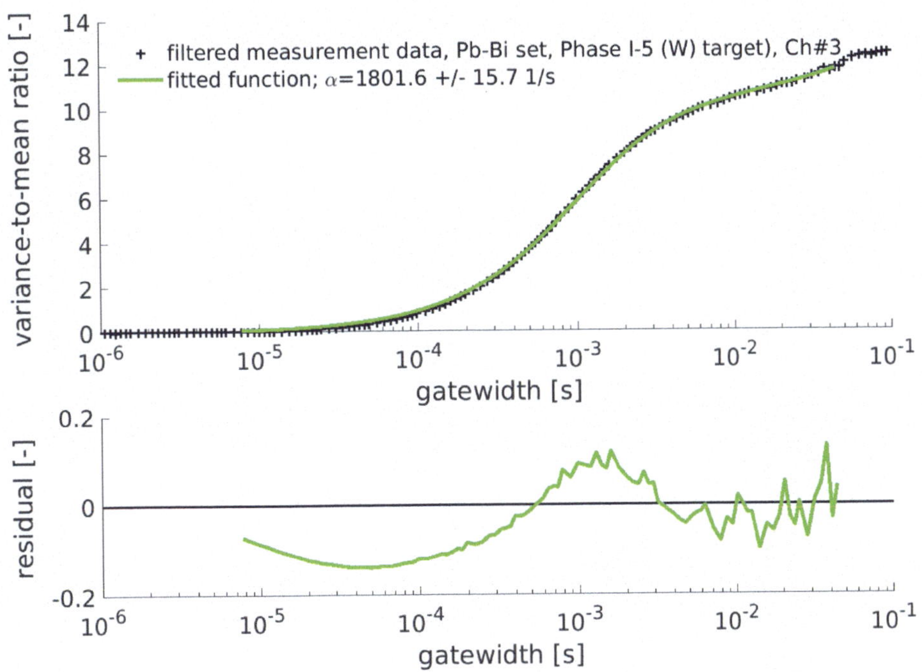

FIG. 5.14. *Evaluated variance-to-mean (Feynman) curve of measurement case Phase I (W target) 5 Ch#3, Pb-Bi benchmark [5.53] , (f_p =20Hz, k_{eff}≈ 0.9, α_{pulsed} = 1965.8 ± 91.7 s^{-1}), and the result of the least-squares fit.*

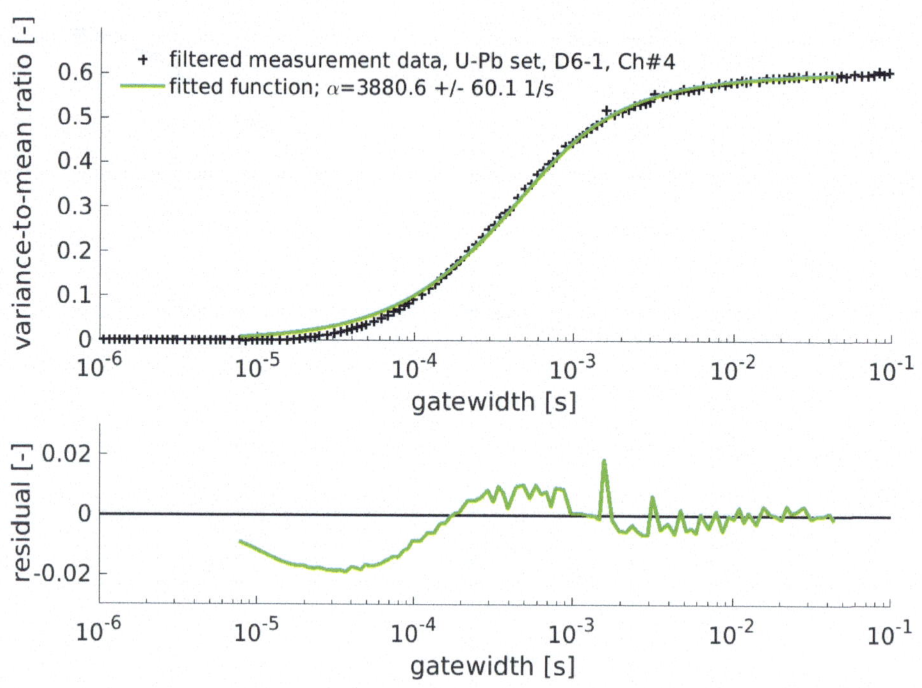

FIG. 5.15. *Evaluated variance-to-mean (Feynman) curve of measurement case D6-1 Ch#4, U–Pb benchmark [5.54], (f_p =20Hz, k_{eff}= 0.93, α_{pulsed} = 4166.6 ± 77.4 s^{-1}) and the result of the least-squares fit.*

TABLE 5.3. RESULTS OF THE LEAST-SQUARES FITTING OF EQ. (5.9) IN CASE OF THE U–Pb BENCHMARK (SECTION 2.2). [A]

Case	Target detector	Reference values		Filtered values		α-difference (%)
		α (s^{-1})	$\Delta\alpha$ (s^{-1})	α (s^{-1})	$\Delta\alpha$ (s^{-1})	
D1-1	Ch#1	423	13	422	27	0,1
D1-1	Ch#2	498	15	358	23	28,1
D1-1	Ch#3	487	11	464	9	4,6
D1-1	Ch#4	562	18	374	34	33,5
D1-2	Ch#1	458	15	365	5	20,4
D1-2	Ch#2	494	14	330	35	33,2
D1-2	Ch#3	482	9	441	17	8,5
D1-2	Ch#4	836	30	321	44	61,6
D1-2	Ch#6	502	147	281	32	44,0
D1-3	Ch#1	434	124	251	10	42,1
D1-3	Ch#2	644	100	222	8	65,6
D1-3	Ch#3	503	92	313	6	37,8
D1-3	Ch#4	1640	132	202	17	87,7
D1-3	Ch#7	538	17	542	6	0,6
D1-4	Ch#1	565	5	503	5	11,0
D1-4	Ch#2	590	5	465	24	21,2
D1-4	Ch#3	555	2	523	5	5,9
D1-4	Ch#6	490	42	217	28	55,7
D1-5	Ch#1	498	7	463	26	7,0
D1-5	Ch#2	489	7	364	45	25,7
D1-5	Ch#3	489	3	464	9	5,0
D1-5	Ch#6	541	128	131	26	75,7
D2-3	Ch#1	250	154	180	1	28,1
D2-3	Ch#2	790	57	185	94	76,6
D2-3	Ch#3	384	38	181	1	52,9
D3-1	Ch#1	669	5	922	12	37,7
D3-1	Ch#2	693	5	940	96	35,6
D3-1	Ch#3	684	2	652	4	4,7
D3-1	Ch#4	1297	35	359	59	72,3
D3-1	Ch#6	704	35	775	80	10,1
D4-1	Ch#1	878	27	906	15	3,2
D4-1	Ch#2	980	28	737	76	24,8
D4-1	Ch#3	997	14	970	7	2,8
D4-1	Ch#4	2346	86	3175	45	35,4
D5-1	Ch#1	1144	18	973	13	15,0
D5-1	Ch#2	1204	15	794	106	34,0
D5-1	Ch#3	1188	6	1085	8	8,7
D5-1	Ch#4	2954	87	3459	44	17,1
D5-2	Ch#1	1094	23	986	13	9,9
D5-2	Ch#2	1193	18	782	155	34,5
D5-2	Ch#3	1138	9	1062	7	6,7
D5-2	Ch#6	824	179	1035	108	25,6
D5-3	Ch#1	960	48	822	21	14,3

TABLE 5.3. RESULTS OF THE LEAST-SQUARES FITTING OF EQ. (5.9) IN CASE OF THE U–Pb BENCHMARK (SECTION 2.2). [A]

Case	Target detector	Reference values		Filtered values		α-difference (%)
		α (s^{-1})	$\Delta\alpha$ (s^{-1})	α (s^{-1})	$\Delta\alpha$ (s^{-1})	
D5-3	Ch#2	1215	44	616	89	49,3
D5-3	Ch#3	1105	16	1006	12	9,0
D5-3	Ch#4	3584	187	3508	76	2,1
D6-1	Ch#1	1493	25	1288	14	13,7
D6-1	Ch#2	1636	42	1234	42	24,6
D6-1	Ch#3	1641	9	1511	11	7,9
D6-1	Ch#4	4167	77	3881	60	6,9
D6-2	Ch#1	1672	80	1451	40	13,2
D6-2	Ch#2	1791	101	973	188	45,7
D6-2	Ch#3	1622	49	1524	16	6,0
D6-2	Ch#4	4289	202	3955	56	7,8
D6-3	Ch#1	1409	46	1178	23	16,4
D6-3	Ch#2	1527	132	1265	88	17,1
D6-3	Ch#3	1641	12	1476	18	10,1
D6-3	Ch#4	4356	139	3852	63	11,6
F2-1	Ch#1	378	4	391	33	3,4
F2-1	Ch#2	369	4	367	9	0,5
F2-1	Ch#3	371	4	353	5	4,9
F2-1	Ch#4	517	10	1497	52	189,2
F2-1	Ch#5	372	4	337	6	9,3
F4-2	Ch#1	950	3	836	8	12,0
F4-2	Ch#2	985	3	902	6	8,4
F4-2	Ch#3	945	5	735	19	22,2
F4-2	Ch#4	990	4	967	6	2,3
F4-2	Ch#6	1126	52	914	39	18,8
F4-2	Ch#7	1413	68	1100	39	22,1
F5-1	Ch#1	1101	4	1012	8	8,0
F5-1	Ch#2	1111	4	962	7	13,4
F5-1	Ch#3	1115	3	1006	8	9,8
F5-1	Ch#4	2633	27	2779	28	5,6
F5-1	Ch#5	1667	4	906	9	45,7
F5-1	Ch#6	1913	153	1192	64	37,7
F5-1	Ch#7	1300	88	1176	40	9,5
F6-1	Ch#1	1148	4	1016	9	11,5
F6-1	Ch#2	1171	4	1054	8	9,9
F6-1	Ch#3	1178	4	1061	9	9,9
F6-1	Ch#4	2811	25	2892	30	2,9
F6-1	Ch#5	1127	4	1049	10	6,9
F6-1	Ch#6	1198	70	1100	30	8,2
F6-1	Ch#7	1305	40	1327	28	1,7
F7-2	Ch#1	1503	7	1297	13	13,7
F7-2	Ch#2	1522	7	1348	12	11,4
F7-2	Ch#4	1562	9	1692	16	8,4

TABLE 5.3. RESULTS OF THE LEAST-SQUARES FITTING OF EQ. (5.9) IN CASE OF THE U–Pb BENCHMARK (SECTION 2.2). [a]

Case	Target detector	Reference values		Filtered values		α-difference (%)
		α (s^{-1})	$\Delta\alpha$ (s^{-1})	α (s^{-1})	$\Delta\alpha$ (s^{-1})	
F7-2	Ch#5	1459	7	1249	15	14,4
F7-2	Ch#6	1582	149	1418	54	10,4
F7-2	Ch#7	1286	105	249	62	80,7

[a] The reference values were obtained from [5.54]. α-difference refers to the absolute value of the difference between the reference α-values and α-values obtained from the filtered experimental data.

Case	Target detector	Reference values		Filtered values		abs_diff (-)
		α (s^{-1})	$\Delta\alpha$ (s^{-1})	α (s^{-1})	$\Delta\alpha$ (s^{-1})	
I-1 (W)	Ch#1	633	7	558	5	11.8
I-1 (W)	Ch#2	665	13	599	5	9.9
I-1 (W)	Ch#3	637	11	741	5	16.4
I-2 (W)	Ch#1	879	22	803	10	8.6
I-2 (W)	Ch#2	869	28	834	7	4.1
I-2 (W)	Ch#3	1090	60	1059	8	2.8
I-3 (W)	Ch#1	1086	22	978	14	9.9
I-3 (W)	Ch#2	1103	27	1036	10	6.1
I-3 (W)	Ch#3	1408	57	1319	12	6.3
I-4 (W)	Ch#1	889	12	831	11	6.6
I-4 (W)	Ch#2	942	21	874	5	7.2
I-4 (W)	Ch#3	1073	24	1060	8	1.2
I-5 (W)	Ch#1	1488	49	1308	19	12.1
I-5 (W)	Ch#2	1452	35	1380	11	4.9
I-5 (W)	Ch#3	1966	92	1802	16	8.4
II-1	Ch#1	401	9	366	2	8.6
II-1	Ch#2	367	6	339	2	7.6
II-1	Ch#3	554	17	533	3	3.8
II-2	Ch#1	498	5	649	6	30.2
II-2	Ch#2	464	49	624	4	34.7
II-2	Ch#3	700	7	859	10	22.7
II-3	Ch#1	655	2	614	4	6.3
II-3	Ch#3	877	5	535	210	39.1
II-3	Ch#1	655	2	650	4	0.7
II-3	Ch#3	877	5	833	60	5.0
II-3	Ch#1	655	2	657	3	0.4
II-3	Ch#2	620	2	676	3	9.0
II-4	Ch#1	815	6	821	8	0.8
II-4	Ch#2	822	5	869	5	5.7
II-4	Ch#3	1029	12	2453	14	138.4
II-5	Ch#1	1365	9	1262	15	7.6
II-5	Ch#2	1400	7	1324	13	5.4
II-5	Ch#3	1669	17	1540	17	7.7
II-6	Ch#1	1557	13	1427	17	8.4
II-6	Ch#2	1636	10	1529	15	6.6
II-6	Ch#3	1917	22	1752	20	8.6

[a] The reference values were obtained from [5.53]. α-difference refers to the absolute value of the difference between the reference α-values and α-values obtained from the filtered experimental data.

The observations concerning the bias on the fitted α-values are similar to the simulations. As it can be seen in Fig. 5.16, where the difference of the fitted α-values with and without filtering are plotted against the dimensionless parameter f_p/α, the relation of the prompt decay constant α and the pulsing frequency f_p clearly determines the applicability of the method. Below $f_p/\alpha \approx 0.06$ the filtering method performs well with lower deviations and in many cases less than 10% which is a good agreement concerning the statistical error of the fits. Above this value, almost all points have deviation much higher than 10%.

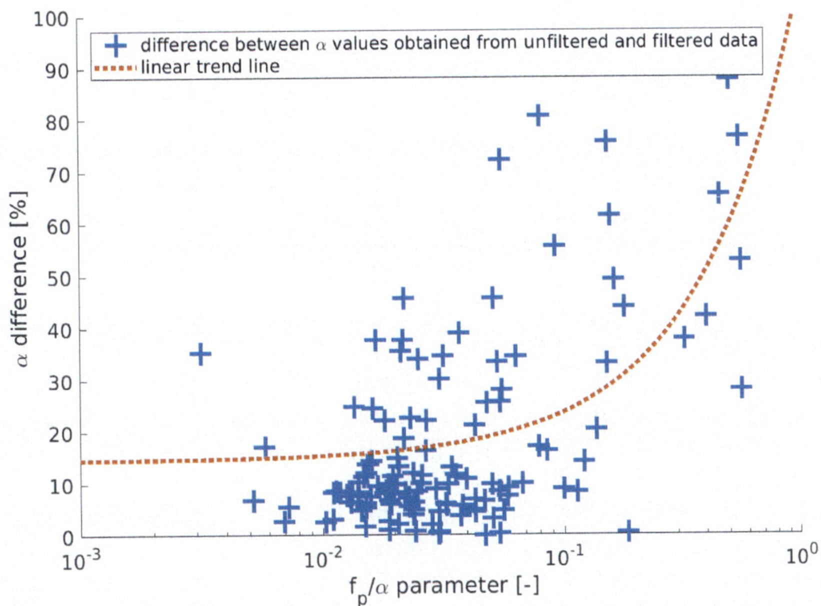

FIG. 5.16. Absolute values of the differences between the reference α-values (obtained from [5.53] and [5.54]) and α-values obtained from the filtered experimental data, plotted against the dimensionless parameter f_p / α.

5.8.5. Conclusions

A new methodology is presented for the evaluation of neutron noise measurements with a pulsed neutron source, which involves the filtering of the correlations introduced by the pulsing by randomizing the pulses. Although the detailed theoretical analysis of the method is not complete yet, the simulations and measurement evaluations presented show promising result when the pulsing frequency is sufficiently lower than the prompt decay constant. This means that the new method can provide simpler and more accurate way to obtain the prompt decay constant and other parameters especially in fast or deeply subcritical system, where higher α is expected, or the determination of higher α-modes.

5.9. DESIGN APPROACH BY A-BAQUS

5.9.1. Introduction

The rationales for designing the core of an ADS have been defined, extending an original work [5.55] based on the optimization of the European Facility for Industrial Transmutation (EFIT). Its application into the activities of this CRP notably involves the applicability of the original models to ternary fuels, thereby including into the scope of the methodology also LEU-based ADS, which were not part of the previous work cited in Ref. [5.55].

The effectiveness of the proposed approach is shown detailing the logical design approach for a hypothetical ADS system that operates in the thorium-uranium cycle while burning minor actinides, thus showing how two apparently alternative applications (among those listed in the scope of the CRP) could be simultaneously targeted.

The A-BAQUS system approaches the basic parameters of the design of an ADS from its envisaged operation. The reactor operation is characterized by the reactivity swing (Δk) and the burning rate (ΔTRU) resulting from the material composition change evolution. The design parameters of an ADS are described within a range of possible values based on physical and mechanical constraints for the design of a nuclear core (e.g., cladding temperature, coolant flow, active height, geometrical buckling, core radius/volume, power rating). The required range of the particle accelerator current is determined by the relationship between core power and current, given a fixed spallation target and particle energy, in spite of the subcriticality level and reactivity swing.

5.9.2. The A-BAQUS entry plot

A multi-entry plot [5.55] is proposed as design guideline for ADS systems, including the utilization of multicomponent fuel, in which thorium is considered the fertile material, to produce the fissile uranium and eliminating transuranics (which includes the plutonium inventory). The plot is organized in three interrelated quadrants, which are described in the following Sections 5.9.2.1 to 5.9.2.3.

5.9.2.1. Quadrant I – Representation of the fuel and identification of the associated key performance

From the fuel composition analysis of the ADS, two key properties are extracted: the enrichment (e) and the waste content (w). These properties mainly determine two key operational parameters: the reactivity swing in the reactor (related to the enrichment), and the burning rate (related to the waste content).

The first quadrant of the A-BAQUS (Fig. 5.17) represents the related reactivity swing and burning rate performances, to catch the implications following from the choice of a specific fuel composition. In the figure, the correspondences between enrichment (e) and reactivity swing (Δk), and between waste content (w) and burning rate (ΔTRU) are shown.

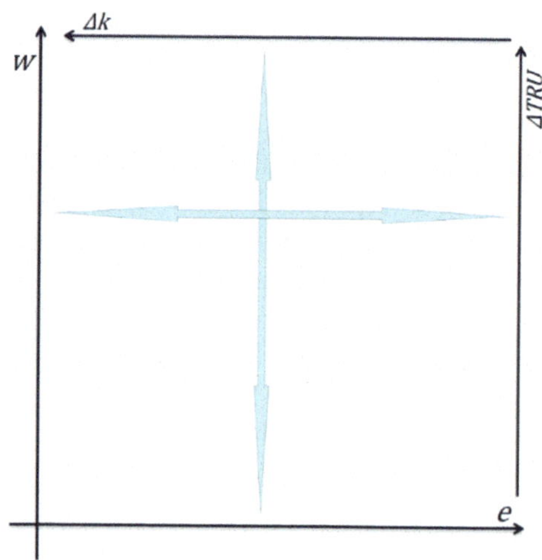

FIG. 5.17. First quadrant of the A-BAQUS multi-entry plot.

5.9.2.2. Quadrant IV – Identification of the (sub-)critical core size and power

The design of the core of a nuclear reactor depends on interrelated parameters that are subject to technological and safety constraints. Limits on the cladding temperature and the coolant flow speed, and objectives on thermodynamic efficiency and average linear power rating, set the diameter of the fuel pin and the area of the associated coolant channel [5.56].

The functional associating a fuel enrichment to the required core size derives from setting the material buckling equal to the geometrical buckling; the former being completely defined by the design of the unit core cell as per considerations above; the latter only depending on the core radius, once the active height is fixed. This is represented in the example shown in Fig. 5.18. The active height is usually fixed along with the unit core cell, being strictly related to some of the same constraining criteria (e.g., safety requirements on natural circulation performance). However, in all cases where the active height can be considered as a degree of freedom, it can be introduced in the above function as parameter for the axial component of the geometrical buckling; in this way, the function returns a family of curves (rather than a single one), each associated to a different active height.

Once the unit core cell and the active height are defined, and the core radius is fixed, the total development of the fuel elements (or the core volume, in a fluid fuel system) is also univocally fixed. Since, however, the average power rating (linear for solid-fuel systems, or volumetric for fluid-fuel ones) is also fixed – along with the unit core cell – in the preliminary considerations above, the core size univocally determines also the total power that can be generated in the system, which can be accordingly represented on an axis in the fourth quadrant that is parallel to the one for the core radius.

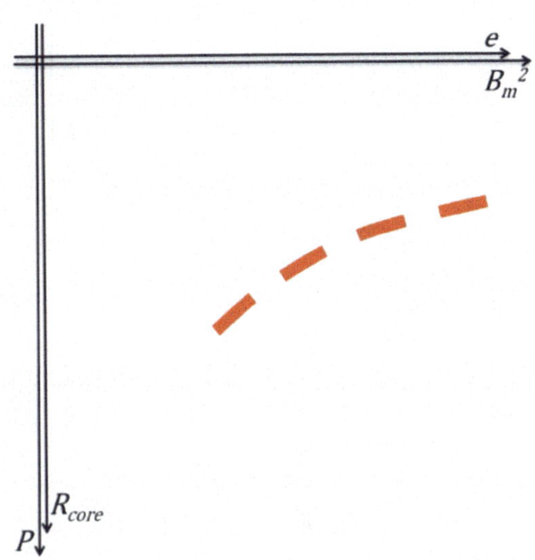

FIG. 5.18. Fourth quadrant of the A-BAQUS multi-entry plot.

5.9.2.3. Quadrant III – Identification of the required accelerator current (range)

In an ADS, the performance required for the accelerator (i.e., the delivered current) is one of the key parameters to be established in defining the design. The current I has indeed to provide – through spallation – the source neutrons required to compensate those lost by the subcritical core at the rated power. For a given spallation target and fixed energy of the injected particles, and for a given subcriticality level, a linear equation relates the core power to the required accelerator current. Since, however, the core might experience a reactivity swing during operation, a current range is identified, considering the two bounding lines referring to the extreme values of k_{eff} experienced during operation (which are retrieved from the first quadrant of the plot), as represented in Fig. 5.19.

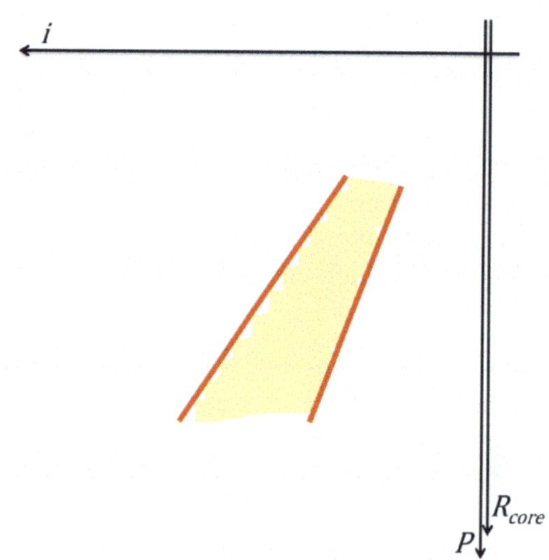

FIG. 5.19. Third quadrant of the A-BAQUS multi-entry plot.

5.9.3. Application to LEU-based ADS

The application of the methodology has been applied to an ADS operating in a Th-U cycle burning minor actinides. The system begins operation with low enriched uranium fuel. The fuel is managed by batches. The following roles are assigned to the fuel components:

- Uranium is the reference fissile, either as ^{235}U in the initial LEU loading or as ^{233}U resulting from the successive breeding from thorium;
- Thorium is therefore the reference fertile;
- Minor actinides are the radioactive waste to be burned. Plutonium has a potential role as fissile. No plutonium handling is foreseen for its practical use; accordingly, plutonium is assimilated to minor actinides as waste for this specific application. Therefore, minor actinides and plutonium will be considered as transuranics.

For this specific case, a feed made by 60% of thorium and 40% of transuranics (namely: all minor actinides) is chosen. This is anticipated to result in burnup figures of approximately 25.2 kg thorium per TWh and 16.8 kg minor actinides per TWh. The resulting fuel composition at equilibrium after burnup is shown in Table 5.5.

TABLE 5.5. COMPOSITION OF THE RESULTING EQUILIBRIUM FUEL

Species	Content
Thorium	62.5%
Uranium	14.7%
Transuranics	22.8%

For the test case considered, a solid-fuelled, lead-cooled core is chosen. The starting point for setting up the core configuration is the design of the fuel pin and associated coolant channel. For this, typical technological limits are used, nominally:

- Average coolant inlet and outlet temperatures of 400 and 480°C, respectively (the former to introduce a margin against lead freezing, the latter for mitigating corrosion issues on the core structures);

- Peak fuel temperature of 2000°C, which translates to a peak linear power rating of ~320 W/cm once uncertainties are accounted for;
- Coolant flow speed below 2 m/s, to both prevent erosion of core structures and promote natural circulation.

With these constraints, a nominal fuel pin is designed, having an outer diameter of 8.6 mm with cladding thickness of 0.6 mm and fuel pellet diameter of 7.1 mm.

The coolant flow area and active height are determined altogether, and along with the actual average coolant flow speed, as both parameters influence the natural circulation; the resulting proposed configuration, complying with a coolant flow speed of 1 m/s, has a lattice pitch of 13.3 mm and 900 mm active height.

With these data, and adopting the equilibrium fuel composition, an infinite multiplication factor k_∞ of ~1.11 is found. Assuming 0.96 as limit for the effective multiplication factor k_{eff}, the core radius can be found as ~1500 mm, corresponding to a power of ~580 MW.

As the core geometry approaches closely the EFIT one [5.57], it was decided to adopt that engineered design for simplicity. By doing so, a three-zones configuration already optimized for flattening the radial power distribution is readily obtained. The final data of the adopted core configuration are listed in Table 5.6. Coherently, also the same accelerator was chosen, providing 800 MeV protons.

TABLE 5.6. MAIN CORE DATA OF THE PROPOSED ADS

Parameter	Inner zone	Middle zone	Outer zone
Pellet radius (mm)	3.55	3.55	4.00
Gap thickness .(mm)	0.16	0.16	0.16
Cladding inner / outer radius .(mm)	3.71 / 4.31	3.71 / 4.31	4.16 / 4.76
Cladding thickness .(mm)	0.60	0.60	0.60
Pins lattice pitch .(mm)	13.63	13.63	13.54
Number of assemblies	42	66	72
Number of pins per assembly	168[a]		
Wrapper inner / outer flat-to-flat (mm)	178 / 186		
Assemblies lattice pitch (mm)	191		

[a] The central position is occupied by a dummy structural pin.

The results of the nuclear criticality analyses on the resulting core show that the effective multiplication factor k_{eff} is equal to 0.96311, if the whole core is loaded with fresh equilibrium fuel and reduces to 0.95100 when the same whole core is burned to 70 MWd/kgHM on average. Considering, however, a multibatch operation (specifically, 5 batches) along the 16 equivalent full-power months of duration, k_{eff} changes from 0.95827 to 0.95584, with a reactivity swing of −243 pcm.

To sustain this criticality behaviour, given that the source efficiency varies between 0.753 to 0.761, the required proton current varies between 19.2 and 20.2 mA over an equilibrium cycle.

The described design approach is visually represented on the A-BAQUS as shown in Fig. 5.20.

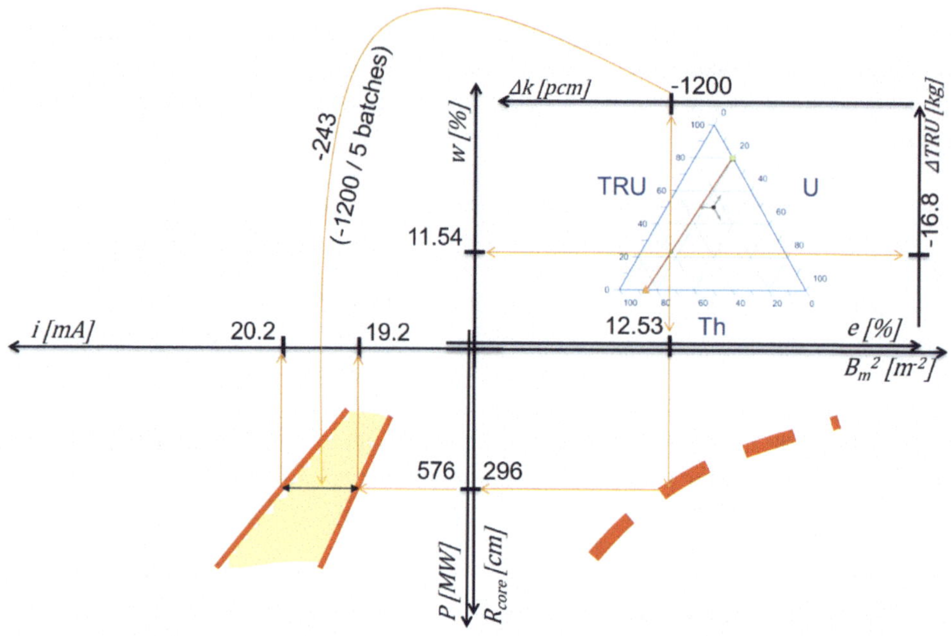

FIG. 5.20. Representation of the design process for the selected ADS case on the A-BAQUS.

5.9.4. Conclusions

ADS are worldwide investigated for their unique capabilities in terms of exploiting exotic fuel cycles, such as deep minor actinides burning or extensive Th–U conversions, deriving from the inherent safety of the subcritical core which relieves all issues typically associated with these unconventional configurations.

Focusing on design approaches that are fuel-cycle-oriented, so as to magnify the benefits of ADS with minimal burden and capitalizing on the physics relating a reactor core to the associated fuel cycle, an extension of the "42-0" approach [5.58] for designing the core of an ADS system (originally introduced for the case of a uranium-free, plutonium-sustained minor actinides burner) has been presented. The extended process provides the rationales for establishing *a priori* the fuel composition that is required to achieve, by design, target fuel cycle performances, also in the case of a ternary fuel (i.e. a fuel made by more than two species). As in the original work [5.55], the present also stands on the A-BAQUS multi-entry plot, meant as a visual aid to the designer in establishing the main correlations among the key core parameters.

Finally, the proposed extended methodology has been applied to a case study, selected to fulfil simultaneously two of the potential applications identified within the scope of this CRP, namely: to operate on the Th–U cycle and to burn minor actinides. Specifying – wherever appropriate – all the assumptions introduced (which do not limit the scope of the proposed methodology, rather orient the final design to a reference objective), the process has been followed step by step up to fixing the design of the fuel pin, coolant channel, core size (both active height and radius), core power and required accelerator current, thereby defining the configuration of the ADS core, permitting to target by design the chosen applications with the defined performance.

5.10. REFERENCES TO CHAPTER 5

[5.1]   PELOWITZ, D., MCNP6TM User's Manual, LA-CP-11-01708, Los Alamos National Laboratory, Los Alamos NM (2011).

[5.2]   LEPPÄNEN, J., Serpent – a Continuous-energy Monte Carlo Reactor Physics Burnup Calculation Code, (2015).

[5.3]   CETNAR, J., et al., User Manual for Monte-Carlo Continuous Energy Burnup (MCB) Code – Version 1C, KTH, Stockholm (2001) 26.

[5.4]   SHIM, H., et al., McCARD: Monte Carlo Code for Advanced Reactor Design and Analysis, Nucl. Eng. Technol. **44** (2012).

[5.5]   REYES-RAMIREZ, R., et al., Comparison of MCNPX-C90 and TRIPOLI-4-D for fuel depletion calculations of a Gas-cooled Fast Reactor, Ann. Nucl. Eng. **37** (2010).

[5.6]   WANG, K., et al., RMC – A Monte Carlo Code for reactor Core Analysis, Ann. Nucl. Eng. **82** (2015).

[5.7]   ROMANO, P., et al., OpenMC: A State-of-the-Art Monte Carlo Code for Research and Development, Ann. Nucl. Eng. **82** (2015).

[5.8]   PARK, H., et al., Monte Carlo burnup and Its Uncertainty Propagation Analyses for VERA Depletion Benchmarks by McCARD, Nucl. Eng. Technology **50** (2018).

[5.9]   WILLIAMS, D., et al., Molten Salt Fuel Cycle Requirements for ADTT Applications, Invited Presentation and Paper at the 3rd International Conference on Accelerator-Driven Transmutation Technologies and Applications (ADTTA '99), Prague, Czech Republic (1999).

[5.10]   GOHAR, Y., et al., Lead-Bismuth Target Design for the Subcritical Multiplier (SCM) of the Accelerator Driven Test Facility (ADTF), ANL/TD/02-01, Argonne National Laboratory, Lemont, IL (2002).

[5.11]   CAO, Y., GOHAR, Y., Monte Carlo Fuel Burnup Analyses of the Accelerator-Driven Subcritical Systems, Ann. Nucl. Eng. **131** (2019).

[5.12]   CUMING, M., et al., GEM*STAR Accelerator-Driven Subcritical System for Improved Safety, Waste Management and Plutonium Disposition, in: Proceedings of NAPAC2016, Chicago, IL, USA (2016).

[5.13]   OECD NUCLEAR ENERGY AGENCY, Physics and Safety of Transmutation Systems: A Status Report, NEA No. 6090, Paris, France (2006).

[5.14]   ABDERRAHIM, H., et. al., Accelerator and Target Technology for Accelerator Driven Transmutation and Energy Production, FERMILAB-FN-0907-DI, LA-UR-10-06754, Los Alamos National Laboratory, Los Alamos, NM (2010).

[5.15]   GOHAR, Y., et al., Accelerator-Driven Subcritical system for Disposing of the U.S. Spent Nuclear Fuel Inventory, ANL-18/07, Argonne National Laboratory, Lemont, IL (2018).

[5.16]   BERMUDEZ, J., et al., Target Design for a Thorium Accelerator-Driven Molten Salt Reactor (AD-MSR), in: Thorium Energy for the world, Proceedings of the Thorium Energy Conference (ThEC13), Geneva, Switzerland (2013). http://www.thoriumenergyworld.com/thec13-geneva.html

[5.17]   BAK, S., et al., Implementation of an LBE Spallation Target in an Accelerator-Driven Molten Salt Subcritical Reactor, J. Nucl. Sci. Tech. **54** [8] (2017).

[5.18]   TSUJIMOTO, K., et al., Neutronics Design for Lead-Bismuth Cooled Accelerator-Driven System for Transmutation of Minor Actinide, J. Nucl. Sci. Tech. **41** (2004) 21–26.

[5.19]   Li, Z., et al., Physics Design of an Accelerator for an Accelerator-Driven Subcritical System, Phys. Rev. Special Topics – Accel. Beams **16** (2013).

[5.20]   FERNADEZ, R., et al., MYRRHA, Technology Development for the Realisation of ADS in Eu: Current Status & Prospects for Realisation, Applications of High Intensity Proton Accelerators, Proceedings of the Workshop, Fermilab, Chicago (2010).

[5.21]   YAMANAKA, M., Private Communication, 2015.

[5.22]   YAMANAKA, M., Private Communication with the developer, 2017.

[5.23]   CHADWICK, M., et al., ENDF/B-VII.1 Nuclear Data for Science and Technology: Cross-Sections, Covariances, Fission Product Yields and Decay Data, Nucl. Data Sheets 112 (2011).

[5.24]   CROFF, A., A User's Manual for the ORIGEN2 Computer Code, ORNL/TM-7157, Oak Ridge National Laboratory, Oak Ridge TN (1980).

[5.25]   ISOTALO, A., WIESELQUIST. A., A Method for Including External Feed in Depletion Calculations with CRAM and implementation into ORIGEN, Ann. Nucl. Eng. 85 (2015) 68–77.

[5.26]   AKCASU, Z., et al., Mathematical methods in nuclear reactor dynamics, Academic Press, New York (1971).

[5.27]   KRILOV, A., The numerical solution of equations determining the frequency of small vibration in material systems in engineering, Izvestiya Akademii Nauk S.S.S.R. (1931) 491–539.

[5.28]   DULLA, S., et al, A method for on-line reactivity monitoring in nuclear reactors, Annals of Nuclear Energy 65 (2014) 433–440.

[5.29]   DULLA, S., et al., A method for the continuous monitoring of reactivity in subcritical source-driven systems, Annals of Nuclear Energy 87 (2015) 1–11.

[5.30]   KALMAN, R., A new approach to linear filtering and prediction problems, Transactions of the ASME Journal of Basic Engineering 82 (1960) 35–45.

[5.31]   SAVITZKY, A., GOLAY, J., Smoothing and differentiation of data by simplified least squares procedures, Analytical Chemistry 36 (1964) 1627–1639.

[5.32]   CHARTRAND, R., Numerical differentiation of noisy, nonsmooth data, ISRN Applied Mathematics 2011, doi:10.5402/2011/164564 (2011).

[5.33]   CARON, D., et al., Assessment of and on-line reactivity monitoring technique, Transactions of the American Nuclear Society 111 (2014) 1185–1187.

[5.34]   DULLA, S., et al., Interpretation of experimental measurements on the SC-1 configuration of the VENUS-F core, Proc. of PHYSOR 2014, Kyoto, Japan (2014).

[5.35]   DORNING, J., Nuclear Reactor Kinetics: 1934–1999 and beyond, Nuclear Computational Science: a Century in Review, Y. Azmy, E. Sartori (Eds.), Springer, New York (2010).

[5.36]   CHERSOLA, D., et al., An alternative observable to estimate k_{eff} in fast ADS, Ann. Nucl. Energy 95 (2016) 42–47.

[5.37]   CARTA, M., D'ANGELO, A., Subcriticality-level evaluation in accelerator-driven systems by harmonic modulation of the external source, Nucl. Sci. Eng. 133 (1999), 282–292.

[5.38]   DELAGE, F., et al., 2011, ADS fuel developments in Europe: results from the EUROTRANS integrated project, Energy Proc. 7 (2011) 303–313.

[5.39]   MANSANI, L., private communications (2014). The data have been deduced under the following hypothesis: the fuel is composed of MOX material; the plutonium isotopic composition is representative of a plutonium vector coming from the reprocessing of PWR spent fuel of 45 GWd/t burn up with an initial enrichment of 4.5% in ^{235}U and a cooling period of 50 years.

[5.40]   SARACCO, P., et al., A useful observable for estimating keff in fast subcritical systems, Proc. of 13th ICRS, 19th RPSD, Paris, France (2016).

[5.41]   BELL, G., GLASSTONE, S., Nuclear reactor theory, Van Nostrand Reinhold, New York (1970).

[5.42]   SINHA, A., et al, BRAHMMA: A compact experimental accelerator driven subcritical facility using D-T/D-D neutron source, Annals of Nuclear Energy 75 (2015) 590–594.

[5.43] 'SINHA, A., et al, Experimental subcritical facility driven by D-D/D-T neutron generator at BARC, India, Nuclear Instruments and Methods in Physics Research B **350** (2015) 66–70.

[5.44] HENRY, A., The application of reactor kinetics to the analysis of experiments, Nucl. Sci. Eng. **3** (1958) 52–70.

[5.45] DULLA, S., et al., Importance weighting of local flux measurements to improve reactivity predictions in nuclear systems. Kerntechnik **80** (2015) 201–207.

[5.46] LEWINS, J., Importance: the adjoint function. Pergamon, Oxford (1965).

[5.47] DULLA, S., et al., Reduction of spatial and spectral effects by adjoint weighting of flux signals in a reactivity reconstruction technique, Transactions of the American Nuclear Society **111** (2014) 1200–1203.

[5.48] DULLA, S., et al., Interpretation of local flux measurements for reactivity prediction, Proc. of PHYSOR 2016, American Nuclear Society, Sun Valley, ID (2016) 2528–2537.

[5.49] DULLA, S., et al., Integral parameters in source-driven systems, Progress in Nuclear Energy **53** (2011) 32–40.

[5.50] INTERNATIONAL ATOMIC ENERGY AGENCY, Pulsed Neutron Research II, IAEA, Vienna, (1965) 239.

[5.51] SHIM, H., et al. Estimation of kinetics parameters by Monte Carlo fixed-source calculations for point kinetic analysis of accelerator-driven system, J. Nucl. Sci. Technol. **57**[2] (2020) 177–186.

[5.52] ORNDOFF, J., Prompt neutron periods of metal critical assemblies, Nucl. Sci. Eng. **2** (1957) 450.

[5.53] KITAMURA, Y., et al., Calculations of the Pulsed Feynman- and Rossi-alpha Formulae with Delayed Neutrons, Ann. Nucl. Energy **32** (2005) 671.

[5.54] CHADWICK, M., et al., ENDF/V-II.0: Next Generation Evaluated Nuclear Data Library for Nuclear Science and Technology, Nucl. Data Sheet **107** (2006) 2931.

[5.55] ARTIOLI, C., A-BAQUS: A multi-entry graph assisting the neutronic design of an ADS. Case study: EFIT, Fifth International Workshop on the Utilisation and Reliability of High Power Proton Accelerator (HPPA 5), Mol, Belgium (2007).

[5.56] ARTIOLI, C., et al., A new paradigm for core design aimed at the sustainability of nuclear energy: The solution of the extended equilibrium state, Ann. Nucl. Energy **37** (2010) 915–922.

[5.57] KNEBEL, J., IP EUROTRANS: A European Research Programme for the Transmutation of High Level Nuclear Waste in an Accelerator Driven System, Proc. of 8th Information Exchange Meeting on Actinide and Fission Product Partitioning & Transmutation (IEMPT8), Las Vegas, NV, USA (2004).

[5.58] ARTIOLI, C., et al., Optimization of the Minor Actinides Transmutation in ADS: The European Facility for Industrial Transmutation EFIT-Pb Concept, Proc. Eighth International Topical Meeting on Nuclear Applications and Utilization of Accelerators (AccApp'07), ISBN: 0-89448-054-5, ANS, Pocatello, ID, USA (2007), CD-ROM.

Chapter 6

GENERAL CONCLUSIONS AND SUGGESTIONS FOR FUTURE WORK

The advancement of Accelerator Driven Systems (ADS) research remains a priority within the international community. This report captures the findings of an IAEA Coordinated Research Project (CRP) conducted from 2016 to 2020 involving 22 institutions from 17 Member States. The CRP followed a previous project (I32006) devoted to understanding the physics of ADS, and focused on the use of Low Enriched Uranium (LEU) fuel and their applications, including experimental, analytical, and modelling aspects.

Chapter 2 provides an in-depth description of experimental configurations and results related to ADS facilities, including experiments carried out at KUCA (Japan), GIACINT (Belarus), QUINTA (the Russian Federation), and BRAHMMA (India) installations. It presents measurements of kinetic parameters, reaction rates, and subcriticality, using various neutron sources and detectors. Notably, the chapter highlights the successful validation of physics models through comparison of experimental data with Monte Carlo and deterministic codes such as MCNP, McCARD, and KIN3D. Key findings include accurate measurements of prompt neutron decay constants, effective delayed neutron fractions, and control rod worths, demonstrating the reliability of the proposed experimental and analysis methodologies. The chapter provides highly valuable and detailed benchmarks crucial for ADS design and safety assessments.

Chapter 3 focuses on ADS applications for nuclear data (cross section) measurement, and the required advanced detectors developments to accommodate a high neutron flux, which in turn enables an enhanced characterization of ADS flux spectra and refined simulations for other applications of ADS such as transmutation. In particular, two types of detectors are covered, namely a diamond detector and a silicon detector, each of which is fully described in terms of performance and operation range. The diamond detector is an innovative compact neutron spectrometer which can be used to improve the characterization of ADS, up from thermal up to 5 MeV neutrons. Limitations on count rate and electronics linearity under high fluxes are being addressed and shall be solved in the most recent prototypes. The silicon detector is shown to be well suited for neutron total cross section measurements, and neutron spectrum measurements.

Chapter 4 offers an extensive overview of advanced accelerator-driven systems (ADS) for nuclear energy and waste transmutation, focusing first on China's ADANES concept and then on Japan's LBE-cooled ADS. The Chinese ADANES system, developed by the Chinese Academy of Sciences, merges fuel breeding, electricity generation, and actinide transmutation in a subcritical reactor driven by a high-power accelerator and granular tungsten spallation target, featuring ceramic materials for enhanced safety and efficiency. Advanced computational tools like GPU-accelerated Monte Carlo simulations are used to optimize the neutron economy and heat management of granular versus monolith targets, aiming for long operational life, high efficiency, and reduced proliferation risk with innovative fuel reprocessing techniques. Japan's ADS project, part of the OMEGA program, emphasizes transmutation of minor actinides using a lead-bismuth eutectic cooled core, detailing benchmark specifications, core and fuel designs, operational cycles, and results of independent modeling efforts, which highlight generally good agreement in predicted reactor performance and isotope evolution. Both approaches focus on safely closing the nuclear fuel cycle, reducing long-lived radioactive waste, and leveraging advanced materials, coolants, and computational modelling for next-generation nuclear systems.

Chapter 5 provides an overview of physics methods developed within this CRP, effectively advancing modelling techniques for LEU-fuelled ADS and their applications. These include:

1. Fuel Burnup Simulation in ADS
Section 5.1 details a two-step method for modelling fuel burnup in subcritical systems. Since traditional Monte Carlo burnup codes (e.g., SERPENT, MCB5) lack high-energy neutron physics, spallation sources are modelled separately (e.g., with MCNP6). The importance of accurate source term definition and power normalization is highlighted, as incorrect treatment can cause >10% errors

in burnup predictions. Power must be adjusted manually based on simulations using coupled proton-neutron-photon models.

2. MAρTA Method for Reactivity Monitoring

MAρTA (Monitoring Algorithm for Reactivity Transient Analysis) is an inverse, passive method to estimate reactivity using neutron flux signals and point kinetic models. It is applicable during both routine operations and transients. It reconstructs the instantaneous time constant (ω) and uses the in-hour equation to derive reactivity, relying on precise knowledge of delayed neutron parameters.

3. Spectral Index for Reactivity Estimation

A semi-empirical spectral index, defined as the ratio of high-energy to total neutron flux, correlates with the effective multiplication factor (k_{eff}) in fast subcritical systems. This index (R) shows linear dependence on k_{eff} and is relatively insensitive to geometric details, making it a practical local observable for real-time monitoring, provided calibration is performed.

4. Correction of Spatial and Spectral Effects

Reactivity measurements using point kinetics are prone to spatial and spectral biases. Correction methods include spatial weighting using neutron importance and Bell-Glasstone correction factors. Combining signals from multiple detectors with optimized weights helps improve global representativeness.

5. Integral Parameters in Source-Driven Systems

Traditional definitions of integral kinetic parameters (e.g., reactivity, neutron lifetime, β_{eff}) need adaptation for ADS due to their subcritical nature and reliance on external sources. These parameters must be defined relative to specific modeling goals, often using adjoint transport equations.

6. Rossi-α Method and β_{eff} Estimation

An advanced formulation of the Rossi-α method is adapted for pulsed neutron source (PNS) experiments to extract the prompt decay constant and β_{eff}, factoring in correlated and uncorrelated neutron detection probabilities. Fitting and correction techniques are applied to separate signal components and accurately estimate kinetics parameters.

7. Filtering Techniques for Pulse Noise Reduction

To overcome complications from pulsed sources in neutron noise analysis, a filtering method randomly redistributes pulse events to simulate isotopic sources. Simulations and experimental validation (e.g., at KUCA) confirm its utility, especially for high decay constants (α), while low-α scenarios show more bias.

8. Structural Design via ABAQUS

In parallel to neutronic studies, ABAQUS was employed as a finite element analysis tool to assist in the mechanical and thermal design of components within the ADS. This includes evaluating stresses, thermal deformation, and material resilience under irradiation and heat loads - particularly in fuel assemblies and spallation targets. ABAQUS enables integration of realistic boundary conditions and supports the coupled multiphysics environment required for safe ADS design.

The work performed under this CRP enhances the body of knowledge on ADS technologies, bridging experimental, analytical, and computational research. In particular, significant progress was made for accurate subcriticality measurements, improved nuclear data usage, and refined simulation techniques that are pivotal for the safe and efficient design, operation, and fuel management of ADS. This integration of comprehensive datasets, innovative methods, and computational tools provides a valuable resource for researchers, designers, and regulatory bodies in the accelerator-driven system field, and paves the way for future developments in nuclear waste transmutation and advanced energy production.

Participants of the CRP suggest that the activities should be continued, as several planned experiments involving low enriched uranium (LEU) fuels were not completed during previous CRPs due to multiple factors.

These include the need for additional facility time, further development of experimental details and analyses, readiness delays of experimental facilities, and the necessity to revisit or extend experimental evaluations.

Looking forward, it would be beneficial to conduct more ADS experiments that complement analytical studies, particularly through the integration of novel ADS analytical methodologies. Experimental facilities worldwide are encouraged to share their data openly to foster collaboration and benchmarking within the international community. Partnerships between institutions possessing experimental capabilities and those focusing on analytical development are vital, and the IAEA CRPs continue to serve as a valuable platform for such cooperation. For example, recent experimental data from the KIPT neutron source facility in Ukraine has proven instrumental in validating analytical models.

The role of fast neutron facilities is increasingly important, especially in the context of spent fuel disposal and production of uranium-233 for future fuel cycles. Design studies, experiments, and neutron spectrum measurements (including those above 10 MeV) are required to understand fast neutron environments and minor actinide transmutation. Additionally, further reactor physics research involving Pb-Bi cooled fast neutron spectrum cores and the neutron characteristics of solid Pb-Bi targets is necessary.

Enhancing methods to measure and monitor the subcriticality of both thermal and fast ADS remains a key priority. Design and analytical efforts concerning ADS spallation targets and the characterization of spallation products should be intensified. Zero-power and small research reactors represent valuable platforms for conducting ADS-related experiments and should be considered in future research initiatives. Furthermore, design studies focused on prospective ADS power applications are strongly encouraged.

A notable challenge is the recruitment and retention of nuclear specialists, particularly for research and operational roles in nuclear facilities and industry. CRPs provide an excellent avenue for training the next generation of nuclear scientists and engineers. Therefore, Member States are urged to actively involve students and young professionals in ADS research and report their engagement to the IAEA as part of CRP contributions.

Suggestions for Future CRP Topics Include:

- Comprehensive studies on ADS performance

- Exploration of ADS power applications, including instrumentation and feedback mechanisms

- Innovative ADS design features and future facility concepts

- Advanced monitoring techniques for new ADS installations

- Investigation of spallation product formation and impacts

- The role of ADS within the nuclear fuel cycle and spent fuel management strategies

Collectively, these efforts will strengthen the global ADS research framework, enhance international collaboration, and support the sustainable development of nuclear technologies.

LIST OF ABBREVIATIONS

ADANES	accelerator driven advanced nuclear energy system
ADCFR	accelerator driven ceramic fast reactor
ADS	accelerator driven system
ANL	Argonne National Laboratory (United States of America)
ASM	amplified source method
BARC	Bhabha Atomic Research Centre (Mumbai, India)
BRAHMMA	Beryllium Oxide Reflected and HDPE Moderated Multiplying Assembly
BOC	beginning of cycle
C/E	Ratio of computed and experimental results
CR	control rod
CRP	Coordinated Research Project
ECCO	European Cell Code
EFIT	European Facility for Industrial Transmutation
EOC	end of cycle
FFAG	fixed-field alternating gradient
GIACINT	Critical facility at the JIPNR-Sosny
GPU	graphics processing unit
HEU	high enriched uranium
IMP	Institute of Modern Physics (Chinese Academy of Sciences, Lanzhou, China)
JINR	Joint Institute for Nuclear Research (Dubna, Russian Federation)
JIPNR	Joint Institute for Power and Nuclear Research (Sosny, Belarus)
KIT	Karlsruhe Institute of Technology (Germany)
KUCA	Kyoto University Critical Assembly
KURRI	Kyoto University Research Reactor Institute (Japan)
LBE	lead–bismuth eutectic
LEU	low enriched uranium
MAρTA	Monitoring Algorithm for Reactivity Transient Analysis
MCNP	Monte Carlo N-Particle
MCNPX	Monte Carlo N-Particle eXtended
MPI	message passing interface
MSM	modified source multiplication
QUINTA	Spallation target facility at the JINR
PNS	pulsed neutron source
SNL	Seoul National University (Republic of Korea)
SuperMC	Super Multi-functional Calculation Program for Nuclear Design and Safety Evaluation
TOF	time of flight

CRP PARTNERS

Country	Name	Affiliation
Argentina	LESZCZYNSKI Francisco	Centro Atómico Bariloche (CAB) - CNEA
Belarus	SIKORIN Svyatoslav	The State Scientific Institution "The Joint Institute for Power and Nuclear Research-Sosny" of the National Academy of Science of Belarus
China	YANG Lei	Chinese Academy of Sciences; Institute of Modern Physics
China	LIU Chao	Institute of Nuclear Energy Safety Technology (INEST); Chinese Academy of Sciences (CAS)
Germany	GABRIELLI Fabrizio	Karlsruhe Institute of Technology (KIT)
Hungary	SZIEBERTH Mate	Budapest University of Technology and Economics (BME); Institute of Nuclear Techniques (INT)
India	SINGH Krishna Prasad	Bhabha Atomic Research Centre
Indonesia	SUDARMONO Sudarmono	Center for Nuclear Reactor Technology and Safety (PTKRN), BATAN
Islamic Republic of Iran	KASESAZ Yaser	Nuclear Science and Technology Research Institute (NSTRI)
Italy	RAVETTO Piero	Politecnico di Torino, Italy
Italy	CARTA Mario	Ente per le Nuove Tecnologie L'Energia e L'Ambiente (ENEA)
Italy	SARACCO Paolo	Istituto Nazionale di Fisica Nucleare (INFN)
Japan	PYEON Cheolho	Research Reactor Institute, Kyoto University
Japan	SUGAWARA Takanori	Japan Atomic Energy Agency (JAEA)
Poland	SZUTA Marcin	National Centre for Nuclear Research
Republic of Korea	SHIM Hyung Jin	SEOUL NATIONAL UNIVERSITY
Russian Federation	TYUTYUNNIKOV Sergey	Joint Institute for Nuclear Research (JINR)
Spain	ABANADES Alberto	Universidad Politécnica de Madrid
Ukraine	PAVLOVYCH Volodymyr	Institute for Safety Problems of Nuclear Power Plants; National Academy of Sciences of Ukraine

Ukraine	KARNAUKHOV Ivan	National Science Center "Kharkov Institute of Physics and Technology" (NSC KIPT)
United States of America	GOHAR Yousry	Argonne National Laboratory
Uzbekistan	YULDASHEV Bekhzod	Institute of Nuclear Physics, Academy of Sciences of the Republic of Uzbekistan

CONTRIBUTORS TO DRAFTING AND REVIEW

Abánades, A.	Universidad Politécnica de Madrid, Spain
Alldred, K.	International Atomic Energy Agency
Álvarez, I.	Universidad Politécnica de Madrid, Spain
Burgio, N.	National Agency for New Technologies, Energy and Sustainable Economic Development (ENEA), Italy
Caron, D.	Politecnico di Torino, Italy
Carta, M.	National Agency for New Technologies, Energy and Sustainable Economic Development (ENEA), Italy
Chakrov, P.	International Atomic Energy Agency
Chao, L	Institute of Nuclear Safety Technology, China
Cruz, P.	Universidad Politécnica de Madrid, Spain
Dewes, J.	International Atomic Energy Agency
Dulla, S.	Politecnico di Torino, Italy
Dunn, K.	International Atomic Energy Agency
Fabrizio, V.	National Agency for New Technologies, Energy and Sustainable Economic Development (ENEA), Italy
Geupel, S.	International Atomic Energy Agency
Gohar, Y.	Argonne National Laboratory, United States of America
Grasso, G.	National Agency for New Technologies, Energy and Sustainable Economic Development (ENEA), Italy
Hoh, S.S.	Politecnico di Torino, Italy
Jinchuk, D.	International Atomic Energy Agency
Mafi, A.	International Atomic Energy Agency
Marana, G.	Politecnico di Torino, Italy
Marshall, F.	International Atomic Energy Agency
Mehdizadeh, S.	Universidad Politécnica de Madrid, Spain
Nervo, M.	Politecnico di Torino, Italy
Osipenko, M.	Istituto Nazionale di Fisica Nucleare (INFN), Genova, Italy

Peluso, V.	National Agency for New Technologies, Energy and Sustainable Economic Development (ENEA), Italy
Pyeon, C.	Research Reactor Institute, Kyoto University Japan
Ravetto, P.	Politecnico di Torino, Italy
Rineiski, A.	Karlsruhe Institute of Technology, Germany
Ripani, M.	Istituto Nazionale di Fisica Nucleare (INFN), Genova, Italy
Sánchez, H.	Universidad Politécnica de Madrid, Spain
Saracco, P.	Istituto Nazionale di Fisica Nucleare (INFN), Genova, Italy
Shim, H.J.	Seoul National University, Republic of Korea
Sikorin, S.	State Scientific Institution "The Joint Institute for Power and Nuclear Research-Sosny" Belarus
Singh, K. P.	Bhabha Atomic Research Centre, India
Swainson, I.	International Atomic Energy Agency
Szieberth, M.	Institute of Nuclear Techniques, Budapest University of Technology, Hungary
Tyutyunnikov, S	Joint Institute for Nuclear Research (JINR) High Energy Particle Laboratory Russian Federation
Yang, L	Institute of Modern Physics, China

Consultants Meetings

Vienna, Austria: 3–6 December 2019, 26–29 July 2021

Research Coordination Meetings

1st RCM: Vienna, Austria, 25–29 July 2016

2nd RCM: Lanzhou, China, 25–29 September 2017

3rd RCM: Budapest, Hungary, 10–14 December 2018

CONTACT IAEA PUBLISHING

Feedback on IAEA publications may be given via the on-line form available at:
www.iaea.org/publications/feedback

This form may also be used to report safety issues or environmental queries concerning IAEA publications.

Alternatively, contact IAEA Publishing:

Publishing Section
International Atomic Energy Agency
Vienna International Centre, PO Box 100, 1400 Vienna, Austria
Telephone: +43 1 2600 22529 or 22530
Email: sales.publications@iaea.org
www.iaea.org/publications

Priced and unpriced IAEA publications may be ordered directly from the IAEA.

ORDERING LOCALLY

Priced IAEA publications may be purchased from regional distributors and from major local booksellers.

Printed and bound by CPI Group (UK) Ltd, Croydon, CR0 4YY

06/07/2026

02160600-0001